Applied and Numerical Partial Differential Equations

Computational Methods in Applied Sciences

Volume 15

Series Editor

E. Oñate
International Center for Numerical Methods in Engineering (CIMNE)
Technical University of Catalonia (UPC)
Edificio C-1, Campus Norte UPC
Gran Capitán, s/n
08034 Barcelona, Spain
onate@cimne.upc.edu
www.cimne.com

For other titles published in this series, go to
www.springer.com/series/6899

Applied and Numerical Partial Differential Equations

Scientific Computing in Simulation, Optimization and Control in a Multidisciplinary Context

Edited by

William Fitzgibbon
University of Houston, TX, USA

Yuri Kuznetsov
University of Houston, TX, USA

Pekka Neittaanmäki
University of Jyväskylä, Finland

Jacques Périaux
University of Jyväskylä, Finland

and

Olivier Pironneau
UPMC, Paris, France

Springer

William Fitzgibbon
Department of Mathematics
University of Houston
USA

Yuri Kuznetsov
Department of Mathematics
University of Houston
USA

Pekka Neittaanmäki
Department of Mathematical
Information Technology
University of Jyväskylä
Finland

Jacques Périaux
Department of Mathematical Information
University of Jyväskylä
40351, Jyväskylä, Finland
and
Numèrics en Enginyeria (CIMNE)
Centre Internacional de Mètodes
C/Gran Capitan s/n, 08034, Barcelona
Spain
jperiaux@gmail.com

Olivier Pironneau
Department of Mathematique
Laboratoire Jacques Louis Lions
UPMC, Paris, France

ISSN 1871-3033
ISBN 978-90-481-3238-6 e-ISBN 978-90-481-3239-3
DOI 10.1007/978-90-481-3239-3
Springer Dordrecht Heidelberg London New York

Library of Congress Control Number: 2009938832

Cover illustration: Done by Chantal Périaux

Cover design: eStudio Calamar S.L.

Printed on acid-free paper

Springer is part of Springer Science+Business Media (www.springer.com)

Dedicated to Prof. R. Glowinski
at the occasion of his 70th birthday.

Preface

The present volume is comprised of contributions solicited from invitees to conferences held at the University of Houston, University of Jyväskylä, and Xi'an Jiaotong University honoring the 70th birthday of Professor Roland Glowinski. Although scientists convened on three different continents, the editors prefer to view the meetings as single event. The three locales signify the fact Roland has friends, collaborators and admirers across the globe. The contents span a wide range of topics in contemporary applied mathematics ranging from population dynamics, to electromagnetics, to fluid mechanics, to the mathematics of finance among others. However, they do not fully reflect the breath and diversity of Roland's scientific interest. His work has always been at the intersection mathematics and scientific computing and their application to mechanics, physics, aeronautics, engineering sciences and more recently biology. He has made seminal contribution in the areas of methods for science computation, fluid mechanics, numerical controls for distributed parameter systems, and solid and structural mechanics as well as shape optimization, stellar motion, electron transport, and semiconductor modeling. Two central themes arise from the corpus of Roland's work. The first is that numerical methods should take advantage of the mathematical properties of the model. They should be portable and computable with computing resources of the foreseeable future as well as with contemporary resources. The second theme is that whenever possible one should validate numerical with experimental data.

The volume is written at an advanced scientific level and no effort has been made to make it self contained. It is intended to be of interested to both the researcher and the practitioner as well to advanced students in computational and applied mathematics, computational science and engineers and engineering.

Many individuals contributed to the success of the celebration honoring Roland's 70th. The scientific coordination of the events was managed by Prof. Tsorng Whay Pan in Houston and Dr. Kirsi Majava in Jyväskylä. Without their dedicated efforts the conferences and this volume would not have existed.

The solicitation and collect of manuscripts was overseen by Ms. Sharon Lahey in Houston and Ms. Marja-Leena Rantalainen in Jyväskylä. The staffs of the Faculty of Information Technology in Jyväskylä and the Departments of Mathematics of the University of Houston and Jiaotong University need to be recognized for their diligent efforts in logistics, local arrangements and support.

Houston, Texas — *William Fitzgibbon*
Houston, Texas — *Yuri Kuznetsov*
Jyväskylä, Finland — *Pekka Neittaanmäki*
Jyväskylä, Finland — *Jacques Périaux*
Paris, France — *Olivier Pironneau*
April 2009

Contents

Roland Glowinski: The Unconventional and Unexpected Path of a Mathematician
William E. Fitzgibbon and Jacques F. Périaux 1

The Scientific Career of Roland Glowinski
Olivier Pironneau 5

On a Class of Partial Differential Equations with Nonlocal Dirichlet Boundary Conditions
Alain Bensoussan and Janos Turi 9

A Unified Discrete–Continuous Sensitivity Analysis Method for Shape Optimization
Martin Berggren 25

A Novel Approach to Modeling Coronary Stents Using a Slender Curved Rod Model: A Comparison Between Fractured Xience-Like and Palmaz-Like Stents
Josip Tambača, Sunčica Čanić, and David Paniagua 41

On the Stochastic Modelling of Interacting Populations. A Multiscale Approach Leading to Hybrid Models
Vincenzo Capasso and Daniela Morale 59

Remarks on the Controllability of Some Parabolic Equations and Systems
Enrique Fernández-Cara 81

Goal Oriented Mesh Adaptivity for Mixed Control-State Constrained Elliptic Optimal Control Problems
Michael Hintermüller and Ronald H.W. Hoppe 97

Feedback Solution and Receding Horizon Control Synthesis for a Class of Quantum Control Problems
Kazufumi Ito and Qin Zhang 113

Fluid Dynamics of Mixtures of Incompressible Miscible Liquids
Daniel D. Joseph 127

Demand Forecasting Method Based on Stochastic Processes and Its Validation Using Real-World Data
Yinggao Zheng, Hiroshi Suito, and Hideo Kawarada 147

Analytic Bounds for Diagonal Elements of Functions of Matrices
Gérard Meurant 161

Numerical Methods for Ferromagnetic Plates
Michel Flück, Thomas Hofer, Ales Janka, and Jacques Rappaz 169

Two-Sided Estimates of the Solution Set for the Reaction–Diffusion Problem with Uncertain Data
Olli Mali and Sergey Repin 183

Guaranteed Error Bounds for Conforming Approximations of a Maxwell Type Problem
Pekka Neittaanmäki and Sergey Repin 199

A Componentwise Splitting Method for Pricing American Options Under the Bates Model
Jari Toivanen 213

Exact Controllability of the Time Discrete Wave Equation: A Multiplier Approach
Xu Zhang, Chuang Zheng, and Enrique Zuazua 229

Index 247

List of Contributors

Alain Bensoussan
International Center for Decision and Risk Analysis, ICDRiA
School of Management
University of Texas at Dallas
Richardson, TX 75083
USA
Alain.Bensoussan@utdallas.edu

Martin Berggren
Department of Computing Science
Umeå University
Sweden
martin.berggren@cs.umu.se

Sunčica Čanić
Department of Mathematics
University of Houston
4800 Calhoun Road
Houston, TX 77204-3476
USA
canic@math.uh.edu

Vincenzo Capasso
Department of Mathematics
University of Milan
IT-20133 Milan
Italy
vincenzo.capasso@unimi.it

Enrique Fernández-Cara
University of Sevilla
Dpto. E.D.A.N., Aptdo. 1160
41080 Sevilla
Spain
cara@us.es

William E. Fitzgibbon
College of Technology
University of Houston
4800 Calhoun Road
Houston, TX 77204
USA
fitz@uh.edu

Michel Flück
École Polytechnique Fédérale de Lausanne
CH-1015 Lausanne
Switzerland
michel.flueck@epfl.ch

Michael Hintermüller
Department of Mathematics
University of Sussex
Brighton BN1 9RH
United Kingdom
M.Hintermueller@sussex.ac.uk

Thomas Hofer
École Polytechnique Fédérale de Lausanne
CH-1015 Lausanne
Switzerland
t.hofer@epfl.ch

Ronald H. W. Hoppe
Department of Mathematics
University of Houston
Houston, TX 77204-3008
USA
rohop@math.uh.edu
and
Institute of Mathematics
University of Augsburg
D-86159 Augsburg
Germany
hoppe@math.uni-augsburg.de

Kazufumi Ito
Department of Mathematics
North Carolina State University
Raleigh, NC 27695-8205
USA
kito@unity.ncsu.edu

Ales Janka
École Polytechnique Fédérale de Lausanne
CH-1015 Lausanne
Switzerland

Daniel D. Joseph
University of Minnesota, Minneapolis
and
University of California
Irvine
USA
joseph@aem.umn.edu

Hideo Kawarada
Applied Analysis and Technology
Japan
kawarada@apantech.com

Olli Mali
Department of Mathematical Information Technology
University of Jyväskylä
P.O. Box 35 (Agora)
FI-40014 University of Jyväskylä
Finland
oljumali@cc.jyu.fi

Gérard Meurant
30 rue de Sergent Bauchat
75012 Paris
France
gerard.meurant@gmail.com

Daniela Morale
Department of Mathematics
University of Milan
IT-20133 Milan
Italy
daniela.morale@unimi.it

Pekka Neittaanmäki
Department of Mathematical Information Technology
University of Jyväskylä
P.O. Box 35 (Agora)
FI-40014 University of Jyväskylä
Finland
pn@mit.jyu.fi

David Paniagua
Baylor College of Medicine
Texas Heart Institute at St Luke's Episcopal Hospital
P.O. Box 20345
Houston, TX 77225-0345
USA
dpaniag@pol.net

Jacques F. Périaux
Department of Mathematical Information
University of Jyväskylä
40351, Jyväskylä, Finland
and
Numèrics en Enginyeria (CIMNE)
Centre Internacional de Mètodes
C/Gran Capitan s/n, 08034
Barcelona, Spain
jperiaux@gmail.com

Olivier Pironneau
Université Paris VI
Laboratoire Jacques-Louis Lions
175 rue du Chevaleret
FR-75013 Paris, France
pironneau@ann.jussieu.fr

Jacques Rappaz
Institute of Analysis and Scientific Computing
École Polytechnique Fédérale de Lausanne
CH-1015 Lausanne
Switzerland
jacques.rappaz@epfl.ch

Sergey Repin
V. A. Steklov Institute
of Mathematics in St. Petersburg
Fontanka 27
191023, St. Petersburg
Russia
and
Department of Applied Mathematics
St. Petersburg State Technical University
195251, St. Petersburg
Russia
repin@pdmi.ras.ru

Hiroshi Suito
Graduate School of Environmental Sciences
Okayama University
3-1-1, Tsushima-naka
Okayama, 700-8530
Japan
suito@ems.okayama-u.ac.jp

Josip Tambača
Department of Mathematics
University of Zagreb
Bijenička 30
HR-10000 Zagreb
Croatia
tambaca@math.hr

Jari Toivanen
Institute for Computational
and Mathematical Engineering
Stanford University
Stanford, CA 94305
USA
toivanen@stanford.edu

Janos Turi
Programs in Mathematical Sciences
University of Texas at Dallas
Richardson, TX 75083
USA
turi@utdallas.edu

Qin Zhang
Department of Mathematics
North Carolina State University
Raleigh, NC 27695-8205
USA

Xu Zhang
Academy of Mathematics
and Systems Sciences
Chinese Academy of Sciences
CN-100080 Beijing
China
xuzhang@amss.ac.cn

Chuang Zheng
School of Mathematical Sciences
Beijing Normal University
CN-100875 Beijing, China
chuang.zheng@bnu.edu.cn

Yinggao Zheng
Graduate School of Environmental Sciences
Okayama University
3-1-1, Tsushima-naka
Okayama, 700-8530
Japan
zheng@s.ems.okayama-u.ac.jp

Enrique Zuazua
Basque Center for Applied Mathematics
Bizkaia Technology Park, Building 500
ES-48160 Derio, Basque Country
Spain
zuazua@bcamath.org
http://www.bcamath.org/zuazua/

Roland Glowinski: The Unconventional and Unexpected Path of a Mathematician

William E. Fitzgibbon[1] and Jacques F. Périaux[2]

[1] College of Technology, University of Houston, 4800 Calhoun Road, Houston, TX 77204, USA, fitz@uh.edu

[2] Department of Mathematical Information, University of Jyväskylä, 40351, Jyväskylä, Finland
and
Numèrics en Enginyeria (CIMNE), Centre Internacional de Mètodes, C/Gran Capitan s/n, 08034, Barcelona, Spain, jperiaux@gmail.com

More than 10 years have elapsed since the conference, "Computational Science for the Twentieth Century", was held in Tours, France. The Tours event honored the 60th birthday of Roland Glowinski. The world has witnessed many changes in the last decade, but Roland and his lovely wife, Angela, seem to barely have changed at all. Indeed, they are like fine French wine or Tennessee whiskey; they improve with age. As we reflect upon the career of Roland, it is important that we not underestimate the role of Angela. Everyone knows the old saying,

Beside every great man stands a great woman.

The quote becomes more complete, and perhaps appropriate, if we include Voltaire's addendum,

a surprised mother in law.

Angela Glowinski is the first lady of Franco American mathematics. She serves as Roland's tireless confidante, supporter, cheerleader, and at times, task master. Roland never expected to become a professor and renowned scientist. Only at the urging of his wife did Roland make the decision to return to the academy and enroll at the Institut Blaise Pascal. For that, the applied mathematical community, as well as Roland, owes Angela a profound debt of gratitude. In fact, we think that if Jacques-Louis Lions had not existed, Angela would have found a Lions.

The event celebrating Roland's 70th birthday was a peripatetic one, taking place at the University of Houston in Houston, Texas; University of Jyväskylä in Jyväskylä, Finland; and Xi'an Jiaotong University in Xi'an, China. Roland has friends and collaborators across the globe. Many old, loyal friends and some new were present at the gatherings. However, the joy at these events was dampened by the sad realization that some old friends were missing and will

W. Fitzgibbon et al. (eds.), *Applied and Numerical Partial Differential Equations*, Computational Methods in Applied Sciences 15, DOI 10.1007/978-90-481-3239-3_1,

not return. In particular, we pay tribute to Professor Jacques-Louis Lions. He was an inspiration and mentor to us all. His torch will now have to be carried by Roland and his fellow *lionceaux*.

We will not deign to give a full discussion or even a list of the research achievements, distinctions, and accolades of Roland Glowinski. We will say that over the course of his career Roland has authored or coauthored over 250 scientific articles, has written or edited about twenty books, has served on a panoply of editorial boards, advised numerous students and post doctoral fellows, and has collaborated with scientists across the globe. Among other honors, he has been elected to the French Academy of Science and is a member of the French Legion d'Honneur at the level *chevalier*.

The eminent German poet and author Johann Wolfgang von Goethe once said,

> Mathematicians are like Frenchmen: whatever you say to them they translate into their own language and forthwith, it is something entirely different.

Although Roland is, and always will be, quintessentially French, the corpus of his work serves as a marked counterexample to the sentiments Goethe expressed. If there is a common thread running through the large and broad corpus of his work, it is his four-step approach:

1. Identification of the model
2. Determination of the structure and mathematical properties of the model
3. Development of numerical methods that take advantage of the model's mathematical properties, while at the same time making optimal use of available computing resources
4. Validation and verification of the numerical results

It has always been Roland's concern to construct portable methods that can readily be adapted by other scientists in different contexts. Today, applied and computational mathematics is in vogue. Academic institutions compete to develop it. This was not the case when Roland began to follow his muse. Pure mathematics was the mode and even applied mathematics tended to be highly theoretical. Roland's decision to engage the applied problems of industry, engineering, and science was both unconventional and bold. His work has always been in mathematics and scientific computing and their application to mechanics, physics, aeronautics, engineering sciences and, more recently, biology. He has made seminal contributions in the areas of methods for science computation, fluid mechanics, numerical controls for distributed parameter systems, and solid and structural mechanics, as well as shape optimization, stellar motion, electron transport, and semiconductor modeling. Indeed, Roland's work demonstrates that Goethe should have paid attention to the words of Leonardo di Vinci,

> Mechanics is the paradise of the mathematical sciences because by means of it one comes to the fruits of mathematics.

Roland's scientific journey began in the late fifties with his education at the elite École Polytechnique. Air France awarded Roland a traineeship at the Boeing Company in Seattle, in 1960. This is significant in two ways; one, it gave Roland an introduction to aviation and aeronautics, and it introduced Roland to the United States and kindled his affection for life in the United States. Roland fulfilled his military obligation in the French Army serving in Algeria during the period of trouble. Roland then worked as a telecommunications engineer for ORTF (the French Broadcasting System) from 1963–1968 and became well grounded in electromagnetism, as well as learning FORTRAN.

Roland's decision at Angela's urging to enroll in Professor J.-L. Lions' Post DEA Course in Numerical Analysis at Institut Blaise Pascal proved to have a major impact upon Roland's subsequent career. This course is made notable by the careers that it launched. The list of those who have benefited from it includes: J. Cea, A. Bensoussan, P. A. Raviart, J. C. Nédélec, G. Chavent, L. Tartar and O. Pironneau. Roland grabbed Lions' attention and came under his influence. In 1967, Professor Lions hired Roland at l'Institut de Recherche en Informatique et en Automatique (IRIA). Roland excelled and rapidly became a Scientific Director in 1971, serving until 1985. On the academic side, Roland was elevated to a professorship at the Université de Pierre et Marie Curie. Following the dictum of Lions, Roland, as did other disciples of Lions, maintained close connections with industry and government agencies (be they French or, more recently, American) in the areas of aeronautics, nuclear energy, space exploration and hydrocarbon recovery.

Roland's career path is both unconventional and unexpected. It is unconventional by virtue of his decision to become involved in the applied problems of industry, engineering, and science at a time when pure mathematics was the mode. His applied orientation is well illustrated by his highly acclaimed collaboration with Dassault Aviation as leader of the Glowinski–Bristeau–Pironneau–Perrier–Periaux–Poirier GB4P team. This effort culminated in the finite element simulation by least squares techniques of the 3-D shocked transonic flow around a complete Falcon 50 business jet geometry.

In the 1980s, Roland made a series of major contributions in the domain decomposition and fictitious domain methods. This work was initially motivated by large scale industrial applications in aeronautics and the oil industry, and extended recently to applied electromagnetics the identification of the signature of coated materials on aircraft, ships, submarines or mobile phones. The latter work applied exact controllability methods derived from the Hilbert Uniqueness Method of J.-L. Lions. Roland's important contributions to the numerical solution using Lagrange multiplier methods are documented in his paper, Augmented Lagrangians and Operator Splitting, which he coauthored with P. Le Tallec. He subsequently, in collaboration with D. Joseph and T. W. Pan, extended this to the theoretical description of the fluidization and the sedimentation of particular flows.

If there is a feature in Roland's background that distinguishes him from most of his contemporaries in applied mathematics, it is probably the fact that he began by obtaining experience as an engineer in aeronautics at Boeing telecommunications ORTF and then undertaking the study applied mathematics and numerical analysis with J.-L. Lions and other distinguished mathematicians. All of this background was coupled with a native intelligence, an open mind, and relentless curiosity. Roland's affable personality, constant good mood, patience, and willingness to listen, together with his innate ability to interact and collaborate with people across a wide spectrum of scientific and engineering disciplines, have enabled his putting together an impressive international network of friends and colleagues.

In the administrative area, Roland served as Director at CERFACS, Toulouse, from 1992 to 1994, a unique experience with EC and Aerospatiale. Although Roland first came to the United States on an internship with Boeing in the early 1960s, the American portion of his career began in 1985, with his assumption of the M. D. Anderson Professorship of Mathematics at the University of Houston. Roland subsequently became the Hugh and Roy Cullen Professor of Mathematics and Yuri Kuznetsov assumed the M. D. Anderson Chair. Roland's presence at the University of Houston has had significant impact on the development of applied mathematics in Houston and in Texas. Under Roland's leadership and guidance, we have developed into a major node on the international applied and computational network. Many well known scientists joined our faculty – to name a few: Mary Wheeler, Yuri Kuznetsov, Tsorng-Whay Pan, Ed Dean, Jiwen He, Ronald Hoppe, Jeffery Morgan, Robert Azencott and Sunčica Čanić. We will not even attempt to list the visitors who have streamed through Houston. It will suffice to say that one can expect to hear French, Russian, German, Chinese, Spanish, and Croatian, as well as the Texas drawl, along the corridors of the mathematics department. It is fair to say that Roland was a bellwether for the State of Texas. Subsequent to his arrival, both the University of Texas and Texas A&M University have emerged as major centers of computational mathematics. Texas can now be known for computational science, as well as horses, cattle, oil and barbecue.

We find ourselves on the threshold of a new era with interesting and challenging problems concerning the areas of medicine, life science, the environment, energy, information technology, communications, and materials science. Now more than ever, we will need scientists like Roland with innovation, vision, and ability to work across both disciplinary and national boundaries.

We conclude with the last stanza of a poem dedicated to Roland by Professor Zhong-Ci Shi of the Institute of Computational Mathematics in Beijing:

> You earned your success and you should feel very confident with yourself for all that you have achieved.

The Scientific Career of Roland Glowinski

Olivier Pironneau

Université Paris VI, Laboratoire Jacques-Louis Lions, 175 rue du Chevaleret, FR-75013 Paris, France, pironneau@ann.jussieu.fr

Roland Glowinski is a former student of one of the best school for mathematics and engineering in France, the Ecole Polytechnique.

After a first employment at the French television company ORTF, he decided to go back to the university for a thesis (thèse d'Etat) and received his degree and a position of professor at the University of Paris VI in 1970. Already a fervent admirer and colleague of Professor Jacques-Louis Lions, his former advisor, he succeeded him as head of the numerical analysis group at IRIA (now INRIA) in 1976.

Soon he became the best known French algorithm designer for solid and fluid mechanics, a talent which will give him many awards and nominations as scientific advisor in hi-tech companies, a temporary teaching position at Ecole Polytechnique and the worldwide reputation of the best scientific advisor for partial differential equations in industry. Nevertheless, as if life was too easy, Roland decided to move to University of Houston, Texas, in the 1980s. At the request of J.-L. Lions he came back for a couple of years to France to lead the CERFACS, at the time of writing the best French lab in Toulouse for scientific computing. Since then he is a full time Professor at University of Houston and Honorary Professor at the University of Jyväskylä.

Roland Glowinski is the author of 7 books and more than 300 articles. His main contributions are in many fields of applied mathematics, simulations and scientific computing; we may order them in eight groups:

1. *Domain decomposition methods.* He is the first to have understood the links between Schwarz algorithms and Lagrange multiplier methods; one of the first domain decomposition method without overlap is his. He is also among the first to have proposed the framework of mixed methods for domain decompositions. Finally, he is a co-founder of the famous DDM conference series. He received the Cray prize for his achievements in this important field of parallel computing.
2. *Fictitious domain methods.* With equal success he applied the framework of Lagrange multipliers to the fictitious domain embedding methods, thereby

W. Fitzgibbon et al. (eds.), *Applied and Numerical Partial Differential Equations*, Computational Methods in Applied Sciences 15, DOI 10.1007/978-90-481-3239-3_2,

establishing convergence of old Russian algorithms and new methods of his. He gave the appropriate functional framework for the variational numerical methods with important applications to time dependant domains. In an article with V. Girault a fundamental error estimates was established which is one of the most cited result of the field also because it contains a compatibility condition which says that the discretization of the physical domain must be coarser than the background mesh.

3. *Robust preconditioners for the Navier–Stokes equations and other nonlinear partial differential equations.* In his book (Springer-Verlag) on nonlinear problems Roland Glowinski proposed several iterative algorithms (conjugate gradient and augmented lagrangian methods) with optimal preconditioners especially for the Stokes equations thereby opening the way to modern scientific computing, a method which everybody use nowadays. He received the Prix Marcel Dassault of the French academy of sciences for his work in this field.
4. *Several iterative algorithms for visco-elastic problems.* With J.-L. Lions a family of methods based on augmented Lagrangian formulations and other penalties were given to solve the variational inequalities of physics, an approach which is still, when possible, the most stable way to find free boundaries, thereby avoiding remeshing of the moving domains.
5. *Algorithms for the biharmonic problem.* R. Glowinski proposed a formulation of the problem which could be discretized with low degree finite elements and which leads to the fastest numerical method to the point that at Dassault Aviation their first Navier–Stokes solver was based on this formulation. It was also the first in a series of scientific "coups" in the fruitful cooperation between Roland, Jacques Périaux and Pierre Perrier at Dassault Aviation.
6. *An algorithm for the transonic equation.* In this industrial cooperation a variational formulation of the transonic equation was tested with an entropy condition based on the potential of the flow, which again brought Dassault Aviation to the front line of scientific computing with the first complete numerical aircraft at transonic speed.
7. *A numerical implementation compatible with the controllability conditions of hyperbolic problems, the famous H.U.M. of J.-L. Lions.* With several collaborators at INRIA and at Dassault Aviation the method was proven to be very efficient for solving the Maxwell equations of electromagnetism in the physical variables yet seeking for periodic solutions and hence avoiding frequency domain reformulations.
8. *A mixed formulation for fluid–structure interactions.* In cooperation with D. Joseph at University of Minnesota (Minneapolis) for pipelines, R. Glowinski and his team at University of Houston solved the very difficult challenge of simulating the fluidized bed problem. This is a 3D flow with thousands of solid balls moving with the flow. This was the problem that lead Roland to use the fictitious domain method, although he

tried also body fitted meshes. The originality of his approach is to have embedded the Newton laws for the balls into the variational formulation of the problem.

Roland Glowinski is world famous for his contributions to parallel computing especially for problems with unstructured meshes, with the finite element method, linear or non-linear, with or without free boundaries and with multi physics. His collected work in book form would amount to perhaps 12 volumes, fairly easy to read and yet to the point, with the right dose of mathematics for beautiful theories pertinent to the applications which Roland never loose sight of what he calls "applied mathematics of good taste".

It would be tedious to list all the prizes and honors he received but let us cite two: the French Academy of Sciences and the von Karman Lecture at the 2004 SIAM meeting.

Books by Roland Glowinski

1. R. Glowinski, J.-L. Lions, and J. He. *Exact and approximate controllability for distributed parameter systems. A numerical approach*, volume 117 of *Encyclopedia of Mathematics and its Applications*. Cambridge University Press, Cambridge, 2008.
2. R. Glowinski. Finite element methods for incompressible viscous flow. In P. G. Ciarlet and J.-L. Lions, editors, *Handbook of Numerical Analysis, Vol. IX*, pages 3–1176. North-Holland, Amsterdam, 2003.
3. R. Glowinski and P. Le Tallec. *Augmented Lagrangian and operator splitting methods in nonlinear mechanics.* SIAM, Philadelphia, PA, 1989.
4. M. Blanc, D. Fontaine, R. Glowinski, and L. Reinhart. *Simulation of electron transport in the earth magneto sphere.* Gordon Breach, 1987.
5. R. Glowinski. *Numerical methods for nonlinear variational problems.* Springer, New York, 1984.
6. M. Fortin and R. Glowinski. *Méthodes de Lagrangien augmenté.* Gauthier-Villars, Paris, 1982. (publié en anglais par North-Holland en 1983).
7. R. Glowinski, J.-L. Lions, and R. Trémolières. *Analyse numérique des inéquations variationnelles. Tome 1. Théorie générale premiéres applications. Tome 2. Applications aux phénomènes stationnaires et d'évolution.* Dunod, Paris, 1976. (et en anglais par North-Holland, 1981 avec 200 pages de plus).

On a Class of Partial Differential Equations with Nonlocal Dirichlet Boundary Conditions

Alain Bensoussan[1] and Janos Turi[2]

[1] International Center for Decision and Risk Analysis, ICDRiA, School of Management, University of Texas at Dallas, Richardson, TX 75083, USA, Alain.Bensoussan@utdallas.edu

[2] Programs in Mathematical Sciences, University of Texas at Dallas, Richardson, TX 75083, USA, turi@utdallas.edu

1 Introduction

We establish existence and uniqueness of solutions of a class of partial differential equations with nonlocal Dirchlet conditions in weighted function spaces. The problem is motivated by the study of the probability distribution of the response of an elasto-plastic oscillator when subjected to white noise excitation (see [1,2] on the derivation of the boundary value problem). Note that the developments in [1,2] are based on an extension of Khasminskii's method (see, e.g. [5]) and in this paper we use a direct approach to achieve our objectives.

We refer the reader to [3, 4, 6, 7] for general background on modeling, theoretical, and computational issues related to elasto-plastic oscillators.

2 Setting of the Problem

2.1 Notation

We set $D = R \times (-Y, Y)$. A point in D is denoted by (y, z). We define the operators

$$A\zeta(y,z) = -\frac{1}{2}\frac{\partial^2\zeta}{\partial y^2} + \frac{\partial\zeta}{\partial y}(c_0 y + kz) - y\frac{\partial\zeta}{\partial z}, \tag{1}$$

$$\begin{aligned} B_+\psi(y) &= -\frac{1}{2}\frac{\partial^2\psi}{\partial y^2} + \frac{\partial\psi}{\partial y}(c_0 y + kY), \\ B_-\psi(y) &= -\frac{1}{2}\frac{\partial^2\psi}{\partial y^2} + \frac{\partial\psi}{\partial y}(c_0 y - kY). \end{aligned} \tag{2}$$

W. Fitzgibbon et al. (eds.), *Applied and Numerical Partial Differential Equations*, Computational Methods in Applied Sciences 15, DOI 10.1007/978-90-481-3239-3_3,

Let $g(y,z)$ be bounded. We define

$$g_+(y) = g(y,Y) \quad \text{and} \quad g_-(y) = g(y,-Y). \tag{3}$$

We also define the weightfunction

$$\rho_m(y) = \frac{1}{(1+y^2)^m}, \quad m \geq 1. \tag{4}$$

2.2 The Problem

Let $\lambda > 0$, we look for a function

$$\begin{aligned} u \in L^\infty(D), \quad \frac{\partial u}{\partial y} &\in L^2_m(D), \quad m \geq 1, \\ y^2\frac{\partial u}{\partial z}, y\frac{\partial^2 u}{\partial y^2} &\in L^2_m(D), \quad m \geq 3, \\ \int_R |y|\rho_m u^2(y,\pm Y)dy &< \infty, \quad m \geq 1, \\ \int_R |y|^3\rho_m \left(\frac{\partial u}{\partial y}\right)^2 (y,\pm Y)dy &< \infty, \quad m \geq 3, \end{aligned} \tag{5}$$

satisfying

$$\lambda u + Au = g \qquad \text{in } D, \tag{6}$$

$$\lambda u + B_+u = g_+, \quad z = Y, \ y > 0, \tag{7}$$

$$\lambda u + B_-u = g_-, \quad z = -Y, \ y < 0. \tag{8}$$

Note that on a subdomain,

$$D_\rho = \{(\rho,\infty) \times (-Y,Y) \cup (-\infty,-\rho) \times (-Y,Y)\}, \quad \rho > 0,$$

the function u is continuous up to the boundary, the derivatives $\frac{\partial u}{\partial z}$, $\frac{\partial u}{\partial y}$, and $\frac{\partial^2 u}{\partial z^2}$ are L^2_{loc} and the equations (6), (7), and (8) are satisfied in the strong sense.

3 The Main Result

Theorem 1. *Assuming that g is bounded and $\lambda > 0$, there exists one and only one solution of (5)–(8).*

3.1 Boundary Conditions

The boundary conditions can be replaced by nonlocal Dirichlet conditions

$$u(y,Y) = u_+\chi_+(y) + \beta_+(y), \quad y > 0, \tag{9}$$
$$u(y,-Y) = u_-\chi_-(y) + \beta_-(y), \quad y < 0, \tag{10}$$

where u_+ and u_- are constants. The functions β_+, χ_+, β_-, and χ_- are defined by

$$\lambda\beta_+ + B_+\beta_+ = g_+, \quad \beta_+(0) = 0,\ \beta_+ \in H^1_m(0,\infty), \tag{11}$$

where

$$H^1_m(0,\infty) = \left\{\psi(y) \mid \int_0^\infty \rho_m(y)\psi^2(y)dy < \infty, \int_0^\infty \rho_m(y)\left(\frac{d\psi}{dy}\right)^2(y)dy < \infty\right\},$$

similarly

$$\lambda\beta_- + B_-\beta_- = g_-, \quad \beta_-(0) = 0,\ \beta_- \in H^1_m(-\infty,0), \tag{12}$$

where

$$H^1_m(-\infty,0) = \left\{\psi(y) \mid \int_{-\infty}^0 \rho_m(y)\psi^2(y)dy < \infty, \int_{-\infty}^0 \rho_m(y)\left(\frac{d\psi}{dy}\right)^2(y)dy < \infty\right\},$$

furthermore,

$$\lambda\chi_+ + B_+\chi_+ = 0, \quad y > 0,\ \chi_+(0) = 1,\ \chi_+ \in H^1_m(0,\infty), \tag{13}$$

and

$$\lambda\chi_- + B_-\chi_- = 0, \quad y < 0,\ \chi_-(0) = 1,\ \chi_- \in H^1_m(-\infty,0). \tag{14}$$

3.2 A Priori Estimates

Consider a solution u of (5)–(8). We test (6) with $u\rho_m$. We get easily

$$\begin{aligned}
\lambda\int_D u^2\rho_m\,dy\,dz + \frac{1}{2}\int_D\left(\frac{\partial u}{\partial y}\right)^2\rho_m\,dy\,dz + \int_D \rho_m\frac{\partial u}{\partial y}u(c_0y+kz)\,dy\,dz \\
- m\int_D \frac{\partial u}{\partial y}u\frac{y}{1+y^2}\rho_m\,dy\,dz - \frac{1}{2}\int_{-\infty}^0 y\rho_m u^2(y,Y)\,dy \\
+ \frac{1}{2}\int_0^\infty y\rho_m u^2(y,-Y)dy = \int_D g\rho_m u\,dy\,dz \\
+ \frac{1}{2}\int_0^\infty y\rho_m(u_+\chi_+ + \beta_+)^2dy - \frac{1}{2}\int_{-\infty}^0 y\rho_m(u_-\chi_- + \beta_-)^2dy.
\end{aligned} \tag{15}$$

We next test (6) with $\frac{\partial u}{\partial z} y^3 \rho_m$. We obtain

$$\lambda \int_D y^4 \rho_m \left(\frac{\partial u}{\partial z}\right)^2 dy\, dz - \frac{\lambda}{2}\int_{-\infty}^0 y^3 \rho_m u^2(y,Y)\, dy + \frac{\lambda}{2}\int_0^{\infty} y^3 \rho_m u^2(y,-Y)\, dy$$
$$- \frac{1}{4}\int_{-\infty}^0 y^3 \rho_m \left(\frac{\partial u}{\partial y}\right)^2 (y,Y)\, dy + \frac{1}{4}\int_0^{\infty} y^3 \rho_m \left(\frac{\partial u}{\partial y}\right)^2 (y,-Y)\, dy$$
$$= \int_D \frac{\partial u}{\partial y}\frac{\partial u}{\partial z} y^2 \rho_m \left[c_0 y^2 + kzy + \frac{3}{2} - \frac{my^2}{1+y^2}\right] dy\, dz - \int_D g y^3 \rho_m \frac{\partial u}{\partial z}\, dy\, dz$$
$$+ \frac{\lambda}{2}\int_0^{\infty} y^3 \rho_m (u_+\chi_+ + \beta_+)^2 dy - \frac{\lambda}{2}\int_{-\infty}^0 y^3 \rho_m (u_-\chi_- + \beta_-)^2 dy$$
$$+ \frac{1}{4}\int_0^{\infty} y^3 \rho_m \left(u_+\frac{d\chi_+}{dy} + \frac{d\beta_+}{dy}\right)^2 dy - \frac{1}{4}\int_{-\infty}^0 y^3 \rho_m \left(u_-\frac{d\chi_-}{dy} + \frac{d\beta_-}{dy}\right)^2 dy. \tag{16}$$

Since the right-hand side of (15) is bounded, a simple application of Hölder's inequality in (16) allows to obtain bounds on the norms of the functions listed in (5), except for the L^∞ norm of u, which does not follow from the energy equalities (15) and (16).

3.3 Further Regularity

Proposition 1. *Assume that* λ *is sufficiently large and* $\frac{\partial g}{\partial z} \in L^2_m(D)$, *for* $m \geq 1$. *Then*

$$\begin{aligned}
&\frac{\partial u}{\partial z} \in L^2_m(D), \quad \frac{\partial^2 u}{\partial z \partial y} \in L^2_m(d), && \textit{for } m \geq 1,\\
&y\frac{\partial^3 u}{\partial z \partial y^2} \in L^2_m(D), \quad y^2\frac{\partial^2 u}{\partial z^2} \in L^2_m(D), && \textit{for } m \geq 3\\
&\int_R |y|\rho_m \left(\frac{\partial u}{\partial z}\right)^2 (y,\pm Y) dy < \infty, && \textit{for } m \geq 1,\\
&\int_R |y|^3\rho_m \left(\frac{\partial^2 u}{\partial y \partial z}\right)^2 (y,\pm Y) dy < \infty, && \textit{for } m \geq 3.
\end{aligned}$$

Also,

$$\frac{\partial^2 u}{\partial y^2} \in L^2_m(D), \quad \textit{for } m \geq 2.$$

Proof. We find the problem for $v = \frac{\partial u}{\partial z}$ by differentiating (6). We get

$$\begin{aligned}
&\lambda v + Av = \frac{\partial g}{\partial z} - k\frac{\partial u}{\partial y} \quad \text{in } D,\\
&v = 0, \quad \text{for } z = Y,\ y > 0,\\
&v = 0, \quad \text{for } z = -Y,\ y < 0.
\end{aligned} \tag{17}$$

We can obtain the analogues of (15) and (16), namely

$$\lambda \int_D v^2 \rho_m \, dy \, dz + \frac{1}{2} \int_D \left(\frac{\partial v}{\partial y} \right)^2 \rho_m \, dy \, dz + \int_D \rho_m \frac{\partial v}{\partial y} v (c_0 y + kz) \, dy \, dz$$
$$- m \int_D \frac{\partial v}{\partial y} v \frac{y}{1+y^2} \rho_m \, dy \, dz - \frac{1}{2} \int_{-\infty}^{0} y \rho_m v^2(y,Y) \, dy + \frac{1}{2} \int_0^{\infty} y \rho_m v^2(y,-Y) \, dy$$
$$= \int_D \left(\frac{\partial g}{\partial z} - k \frac{\partial u}{\partial y} \right) \rho_m v \, dy \, dz, \tag{18}$$

and

$$\lambda \int_D y^4 \rho_m \left(\frac{\partial v}{\partial z} \right)^2 dy \, dz - \frac{\lambda}{2} \int_{-\infty}^{0} y^3 \rho_m v^2(y,Y) \, dy + \frac{\lambda}{2} \int_0^{\infty} y^3 \rho_m v^2(y,-Y) \, dy$$
$$- \frac{1}{4} \int_{-\infty}^{0} y^3 \rho_m \left(\frac{\partial v}{\partial y} \right)^2 (y,Y) \, dy + \frac{1}{4} \int_0^{\infty} y^3 \rho_m \left(\frac{\partial v}{\partial y} \right)^2 (y,-Y) \, dy$$
$$= \int_D \frac{\partial v}{\partial y} \frac{\partial v}{\partial z} y^2 \rho_m \left[c_0 y^2 + kzy + \frac{3}{2} - \frac{my^2}{1+y^2} \right] dy \, dz$$
$$- \int_D \left(\frac{\partial g}{\partial z} - k \frac{\partial u}{\partial y} \right) y^3 \rho_m \frac{\partial v}{\partial z} \, dy \, dz. \tag{19}$$

Since λ can be taken sufficiently large, these relations prove the properties stated. Note that the last one follows from (6) itself and already proven properties. □

Since

$$\int_R |y| \rho_m \left(\frac{\partial u}{\partial z} \right)^2 (y, \pm Y) dy < \infty, \quad \text{for } m \geq 1,$$

the function $u(y, \pm Y)$ satisfies the differential equation

$$\lambda u - \frac{1}{2} \frac{\partial^2 u}{\partial y^2} + \frac{\partial u}{\partial y} (c_0 y \pm kY) = y \frac{\partial u}{\partial z} (y, \pm Y) + g(y, \pm Y) \tag{20}$$

and we can consider the right-hand side as a given function in $L^2_m(R)$, $m \geq 2$. Note that

$$\frac{\partial u}{\partial z}(y,Y) = 0, \quad \text{if } y > 0 \quad \text{and} \quad \frac{\partial u}{\partial z}(y,-Y) = 0, \quad \text{if } y < 0.$$

From this relation, using also the fact that u is bounded, we deduce easily

$$\frac{\partial^2 u}{\partial y^2}(y, \pm Y) \in L^2_m(R), \quad \text{for } m \geq 2. \tag{21}$$

Hence, in particular,

$$u(y, \pm Y) \text{ is } C^1(r). \tag{22}$$

4 Proof of Theorem 1

4.1 Proof of Uniqueness

Proof. We will prove that if λ is sufficiently large, then the solution of (5)–(8) is unique. Moreover, we will prove that for $\lambda > 0$, there exists a solution of (5)–(8) such that

$$\|u\|_{L^\infty} \leq \frac{\|g\|_{L^\infty}}{\lambda}.$$

So, for λ sufficiently large there exists one and only one solution of (5)–(8) and it satisfies (4.1).

But we may then consider the map T_α defined on $L^\infty(D)$ by

$$u = T_\alpha w,$$

where

$$(\lambda+\alpha)u + Au = g + \alpha w \qquad \text{in } D, \tag{23}$$
$$(\lambda+\alpha)u + B_+u = g_+ + \alpha w_+, \quad z = Y,\ y > 0, \tag{24}$$
$$(\lambda+\alpha)u + B_-u = g_- + \alpha w_-, \quad z = -Y,\ y < 0. \tag{25}$$

A solution of (5)–(8) is a fixed point of T_α.

Next we show that T_α is a contraction on L^∞. Indeed, if $w_1, w_2 \in L^\infty(D)$ and $u_1 = T_\alpha w_1$, $u_2 = T_\alpha w_2$, then $u_1 - u_2$ is the solution of

$$(\lambda+\alpha)(u_1-u_2) + A(u_1-u_2) = \alpha(w_1-w_2) \qquad \text{in } D, \tag{26}$$
$$(\lambda+\alpha)(u_1-u_2) + B_+(u_1-u_2) = \alpha(w_1-w_2)_+, \quad z = Y,\ y > 0, \tag{27}$$
$$(\lambda+\alpha)(u_1-u_2) + B_-(u_1-u_2) = \alpha(w_1-w_2)_-, \quad z = -Y,\ y < 0. \tag{28}$$

Since $\lambda + \alpha$ is large, $u_1 - u_2$ is uniquely defined and

$$\|u_1 - u_2\|_{L^\infty} \leq \frac{\alpha\|w_1 - w_2\|_{L^\infty}}{\alpha + \lambda}.$$

It follows that T_α has a unique fixed point, i.e. the solution of (5)–(8).

We shall assume that λ is sufficiently large to prove uniqueness. We must prove that if u satisfies (5), and (6)–(8) with $g = 0$, then $u = 0$. Since $g = 0$, the regularity result of Proposition 1 applies and we may assume that $u \in C^1(D)$. If we exclude a strip $(-\rho, \rho) \times (-Y, Y)$, $\rho > 0$, then $u \in C^2$ and we may apply strong maximum principle considerations.

Define the function

$$\chi(x) = \begin{cases} \log|x| + 1, & \text{if } |x| > 1 \\ \exp\left(-\frac{3}{8}x^4 + \frac{5}{4}x^2 - \frac{7}{8}\right), & \text{if } |x| \leq 1. \end{cases}$$

Then $\chi \in C^2$, $\chi(\pm 1) = 1$, $\chi'(\pm 1) = \pm 1$, $\chi''(\pm 1) = -1$, and $\chi(x) \geq \chi(0) = e^{-\frac{7}{8}}$.

We set

$$\Psi(y) = \chi\left(\frac{y}{\bar{y}}\right),$$

where $\bar{y}$ is a constant, and let

$$w(y,z) = \frac{u(y,z)}{\Psi(y)}.$$

Then w satisfies

$$\begin{aligned}
Aw + \lambda w - \frac{\Psi'}{\Psi}\frac{\partial w}{\partial y} + \frac{w}{\Psi}\left(-\frac{1}{2}\Psi'' + \Psi'(c_0 y + kz)\right) &= 0, \quad -Y<z<Y,\ y\in\mathbb{R},\\
B_+ w + \lambda w - \frac{\Psi'}{\Psi}\frac{\partial w}{\partial y} + \frac{w}{\Psi}\left(-\frac{1}{2}\Psi'' + \Psi'(c_0 y + kY)\right) &= 0, \quad y>0,\ z=Y,\\
B_- w + \lambda w - \frac{\Psi'}{\Psi}\frac{\partial w}{\partial y} + \frac{w}{\Psi}\left(-\frac{1}{2}\Psi'' + \Psi'(c_0 y - kY)\right) &= 0, \quad y<0,\ z=-Y.
\end{aligned} \tag{29}$$

Using the definition of Ψ, we obtain

$$\Psi' = \frac{1}{\bar{y}}\chi'\left(\frac{y}{\bar{y}}\right) = \begin{cases} \dfrac{1}{y}, & |y| > \bar{y},\\ \dfrac{\Psi(y)}{2\bar{y}^2}\, y\left(-3\dfrac{y^2}{\bar{y}^2} + 5\right), & |y| < \bar{y}, \end{cases}$$

and

$$\Psi'' = \frac{1}{\bar{y}^2}\chi''\left(\frac{y}{\bar{y}}\right) = \begin{cases} -\dfrac{1}{y^2}, & |y| > \bar{y},\\ \dfrac{\Psi(y)}{2\bar{y}^2}\left[\dfrac{9}{2}\left(\dfrac{y}{\bar{y}}\right)^6 - 15\left(\dfrac{y}{\bar{y}}\right)^4 + \dfrac{7}{2}\left(\dfrac{y}{\bar{y}}\right)^2 + 5\right], & |y| < \bar{y}. \end{cases}$$

Combining these relations, we get

$$\Psi'(c_0 y + kz) - \frac{1}{2}\Psi'' = \begin{cases} \dfrac{1}{y}(c_0 y + kz) + \dfrac{1}{2y^2}, & |y| > \bar{y},\\ \dfrac{\Psi(y)}{2\bar{y}^2}\left[-\dfrac{9}{4}\left(\dfrac{y}{\bar{y}}\right)^6 - \dfrac{y^4}{\bar{y}^2}\left(3c_0 - \dfrac{15}{2\bar{y}^2}\right) - 3kz\dfrac{y^3}{\bar{y}^2}\right.\\ \left. -y^2\left(-5c_0 + \dfrac{7}{4\bar{y}^2}\right) + 5kzy - \dfrac{5}{2}\right], & |y| < \bar{y}. \end{cases}$$

Choosing $\bar{y} > \frac{kY}{c_0}$, we have

$$\frac{1}{\Psi}\left(\Psi'(c_0 y + kz) - \frac{1}{2}\Psi''\right) > 0, \quad \text{for } |y| > \bar{y}.$$

In particular,

$$\frac{1}{\Psi}\left(\Psi'(c_0 y + kz) - \frac{1}{2}\Psi''\right) > -\frac{1}{\bar{y}}\left[\frac{13}{4\bar{y}} + 4kY\right]$$

and we may choose $\bar{y}$ sufficiently large so that

$$\lambda - \frac{1}{\bar{y}}\left[\frac{13}{4\bar{y}} + 4kY\right] > 0.$$

Note that $w(y,z) \to 0$ as $|y| \to \infty$ uniformly in z.

If it has a positive maximum, it must be attained at finite distance, say at (y^*, z^*). We cannot have $-Y < z^* < Y$ since the coefficient of $w(y^*, z^*)$ is $\lambda + \frac{1}{\Psi}(\Psi'(c_0 y + kz) - \frac{1}{2}\Psi'') > 0$ and $\frac{\partial w}{\partial y} = \frac{\partial w}{\partial z} = 0$ at (y^*, z^*) while $\frac{\partial^2 w}{\partial y^2}(y^*, z^*) < 0$.

Suppose $z^* = -Y$, $y^* < 0$, then the same conclusion follows from the boundary condition. If $z^* = -Y$, $y^* > 0$, then $\frac{\partial w}{\partial z}(y^*, z^*) < 0$ and $y^* \frac{\partial w}{\partial z}(y^*, z^*) < 0$ and the same conclusion follows from the inner equation.

If $y^* = 0$, $z^* = -Y$, then we have $\frac{\partial u}{\partial y}(0, -Y) = 0$ and hence

$$\lambda u(0, -Y) - \frac{1}{2}\frac{\partial^2 U}{\partial y^2}(0, -Y) = 0.$$

Therefore,

$$\frac{\partial^2 u}{\partial y^2}(0, -Y) > 0,$$

which is a contradiction with the fact that $u(y, -Y)$ attains its maximum at 0.

By a symmetric reasoning, we cannot have a positive maximum with $z^* = Y$. So we cannot have a positive maximum. A similar argument shows that we cannot have a negative minimum. Hence $w = 0$, which implies $u = 0$. □

4.2 Approximation

We study (6)–(8) by a *regularization method* as follows. Define

$$A^\varepsilon = A - \frac{\varepsilon}{2}\frac{\partial^2}{\partial z^2}.$$

We approximate (6)–(8) by

$$\lambda u^\varepsilon + A^\varepsilon u^\varepsilon = g \quad \text{in } D, \tag{30}$$

$$\lambda u^\varepsilon + B_+ u^\varepsilon = g_+, \quad y > 0,\ z = Y, \qquad \frac{\partial u^\varepsilon}{\partial z} = 0, \quad y < 0,\ z = Y, \tag{31}$$

$$\lambda u^\varepsilon + B_- u^\varepsilon = g_-, \quad y < 0,\ z = -Y, \qquad \frac{\partial u^\varepsilon}{\partial z} = 0, \quad y > 0,\ z = -Y. \tag{32}$$

Consider the spaces

$$L^2_m(R) = \left\{ \Psi(y) \mid \int_{-\infty}^{\infty} \rho_m(y)\Psi^2(y)dy < \infty \right\},$$

$$H^1_m(R) = \left\{ \Psi \in L^2_m(R),\ \frac{d\Psi}{dy} \in L^2_m(R) \right\},$$

$$L^2_m(D) = \left\{ \Psi(y,z) \mid \int_D \rho_m(y)\Psi^2(y,z)dydz < \infty \right\},$$

and

$$H^1_m(D) = \left\{ \Psi \in L^2_m(D),\ \frac{\partial\Psi}{\partial y} \in L^2_m(D),\ \frac{\partial\Psi}{\partial z} \in L^2_m(D) \right\}.$$

If $\Psi \in H^1_m(D)$, then $\Psi(y, \pm Y) \in H^{\frac{1}{2}}_m(r)$. We will need a slight modification of $H^1_m(D)$, defined as follows:

$$\tilde{H}^1_m(D) = \left\{ \Psi \mid \int_D \rho_m(y^2+1)\Psi^2 dydz + \int_D \rho_m \left(\frac{\partial\Psi}{\partial y}\right)^2 dydz \right.$$
$$\left. + \int_D \rho_m \left(\frac{\partial\Psi}{\partial z}\right)^2 dydz < \infty \right\}. \quad (33)$$

Note that $H^1_m(D) \cap L^\infty(D) \subset \tilde{H}^1_m(D)$, if $m \geq 2$.

Introduce the set

$$K = \left\{ \Psi \in \tilde{H}^1_m(D) \mid \Psi(y,Y) = \Psi_+\chi_+(y) + \beta_+(y),\ y \geq 0, \right.$$
$$\left. \Psi(y,-Y) = \Psi_-\chi_-(y) + \beta_-(y),\ y \leq 0,\ \|\Psi_\pm\| \leq \frac{\|g\|}{\lambda} \right\}. \quad (34)$$

Lemma 1. *The set K is a convex closed not empty subset of $\tilde{H}^1_m(D)$.*

Proof. The fact that K is convex and closed is clear. To show that it is not empty, we pick

$$\Psi(y,z) = z\frac{\beta_+(y)1\!\!1_{y>0} - \beta_-(y)1\!\!1_{y<0}}{2Y} + \frac{\beta_+(y)1\!\!1_{y>0} - \beta_-(y)1\!\!1_{y<0}}{2}$$

which belongs to K. □

We will give a variational formulation of (30)–(32). We define for $\xi, \eta \in \tilde{H}^1_m(D)$ the bilinear form

$$\begin{aligned} a_{m,\varepsilon}(\xi,\eta) = \frac{\varepsilon}{2}\int_D \rho_m \frac{\partial \xi}{\partial z}\frac{\partial \eta}{\partial z} dydz + \frac{1}{2}\int_D \rho_m \frac{\partial \xi}{\partial y}\frac{\partial \eta}{\partial y} dydz \\ + \lambda \int_D \rho_m \xi \eta dydz - m \int_D \rho_m \frac{\partial \xi}{\partial y} \eta \frac{y}{1+y^2} dydz \\ + \int_D \rho_m \frac{\partial \xi}{\partial y} \eta (c_0 y + kz) dydz - \int_D \rho_m \frac{\partial \xi}{\partial z} \eta y dydz. \end{aligned} \tag{35}$$

To simplify the notation, we drop, when necessary, the indices m, ε. The bilinear form is continuous on $\tilde{H}^1_m(D)$.

If we consider the modified form

$$a(\xi,\eta) + \alpha \int_D (y^2+1)\rho_m \xi \eta dydz, \tag{36}$$

then for α sufficiently large, depending on ε, the modified form is coercive.

If $f \in L^\infty(D)$, then there exists one and only one solution of the variational inequality

$$\begin{aligned} a(\xi, \xi - \eta) + \alpha \int_D \rho_m (y^2+1)\xi(\xi-\eta) dydz \\ \geq \int_D f \rho_m (\xi - \eta) dydz, \quad \forall \eta \in K,\ \xi \in K. \end{aligned} \tag{37}$$

We proceed by defining a map $u = T_\alpha v$. If $v \in L^\infty(D)$, u is the solution of the variational inequality

$$\begin{aligned} a(u, \eta - u) + \alpha \int_D \rho_m (y^2+1) u(\eta - u) dydz \\ \geq \int_D \rho_m (g + \alpha(y^2+1)v)(\eta - u) dydz, \quad \forall \eta \in K,\ u \in K. \end{aligned} \tag{38}$$

Lemma 2. *If* $\|v\|_{L^\infty} \leq \frac{\|g\|_{L^\infty}}{\lambda}$, *then* $\|u\|_{L^\infty} \leq \frac{\|g\|_{L^\infty}}{\lambda}$.

Proof. Let $\gamma = \frac{\|g\|_{L^\infty}}{\lambda}$. We first notice that if $\Psi \in K$, then $|\Psi(y, \pm Y)| \leq \gamma$. Therefore, $(u-\gamma)^+(y, \pm Y) = 0$. We can take $\eta = u - (u-\gamma)^+$. We obtain

$$\begin{aligned} -a(u, (u-\gamma)^+) - \alpha \int_D \rho_m (y^2+1) u (u-\gamma)^+ dydz \\ \geq - \int_D \rho_m (g + \alpha(y^2+1)v)(u-\gamma)^+ dydz; \end{aligned}$$

hence

$$\begin{aligned} a(u, (u-\gamma)^+) + \alpha \int_D \rho_m (y^2+1) u (u-\gamma)^+ dydz \\ \leq \int_D \rho_m (g + \alpha(y^2+1)v)(u-\gamma)^+ dudz. \end{aligned}$$

It follows from the definition of the bilinear form that

$$a(u-\gamma,(u-\gamma)^+)+\gamma\lambda\int_D \rho_m(u-\gamma)^+dydz$$
$$+\alpha\int_D \rho_m(y^2+1)(u-\gamma)(u-\gamma)^+dydz+\alpha\gamma\int_D \rho_m(y^2+1)(u-\gamma)^+dydz$$
$$\leq \int_D \rho_m(g+\alpha(y^2+1)v)(u-\gamma)^+dydz.$$

Since $v \leq \gamma$ and $\gamma\lambda = \|g\|_{L^\infty}$, we deduce

$$a((u-\gamma)^+,(u-\gamma)^+)+\alpha\int_D \rho_m(y^2+1)((u-\gamma)^+)^2dydz \leq 0,$$

which implies $(u-\gamma)^+ = 0$. Similarly $(u-\gamma)^- = 0$ and the proof has been completed. □

Taking $\eta = \eta_0 \in K$ in (36), we obtain

$$a(u,u)+\alpha\int_D (y^2+1)\rho_m u^2dydz \leq a(u,\eta_0)$$
$$+\alpha\int_D (y^2+1)\rho_m u\eta_0dydz-\int_D \rho_m(g+\alpha(y^2+1)v)(\eta_0-u)dydz. \quad (39)$$

Since $\|v\|_{L^\infty} \leq \gamma$, we deduce easily from (33) and the coercivity that there exists a number M depending only on the $\tilde{H}^1_m(D)$ norm of η_0 and of γ (but not on the specific v).

We define the following subset of K:

$$\bar{K} = \left\{ v \in K \mid \|v\|_{L^\infty} \leq \frac{\|g\|_{L^\infty}}{\lambda},\ \|v\|_{\tilde{H}^1_m(D)} \leq M \right\} \quad (40)$$

which is also closed and convex in $\tilde{H}^1_m(D)$ and not empty. Indeed, the function picked in Lemma 1 belongs to $\bar{K}$ if M is sufficiently large. The map T_α transforms $\bar{K}$ into itself. The set $\bar{K}$ is a compact subset of $L^2_m(d)$ and T_α is continuous. Hence T_α has a fixed point.

The fixed point satisfies

$$a(u,\eta-u) \geq \int_D \rho_m g(\eta-u)dydz, \quad \forall\eta \in K,\ u \in K. \quad (41)$$

Take $\Psi \in H^1_m(d) \cap L^\infty(D)$ such that

$$\Psi(y,Y) = 0, \quad \text{if } y > 0, \qquad \Psi(y,-Y) = 0, \quad \text{if } Y < 0. \quad (42)$$

then $\eta = u \pm \Psi \in K$. Hence

$$a(u,\Psi) = \int_D \rho_m g \Psi dydz, \quad \forall \Psi \tag{43}$$

such that (39) is satisfied.

It follows that in the sense of distributions

$$\lambda u + A^{\varepsilon} u = g, \quad \text{in } D. \tag{44}$$

By the definition of K the Dirichlet parts of the boundary conditions in (31)–(32) are satisfied.

Now considering Ψ as in (43) and testing (44) with $\Psi\rho_m$, we obtain

$$\int_{-\infty}^{0} \rho_m \frac{\partial u}{\partial z} \Psi(y,Y)dy - \int_{0}^{\infty} \rho_m \frac{\partial u}{\partial z} \Psi(y,-Y)dy = 0.$$

Since the values of $\Psi(y,Y)$ and $\Psi(y,-Y)$ are arbitrary, we get

$$\frac{\partial u}{\partial z} = 0, \quad \text{for } y < 0,\ z = Y$$

and

$$\frac{\partial u}{\partial z} = 0, \quad \text{for } y > 0,\ z = -Y.$$

Therefore, the fixed point of T_α, solution of (38), is a solution of (30)–(32).

4.3 Estimates

We are going to obtain estimates similar to (15)–(16). Writing (41) explicitly, we obtain

$$\begin{aligned}
&\frac{\varepsilon}{2} \int_D \rho_m \frac{\partial u^\varepsilon}{\partial z} \left(\frac{\partial \eta}{\partial z} - \frac{\partial u^\varepsilon}{\partial z} \right) dydz + \frac{1}{2} \int_D \rho_m \frac{\partial u^\varepsilon}{\partial y} \left(\frac{\partial \eta}{\partial y} - \frac{\partial u^\varepsilon}{\partial y} \right) dydz \\
&+ \lambda \int_D \rho_m u^\varepsilon (\eta - u^\varepsilon) dydz - m \int_D \rho_m \frac{\partial u^\varepsilon}{\partial y} (\eta - u^\varepsilon) \frac{y}{1+y^2} dydz \\
&+ \int_D \rho_m \frac{\partial u^\varepsilon}{\partial y} (\eta - u^\varepsilon)(c_0 y + kz) dydz - \int_D \rho_m \frac{\partial u^\varepsilon}{\partial z} (\eta - u^\varepsilon) y dydz \\
&\geq \int_D \rho_m g (\eta - u^\varepsilon) dydz, \quad \forall \eta \in K.
\end{aligned}$$

Therefore,

$$
\begin{aligned}
&\frac{\varepsilon}{2}\int_D \rho_m \left(\frac{\partial u^\varepsilon}{\partial z}\right)^2 dydz + \frac{1}{2}\int_D \rho_m \left(\frac{\partial u^\varepsilon}{\partial y}\right)^2 dydz + \lambda\int_D \rho_m (u^\varepsilon)^2 dydz \\
&- m\int_D \rho_m \frac{\partial u^\varepsilon}{\partial y} u^\varepsilon \frac{y}{1+y^2} dydz + \int_D \rho_m \frac{\partial u^\varepsilon}{\partial y} u^\varepsilon (c_0 y + kz) dydz \\
&- \frac{1}{2}\int_{-\infty}^0 y\rho_m (u^\varepsilon)^2 (y, Y) dy + \frac{1}{2}\int_0^\infty y\rho_m (u^\varepsilon)^2 (y, -Y) dy \\
&\leq \int_D rho_m g u^\varepsilon dydz + \frac{1}{2}\int_0^\infty y\rho_m (u_+^\varepsilon \chi_+ + \beta_+)^2 dy \\
&\quad - \frac{1}{2}\int_{-\infty}^0 y\rho_m (u_-^\varepsilon \chi_- + \beta_-)^2 dy \\
&\quad + \frac{\varepsilon}{2}\int_D \rho_m \frac{\partial u^\varepsilon}{\partial z}\frac{\partial \eta}{\partial z} dydz + \frac{1}{2}\int_D \rho_m \frac{\partial u^\varepsilon}{\partial z}\frac{\partial \eta}{\partial y} dydz \\
&\quad + \lambda\int_D \rho_m u^\varepsilon \eta dydz - m\int_D \rho_m \frac{\partial u^\varepsilon}{\partial y}\eta \frac{y}{1+y^2} dydz \\
&\quad + \int_D \rho_m \frac{\partial u^\varepsilon}{\partial y}\eta dydz - \int_D \rho_m \frac{\partial u^\varepsilon}{\partial z}\eta y dydz - \int_D \rho_m g\eta dydz.
\end{aligned}
$$

Recalling that u^ε is bounded, we deduce

$$
\begin{aligned}
&\varepsilon\int_D \rho_m \left(\frac{\partial u^\varepsilon}{\partial z}\right)^2 dydz < C, \quad \int_D \rho_m \left(\frac{\partial u^\varepsilon}{\partial y}\right)^2 dydz < C, \\
&\text{and} \int_R |y|\rho_m (u^\varepsilon)^2 (y, \pm Y) dy < C, \quad m \geq 1.
\end{aligned} \tag{45}
$$

Next, considering (30) and testing with $\frac{\partial u^\varepsilon}{\partial z} y^3 \rho_m$, we obtain (cf. (16))

$$
\begin{aligned}
&\int_D \rho_m y^4 \left(\frac{\partial u^\varepsilon}{\partial z}\right)^2 dydz + \frac{\varepsilon}{4}\int_0^\infty \left(\frac{\partial u^\varepsilon}{\partial z}\right)^2 (y, Y) y^3 \rho_m dy \\
&- \frac{\varepsilon}{4}\int_{-\infty}^0 \left(\frac{\partial u^\varepsilon}{\partial z}\right)^2 (y, -Y) y^3 \rho_m dy - \frac{\lambda}{2}\int_{-\infty}^0 \rho_m y^3 (u^\varepsilon(y, Y))^2 dy \\
&+ \frac{\lambda}{2}\int_0^\infty \rho_m y^3 (u^\varepsilon(y, -Y))^2 dy - \frac{1}{4}\int_{-\infty}^0 \rho_m y^3 \left(\frac{\partial u^\varepsilon}{\partial y}\right)^2 (y, Y) dy \\
&+ \frac{1}{4}\int_0^\infty \rho_m y^3 \left(\frac{\partial u^\varepsilon}{\partial y}\right)^2 (y, -Y) dy \\
&= \int_D \rho_m \frac{\partial u^\varepsilon}{\partial y}\frac{\partial u^\varepsilon}{\partial z} y^2 \left[c_0 y^2 + kzy + \frac{3}{2} - \frac{my^2}{1+y^2}\right] dydz - \int_D \rho_m g y^3 \frac{\partial u^\varepsilon}{\partial z} dydz \\
&\quad + \frac{\lambda}{2}\int_0^\infty \rho_m y^3 (u_+^\varepsilon \chi_+ + \beta_+)^2 dy - \frac{\lambda}{2}\int_{-\infty}^0 \rho_m y^3 (u_-^\varepsilon \chi_- + \beta_-)^2 dy \\
&\quad + \frac{1}{4}\int_0^\infty \rho_m y^3 \left(u_+^\varepsilon \frac{d\chi_+}{dy} + \frac{d\beta_+}{dy}\right)^2 dy - \frac{1}{4}\int_{-\infty}^0 \rho_m y^3 \left(u_-^\varepsilon \frac{d\chi_-}{dy} + \frac{d\beta_-}{dy}\right)^2 dy.
\end{aligned} \tag{46}
$$

From this relation we deduce

$$\int_D \rho_m y^4 \left(\frac{\partial u^\varepsilon}{\partial z}\right)^2 dydz \leq C, \quad \varepsilon \int_R \rho_m |y|^3 \left(\frac{\partial u^\varepsilon}{\partial z}\right)^2 (y, \pm Y)dy \leq C,$$
$$\text{and} \int_R |y|^3 \rho_m \left(\frac{\partial u^\varepsilon}{\partial y}\right)^2 (y, \pm Y)dy \leq C, \quad m \geq 3. \tag{47}$$

We next obtain the analogue of (18). Assuming $\frac{\partial g}{\partial z} \in L_m^2(D)$, $m \geq 1$, we differentiate (30)–(32) in z and set

$$v^\varepsilon = \frac{\partial u^\varepsilon}{\partial z}.$$

We obtain

$$\begin{aligned}
&\lambda v^\varepsilon + A^\varepsilon v^\varepsilon = \frac{\partial g}{\partial z} - k\frac{\partial u^\varepsilon}{\partial y} \quad \text{in } D,\\
&\frac{\varepsilon}{2}\frac{\partial v^\varepsilon}{\partial z} + yv^\varepsilon = 0, \quad y > 0,\ z = Y, \qquad v^\varepsilon(y, Y) = 0, \quad y < 0,\\
&\lambda\frac{\varepsilon}{2}\frac{\partial v^\varepsilon}{\partial z} + yv^\varepsilon = 0, \quad y < 0,\ z = -Y, \qquad v^\varepsilon(y, -Y) = 0, \quad y > 0.
\end{aligned} \tag{48}$$

We test with $v^\varepsilon \rho_m$ and obtain

$$\begin{aligned}
&\lambda \int_D \rho_m (v^\varepsilon)^2 dydz + \frac{\varepsilon}{2} \int_D \rho_m \left(\frac{\partial v^\varepsilon}{\partial z}\right)^2 dydz\\
&\frac{1}{2}\int_D \rho_m \left(\frac{\partial v^\varepsilon}{\partial y}\right)^2 dydz + \int_D \rho_m \frac{\partial v^\varepsilon}{\partial y} v^\varepsilon (c_0 y + kz)dydz\\
&- m \int_D \rho_m \frac{\partial v^\varepsilon}{\partial y} v^\varepsilon \frac{y}{1+y^2} dydz + \frac{1}{2}\int_0^\infty \rho_m y (v^\varepsilon(y, Y))^2 dy\\
&- \frac{1}{2}\int_{-\infty}^0 \rho_m y (v^\varepsilon(y, -Y))^2 dy\\
&= \int_D \rho_m \left(\frac{\partial g}{\partial z} - k\frac{\partial u^\varepsilon}{\partial y} v^\varepsilon\right) dydz.
\end{aligned} \tag{49}$$

Again if λ is sufficiently large and $\frac{\partial g}{\partial z} \in L_m^2(D)$, $m \geq 1$ we have

$$\int_D \rho_m \left(\frac{\partial u^\varepsilon}{\partial z}\right)^2 dydz \leq C, \quad \int_D \rho_m \left(\frac{\partial^2 u^\varepsilon}{\partial y \partial z}\right)^2 dydz \leq C,$$
$$\text{and} \int_R \rho_m |y| \left(\frac{\partial u^\varepsilon}{\partial z}\right)^2 (y, \pm Y)dy \leq C, m \geq 1. \tag{50}$$

4.4 Proof of Existence

Proof. If λ is sufficiently large and $\frac{\partial g}{\partial z} \in L_m^2(D)$ we can use the estimates (45) and (50) and pass to the limit in the variational inequality (41).

We obtain

$$\begin{aligned}
&\frac{1}{2}\int_D \rho_m \frac{\partial u}{\partial y}\left(\frac{\partial \eta}{\partial y}-\frac{\partial u}{\partial y}\right)dydz+\lambda\int_D \rho_m u(\eta-u)dydz\\
&-m\int_D \rho_m \frac{\partial u}{\partial y}(\eta-u)\frac{y}{1+y^2}dydz+\int_D \rho_m \frac{\partial u}{\partial y}(\eta-u)(c_0y+kz)dydz\\
&-\int_D \rho_m \frac{\partial u}{\partial z}(\eta-u)ydydz\\
&\geq \int_D \rho_m g(\eta-u)dydz, \quad \forall \eta\in K,\ u\in K. \qquad (51)
\end{aligned}$$

In general, we cannot use the regularity properties of Proposition 1, and thus we cannot write the variational inequality (51).

However, the weak limit of u^ε in the sense of the bounds (45), (47) will satisfy (6) in the sense of distributions. The boundary conditions (7), (8) are obtained easily by considering the limits of

$$u_+^\varepsilon \chi_+(y)+\beta_+(y) \quad \text{and} \quad u_-^\varepsilon \chi_-(y)+\beta(y)$$

which amount to considering the limits of the numbers u_+^ε, u_-^ε which are bounded by $\frac{\|g\|}{\lambda}$.

So we can extract converging subsequences. This is also the trace of a converging subsequence of u^ε. The proof has been completed.□

Acknowledgement. This research was partially supported by a grant from CEA, Commissariat à l'énergie atomique and by the National Science Foundation under grant DMS-0705247.

References

1. A. Bensoussan and J. Turi. Stochastic variational inequalities for elasto-plastic oscillators. *C. R. Math. Acad. Sci. Paris*, 343(6):399–406, 2006.
2. A. Bensoussan and J. Turi. Degenerate Dirichlet problems related to the invariant measure of elasto-plastic oscillators. *Appl. Math. Optim.*, 58(1):1–27, 2007. DOI 10.1007/s00245-007-9027-4 (available online).
3. C. Féau. *Les méthodes probabilistes en mécanique sismique. Application aux calculs de tuyauteries fissurées.* Thèse, Université d'Evry, 2003.
4. D. Karnopp and T. D. Scharton. Plastic deformation in random vibration. *J. Acoust. Soc. Amer.*, 39:1154–1161, 1966.
5. R. Z. Khasminskii. *Stochastic stability of differential equations.* Sijthoff and Noordhoff, Alphen aan den Rijn, 1980.
6. A. Preumont. *Random vibration and spectral analysis.* Kluwer Academic Publ., Dordrecht, 2nd edition, 1994.
7. J. B. Roberts and P.-T. D. Spanos. *Random vibration and statistical linearization.* John Wiley & Sons, Chichester, 1990.

A Unified Discrete–Continuous Sensitivity Analysis Method for Shape Optimization

Martin Berggren

Department of Computing Science, Umeå University, Sweden,
martin.berggren@cs.umu.se

Summary. Boundary shape optimization problems for systems governed by partial differential equations involve a calculus of variation with respect to boundary modifications. As typically presented in the literature, the first-order necessary conditions of optimality are derived in a quite different manner for the problems before and after discretization, and the final directional-derivative expressions look very different. However, a systematic use of the material-derivative concept allows a unified treatment of the cases before and after discretization. The final expression when performing such a derivation includes the classical before-discretization ("continuous") expression, which contains objects solely restricted to the design boundary, plus a number of "correction" terms that involve field variables inside the domain. Some or all of the correction terms vanish when the associated state and adjoint variables are smooth enough.

1 Introduction

Computer simulations of systems in science and engineering provide an efficient and cost effective tool to explore how performance depends on geometric features of the system components. An attractive alternative to trial-and-error testing is numerical design optimization, in which we introduce a parametrization of the geometry and let a numerical optimization algorithm interact with the simulation software in order to explore the parameter space. *Boundary shape optimization* is a strategy for design optimization that examines displacements of the boundary to a given domain. Such optimization is a powerful tool for final design, in order to put the final touch to a given configuration. Numerical boundary shape optimization typically uses body-fitted meshes, which makes the method suitable for problem exhibiting boundary layers or other phenomena with high sensitivity to boundary smoothness.

Besides boundary shape optimization, there are other, conceptually different techniques for design optimization that can handle much more general geometries than those generated by displacements of a given boundary; the term *topology optimization* is often used to highlight the generality. In the

W. Fitzgibbon et al. (eds.), *Applied and Numerical Partial Differential Equations*, Computational Methods in Applied Sciences 15, DOI 10.1007/978-90-481-3239-3_4,

so-called *material distribution method* for topology optimization, it is coefficients of the governing partial differential equations discretized on a fixed mesh that are subject to optimization [2]. Such methods can generate arbitrarily complex geometries and are therefore suitable for preliminary design studies. The price for the generality is the limited resolution of the boundary geometry: typically, the boundary is represented using a staircase approximation, which is likely to cause problems in connection with boundary layers, for instance.

Conceptually, boundary shape optimization is a calculus of variation with respect to boundary modifications and traces its historical roots back to the works by, for instance, Newton, Lagrange, and Hadamard. The modern development was initiated in the early 1970s, mainly by the French school of numerical analysis, through researchers like Cea, Glowinski, and Pironneau. Although the field has developed and matured over the years, it is perhaps fair to say that the impact on science and engineering practice has been limited.

In contrast, the technique of optimal layout of a linearly elastic structures using the material distribution method for topology optimization has, indeed, had a noticeable impact on the design of mechanical components. There are commercial software packages available, for instance, from Altair Engineering and FE-design, which are increasingly used for the design of mechanical components, particularly in the vehicle and aerospace industries. Boundary shape optimization is then used as a post processing tool for the layout obtained by topology optimization. However, boundary shape optimization is not much used for practical engineering design outside of such structural "sizing". One reason for the limited impact can be the complexity of managing a system for shape optimization: software for parametrization of shapes, mesh deformation, solvers, sensitivity analysis, and optimization needs to be developed and interfaced in an intricate way. Another reason is computational: solving a shape optimization problem takes often at least an order of magnitude longer time than a pure simulation. Because of the explosive development of hard- and software resources, these hurdles are likely to be overcome eventually. The recent appearance of several monograph dedicated to shape optimization [4, 6, 8, 10–12] is hopefully indicative of a revival.

The key to be able to treat shape optimization problems with a large number of design variables lies the use of gradient-based optimization methods and, in particular, in the use of adjoint equations to extract the directional derivatives. The experience collected through my own involvement in boundary shape optimization strongly indicates that the sensitivity information – directional derivatives of objective functions and constraints – needs to be very accurately computed in order for the optimization algorithms to fully converge. As was early on recognized, not the least by Roland Glowinski and his colleagues when developing shape optimization techniques, the processes of discretization and differentiation do not commute, in general. That is, a discretization of the necessary conditions of optimality (differentiate-then-discretize, or the "continuous" approach) does not generally lead to the same expressions as when deriving the necessary conditions for the discretized optimization problem (discretize-then-differentiate, or the "discrete" approach). The latter strategy is more reliable in my experience, but may be difficult to effectuate in practice for complicated problems. Glowinski and He [7] and Gunzburger [8, §2.9], among many others, discuss and offer perspectives on this somewhat controversial issue.

A disturbing fact is that the two approaches often appear to be unrelated: the procedure for deriving the first-order necessary conditions in the undiscretized case is typically different from the one used in the discrete case, and the final expressions look very different. These problems may have contributed to the reason why there are very attempts to perform analysis of convergence and approximation errors for shape optimization problems. One of the few attempts reported in the literature are by Di Cesare et al. [5], [12, Chapter 6].

The present article shows that a systematic use of the material derivative allows a unified sensitivity analysis in the undiscretized and discretized cases. To minimize technical issues, the derivation will be made for a model elliptic problem and will be largely formal (without existence proofs, for instance). However, the derivation will be made in a way that does not violate the regularity properties of the discrete problem. The final directional-derivative expression (45) (which appears to be new) contains the "continuous" expression plus a number of correction terms that are generally nonzero in the discrete case, but that vanish when the state and adjoint solutions are regular enough.

2 A Potential Flow Model Problem

We consider the flow of an incompressible fluid in a bounded domain $\Omega \subset \mathbb{R}^d$, $d = 2, 3$ with a Lipschitz boundary $\partial\Omega$ (Figure 1). Fluid is flowing in and out through $\Gamma_{\text{io}} \subset \partial\Omega$; otherwise there are impenetrable walls at the boundary. Let $\Gamma_{\text{d}} \in \partial\Omega \setminus \Gamma_{\text{io}}$ be a part of the boundary. We wish to manipulate the shape of the *design boundary* Γ_{d} in order to affect the velocity field in a desired way. Let $\mathscr{U}$ be the set of admissible design boundaries, whose definition may provide conditions such as bounds on curvature, bounds on displacements from

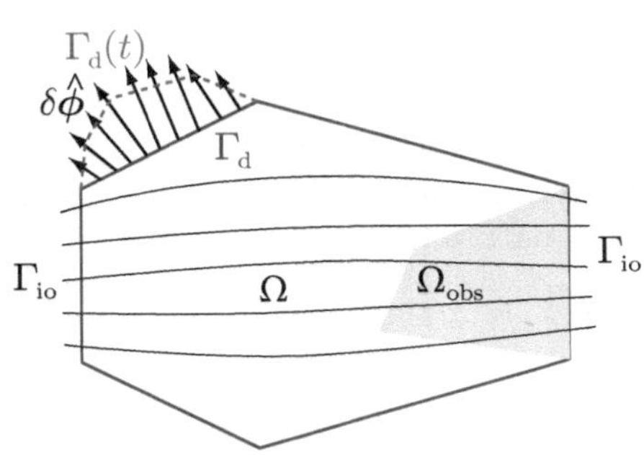

Fig. 1. An example domain for the model shape optimization problem.

a reference configuration, or requirements such as convexity of the domain. In order to perform a calculus of variation on $\Gamma_{\rm d}$, we introduce a *design variation* $\delta\hat{\boldsymbol{\phi}} : \Gamma_{\rm d} \to \mathbb{R}^d$ that generates a family of deformed design boundaries $\Gamma_{\rm d}(t) \in \mathscr{U}$ in the following way: for each $\mathbf{x} \in \Gamma_{\rm d}$, there is an $\mathbf{x}(t) \in \Gamma_{\rm d}(t)$ such that

$$\mathbf{x}(t) = \mathbf{x} + t\delta\hat{\boldsymbol{\phi}}(\mathbf{x}), \quad t \in [0, \alpha]. \tag{1}$$

In order to generate for the formula (1) Lipschitz design boundaries that are connected to the rest of the boundary, any feasible design variation needs to be Lipschitz continuous and vanishing on $\partial\Gamma_{\rm d}$. Any admissible $\delta\hat{\boldsymbol{\phi}}$ should also, of course, be compatible with the definition of $\mathscr{U}$. Further smoothness requirements on $\delta\hat{\boldsymbol{\phi}}$ will be introduced in Section 4 to allow differentiation. We assume that $\alpha > 0$ is small enough so that the mapping between $\Gamma_{\rm d}$ and $\Gamma_{\rm d}(t)$ is bijective for each $t \in [0, \alpha]$.

The displaced design boundaries $\Gamma_{\rm d}(t)$ generate a family of domains $\Omega(t)$ with Lipschitz boundaries. We consider the following potential-flow model defined on $\Omega(t)$:

$$\begin{aligned} -\Delta u + \varepsilon u &= 0 \quad \text{in } \Omega(t), \\ \frac{\partial u}{\partial n} &= g \quad \text{on } \Gamma_{\rm io}, \\ \frac{\partial u}{\partial n} &= 0 \quad \text{on } \partial\Omega(t) \setminus \Gamma_{\rm io}, \end{aligned} \tag{2}$$

where $\varepsilon > 0$ is a small "regularization" parameter introduced to avoid the singularity of the pure Neumann problem. The standard variational form of the state equation (2) is

Find $u(t) \in H^1(\Omega(t))$ such that

$$\int_{\Omega(t)} \nabla v \cdot \nabla u(t) + \varepsilon \int_{\Omega(t)} v u(t) = \int_{\Gamma_{\rm io}} v g \quad \forall v \in H^1(\Omega(t)), \tag{3}$$

where the notation $u(t)$ indicates the dependency on t.

Remark 1. Throughout this article, we will leave out symbols for volume and surface measure in the integrals, since the appropriate measures will be clear from the context.

Now introduce an *observation domain* $\Omega_{\rm obs}$ that does not intersect with the design boundary; that is, $\Omega_{\rm obs} \subset \Omega$ such that $\overline{\Omega}_{\rm obs} \cap \overline{\Gamma}_{\rm d}(t) = \emptyset$. We wish to manipulate the shape of $\Gamma_{\rm d}$ such that the velocity field within the observation domain coincides as closely as possible with a given velocity field $\mathbf{u}_{\rm obs}$, a requirement that naturally leads to the objective function

$$J(\delta\hat{\boldsymbol{\phi}}; t) = \frac{1}{2} \int_{\Omega_{\rm obs}} |\nabla u(t) - \mathbf{u}_{\rm obs}|^2. \tag{4}$$

Some variation of the above problem is a common model problem for shape optimization in the context of fluid flow; Cesare et al. [5] consider essentially the same problem, for instance.

3 Sensitivity Analysis

Here, we present the well-known formulas resulting from a sensitivity analysis of the objective function (4), as described in the book by Pironneau [13], for instance. Section 3.1 gives the expressions before discretization, whereas corresponding expressions obtained after a finite-element discretization are reported in Section 3.2.

3.1 Before Discretization

A sensitivity analysis of the objective function (4) and the state equation (3) concerns the calculation of one-sided directional derivatives of the objective function with respect to design variation $\delta\hat{\phi}$; we will use the notation

$$\delta J(\delta\hat{\phi}) = \frac{\mathrm{d}^+}{\mathrm{d}t} J(\delta\hat{\phi};t)\big|_{t=0} = \lim_{t\to 0^+} \frac{J(\delta\hat{\phi};t) - J(\delta\hat{\phi};0)}{t}. \tag{5}$$

The use of the *one-sided* derivative is essential when performing sensitivity analysis around admissible designs for which geometry constraints are active.

The classical expression for the directional derivative is

$$\delta J(\delta\hat{\phi}) = -\int_{\Gamma_\mathrm{d}} \mathbf{n}\cdot\delta\hat{\phi}\,\nabla u\cdot\nabla u^* - \varepsilon\int_{\Gamma_\mathrm{d}} \mathbf{n}\cdot\delta\hat{\phi}\,uu^*, \tag{6}$$

where $u^* \in H^1(\Omega)$ satisfies the adjoint equation

$$\int_{\Omega} \nabla w\cdot\nabla u^* + \varepsilon\int_{\Omega} wu^* = \int_{\Omega_\mathrm{obs}} \nabla w\cdot(\nabla u - \mathbf{u}_\mathrm{obs}) \quad \forall w\in H^1(\Omega). \tag{7}$$

Note the advantage of introducing the adjoint equation: the directional derivative for each feasible design variation $\delta\hat{\phi}$ can be computed by repeated evaluation of the integral (6) without solving any more equations.

The expression (6) is typically derived through a change of variables involving a smooth bijection between Ω and $\Omega(t)$. Such a mapping can be constructed by extending the boundary variation $\delta\hat{\phi}$ to a domain variation $\delta\phi : \overline{\Omega} \to \mathbb{R}^d$ such that for each point $\mathbf{x}\in\Omega$, there is a unique point $\mathbf{x}(t)\in\Omega(t)$ given by

$$\mathbf{x}(t) = \mathbf{x} + t\delta\phi(\mathbf{x}), \tag{8}$$

and such that $\delta\phi|_{\Gamma_\mathrm{d}} = \delta\hat{\phi}$.

Although the extended mapping is used for the derivation, under certain smoothness assumptions of $\delta\phi$ together with regularity properties that will be made explicit in Section 6, it holds that the final expression (6) is independent of the particular choice of extension.

3.2 After Finite-Element Discretization

In the discrete case, it is natural to use the locations of the mesh vertices at the design boundary Γ_{d} as design variables. However, in order to retain mesh quality, it is, in general, necessary to modify the mesh inside the domain as well. Thus, for generality, associate with each mesh vertex a vector $\delta\mathbf{x}_k \in \mathbb{R}^d$ that indicates a feasible direction of movement for vertex k. Associated with *mesh vertex variation* $\delta\mathbf{x}_k$, it is convenient to define the domain variation $\delta\boldsymbol{\phi}_k = N_k^1 \delta\mathbf{x}_k$, where N_k^1 is the continuous piecewise-linear finite-element basis function at vertex k. Subject to variation $\delta\mathbf{x}_k$, each $\mathbf{x}(t)$ in the deformed domain $\Omega(t)$ is then given by

$$\mathbf{x}(t) = \mathbf{x} + t\,\delta\mathbf{x}_k N_k^1(\mathbf{x}) = \mathbf{x} + t\,\delta\boldsymbol{\phi}_k(\mathbf{x}), \tag{9}$$

and $\Omega(0) = \Omega$. The formula (9) interpolates deformation $t\,\delta\mathbf{x}_k$ at vertex k on the support of N_k^1. Note that the use of piecewise-linear basis functions implies that planar mesh surfaces and edges will remain planar under the deformation. Figure 2 illustrates the deformation (9) in two cases. If the mesh on domain Ω is nondegenerate, then for each mesh vertex k, there is an $\alpha_k > 0$ such that the mesh associated with the deformation (9) will also be nondegenerate for all $t \in [0, \alpha_k]$.

Now discretize the equation (3) on the domain Ω using a conforming finite-element discretization in a subspace $V_h \subset H^1(\Omega)$. Given the deformation (9) associated with an arbitrary mesh vertex k, we may then define a family of discrete solutions

$$\begin{aligned} &u_h(t) \in V_h(t) \subset H^1(\Omega(t)) \text{ such that} \\ &\int_{\Omega(t)} \nabla v_h \cdot \nabla u_h(t) + \varepsilon \int_{\Omega(t)} v_h u_h(t) = \int_{\Gamma_{\mathrm{io}}} v_h g \quad \forall v_h \in V_h(t), \end{aligned} \tag{10}$$

and consider the discrete objective function

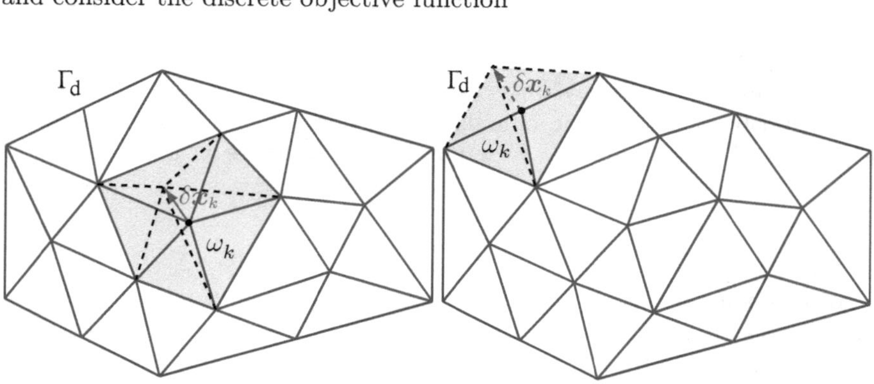

Fig. 2. Each mesh vertex displacement $t\,\delta\mathbf{x}_k$ is interpolated onto the support ω_k of the continuous piecewise-linear finite-element basis functions N_k^1.

$$J_h(\delta\phi_k;t) = \frac{1}{2}\int_{\Omega_{\text{obs}}} |\nabla u_h(t) - \mathbf{u}_{\text{obs}}|^2. \tag{11}$$

The following classical expression (e.g. [12, §6.5]) holds for the directional derivative of J_h:

$$\begin{aligned}\delta J_h = -\delta\mathbf{x}_k \cdot \int_\Omega \nabla N_k^1(\nabla u_h \cdot \nabla u_h^*) + \delta\mathbf{x}_k \cdot \int_\Omega \nabla u_h(\nabla u_h^* \cdot \nabla N_k^1) \\ + \delta\mathbf{x}_k \cdot \int_\Omega \nabla u_h^*(\nabla u_h \cdot \nabla N_k^1) - \varepsilon\, \delta\mathbf{x}_k \cdot \int_\Omega u_h u_h^* \nabla N_k^1,\end{aligned} \tag{12}$$

where $u_h^* \in V_h$ such that

$$\int_\Omega \nabla w_h \cdot \nabla u_h^* + \varepsilon \int_\Omega w_h u_h^* = \int_{\Omega_{\text{obs}}} \nabla w_h \cdot (\nabla u_h - \mathbf{u}_{\text{obs}}) \quad \forall w_h \in V_h. \tag{13}$$

The expression (12) reveals expressions for the derivative of J_h with respect to variations of each mesh vertex in all coordinate directions (note that the integrals are vectors with d components). Once the state u_h and adjoint state u_h^* are known, all these derivatives can be computed by a single assembly loop over all elements. The derivatives can, for instance, be stored in a vector $\mathrm{D}J_h$ of dimension dn, where n is the total number of mesh vertices. Elements dk, $dk+1, \ldots, dk+d-1$ of $\mathrm{D}J_h$ then contains the d components of the integrals in the expression (12).

However, in shape optimization, it does not make much sense to optimize the position of each mesh points independently. A good strategy is to modify the locations of the mesh vertices on Γ_{d} explicitly using updates from the optimization algorithm, and employ a mesh deformation strategy to move the rest of the mesh vertices indirectly in order to preserve mesh quality. In simple geometries, such a mesh deformation can be defined by an explicit formula based on the distance to Γ_{d}. A more general strategy, however, is to use a numerical deformation strategy, for instance, based on elliptic smoothing [12, §5.3]. To describe the role of the mesh deformation in the derivative calculations, consider the spaces of discrete boundary and domain variations, $\widehat{\mathcal{U}}_h = \text{span}(\delta\phi_k)_{k\in\mathcal{V}(\Gamma_{\text{d}})}$ and $\mathcal{U}_h = \text{span}(\delta\phi_k)_{k\in\mathcal{V}(\overline{\Omega})}$, where $\mathcal{V}(\gamma)$ denotes the set of mesh vertices located in the subdomain γ. A mesh deformation strategy defines a mapping $a : \widehat{\mathcal{U}} \to \mathcal{U}$, and the objective function that is in reality used for optimization when employing a mesh deformation is the composition $\widehat{J}_h = J_h \circ a$. By the chain rule, the derivative of mapping $\widehat{J}_h$ will be

$$\mathrm{D}\widehat{J}_h = \mathcal{A}^{\mathrm{T}}\,\mathrm{D}J_h, \tag{14}$$

where $\mathcal{A}$ is a matrix representation of the Jacobian of the mesh deformation mapping a.

Note that the discrete adjoint equation (13) constitutes a finite-element discretization of the adjoint equation (7). However, the discrete directional derivative expressions (14), (12) carry no obvious resemblance to the expression (6).

4 Shape and Material Derivatives of Functions

In Section 6, we will perform the sensitivity analysis in a way that simultaneously provides the seemingly quite different expressions (6) and (12). A main component of the derivation is the differentiation of the state equations (3) and (10). There are two fundamentally different ways in which a function can be differentiated with respect to variations in the domain on which it is defined: as a *material* or as a *shape* derivative. These concepts, shortly reviewed below, are analogues to the material and spatial derivatives in continuous mechanics [9, §8]. For a thorough treatment of these concepts in the framework of shape optimization, see the monograph by Sokolowski and Zolésio [14]. This section aims to demonstrate a fact that seems curiously underappreciated in the shape-optimization literature: the material derivative is better suited, due to its favorable regularity properties, than the shape derivative for use in the sensitivity analysis.

We start by introducing the notation $\Omega(t) = \boldsymbol{\tau}_t(\Omega)$, where, for $\mathbf{x} \in \Omega$,

$$\boldsymbol{\tau}_t(\mathbf{x}) = \mathbf{x} + t\,\delta\boldsymbol{\phi}(\mathbf{x}), \quad t \in [0, \alpha]. \tag{15}$$

For simplicity, we assume that the domain variation $\delta\boldsymbol{\phi}$ vanishes on Ω_{obs} and $\partial\Omega \backslash \Gamma_{\text{d}}$. For the problem before discretization, $\delta\boldsymbol{\phi}$ is an extension of $\delta\hat{\boldsymbol{\phi}}$ (which was defined solely on Γ_{d}) into a mapping from $\overline{\Omega}$ into $\mathbb{R}^d$. We require that the extended mapping is smooth enough so that the components of the second-order tensor $\nabla\delta\boldsymbol{\phi}$ are in $L^\infty(\Omega)$. In the discrete case, $\delta\boldsymbol{\phi}(\mathbf{x}) = \delta\boldsymbol{\phi}_k(\mathbf{x})$, where $\delta\boldsymbol{\phi}_k$ is given by the expression (9) (here, $\delta\boldsymbol{\phi}$ can be made to vanish on Ω_{obs} and $\partial\Omega \setminus \Gamma_{\text{d}}$ by simply not considering any k for which corresponding mesh vertices are in Ω_{obs} or $\partial\Omega \setminus \Gamma_{\text{d}}$). By the definition (9), it follows that the components of $\nabla\delta\boldsymbol{\phi}$ are in $L^\infty(\Omega)$ in the discrete case.

Now consider functions $p : \Omega(t) \times \mathbb{R} \to \mathbb{R}$. We will use $p(t)$ as a shorthand notation the for function $\mathbf{x} \mapsto p(\mathbf{x}, t)$.

Definition 1. *The* material derivative *of p with respect to domain variation $\delta\boldsymbol{\phi}$ at point $\mathbf{x} \in \Omega$ is*

$$\delta_m p(\mathbf{x}; \delta\boldsymbol{\phi}) = \lim_{t\to 0^+} \frac{p(\boldsymbol{\tau}_t(\mathbf{x}), t) - p(\mathbf{x}, 0)}{t} = \frac{\mathrm{d}^+}{\mathrm{d}t} p(\boldsymbol{\tau}_t(\mathbf{x}), t)\big|_{t=0},$$

provided that the point-wise limit exists.

(Whenever there is no risk for confusion, we will suppress the second argument and just use the notation $\delta_{\text{m}} p(\mathbf{x})$.) The material derivative is thus a (one-sided) derivative of the *compound* function $t \mapsto p(t) \circ \boldsymbol{\tau}_t$ (the "total" derivative). For $p(t)$ in a Banach space W, Definition 1 is easily extended to

$$\delta_{\text{m}} p = \frac{\mathrm{d}^+}{\mathrm{d}t} p(t) \circ \boldsymbol{\tau}_t\big|_{t=0} \tag{16}$$

with the limit in a Banach space $X \supset W$.

Definition 2. *The* shape derivative *of p with respect to design variation $\delta\boldsymbol{\phi}$ is the function*

$$\delta p = \delta_m p - \delta\boldsymbol{\phi} \cdot \nabla p(0). \tag{17}$$

Remark 2. Definition 2 imposes an a priori regularity difference between $\delta_{\mathrm{m}}p$ and δp due to the right-side gradient in the expression (17). This difference is consistent with the typical behavior when differentiating the state variable in a shape optimization problem. As illustrated in Examples 3 and 4 below, the material derivative of the state variable can typically be defined in the same space as the state variable itself, whereas the shape derivative typically cannot. An alternative definition of the shape derivative is as the partial derivative

$$\delta p(\mathbf{x}) = \lim_{t\to 0^+} \frac{p(\mathbf{x},t) - p(\mathbf{x},0)}{t} \tag{18}$$

from which the expression (17) follows by the chain rule applied on $\delta_{\mathrm{m}}p$. However, a complicating factor with the definition (18) is that the two terms on the right side has different domains of definition, $\Omega(t)$ and Ω, respectively.

Following four examples highlight the different properties of the material and shape derivative.

Example 1. Let $g : \Omega \to \mathbb{R}$ be given. Define $p(t) = g \circ \boldsymbol{\tau}_t^{-1}$; that is, $p(\mathbf{x},t)$ is defined by mapping back $\mathbf{x} \in \Omega(t)$ to corresponding point in Ω and evaluating g at the mapped-back point. Then

$$\begin{aligned} \delta_{\mathrm{m}}p &= \frac{\mathrm{d}^+}{\mathrm{d}t}\left(p(t)\circ\boldsymbol{\tau}_t\right)\big|_{t=0} = \frac{\mathrm{d}^+}{\mathrm{d}t}\left(g\circ\boldsymbol{\tau}_t^{-1}\circ\boldsymbol{\tau}_t\right)\big|_{t=0} = 0,\\ \delta p &= \delta_{\mathrm{m}}p - \delta\boldsymbol{\phi}\cdot\nabla p(0) = -\delta\boldsymbol{\phi}\cdot\nabla p(0). \end{aligned} \tag{19}$$

Thus, when $p(t)$ is "moving along" with the deformation, the material derivative vanishes. Next example illustrates the opposite situation.

Example 2. Let $f : \mathbb{R}^d \to \mathbb{R}$. Define $p(t) = f|_{\Omega(t)}$. Then

$$\begin{aligned} \delta_{\mathrm{m}}p &= \frac{\mathrm{d}^+}{\mathrm{d}t}\left(p(t)\circ\boldsymbol{\tau}_t\right)\big|_{t=0} = \frac{\mathrm{d}^+}{\mathrm{d}t}\left(f|_{\Omega(t)}\circ\boldsymbol{\tau}_t\right)\big|_{t=0}\\ &= \nabla f|_{\Omega(0)}\cdot\left(\frac{\mathrm{d}^+}{\mathrm{d}t}\boldsymbol{\tau}_t\right)\bigg|_{t=0} = \nabla f|_{\Omega(0)}\cdot\delta\boldsymbol{\phi} = \delta\boldsymbol{\phi}\cdot\nabla p(0),\\ \delta p &= \delta_{\mathrm{m}}p - \delta\boldsymbol{\phi}\cdot\nabla p(0) = 0. \end{aligned} \tag{20}$$

Thus, a function that is "fixed" with respect to the deformation yields a vanishing shape derivative, a property that is consistent with the interpretation (18) of the shape derivative as a partial derivative.

Example 3. Let g belong to a finite element space $V_h \subset H^1(\Omega)$ such that $g(\mathbf{x}) = \sum_{k=1}^N g_k N_k^p(\mathbf{x})$, where N_k^p is a finite-element basis function that is globally continuous and whose restriction on each triangular or tetrahedral element is a polynomial of degree p. Define basis functions on the deformed domain $\Omega(t)$ by the expression $N_k^p(\mathbf{x},t) = N_k^p(\boldsymbol{\tau}_t^{-1}(\mathbf{x}))$, as in Example 1. The span of the functions $N_k^p(t)$ defines a family of finite-element spaces $V(t)$ on $\Omega(t)$. Each $p(t) \in V_h(t)$ may then be written

$$p(\mathbf{x},t) = \sum_{k=1}^{N} p_k(t) N_k^p(\mathbf{x},t). \tag{21}$$

As in Example 1, we find that $\delta_{\mathrm{m}} N_k^p = 0$, $\delta N_k^p = -\delta\boldsymbol{\phi}\cdot\nabla N_k^p$ and thus

$$\begin{aligned}\delta_{\mathrm{m}} p &= \sum_{k=1}^{N} \delta p_k \, N_k^p, \\ \delta p = \delta_{\mathrm{m}} p - \delta\boldsymbol{\phi}\cdot\nabla p &= \sum_{k=1}^{N} \left(\delta p_k \, N_k^p - p_k \, \delta\boldsymbol{\phi}\cdot\nabla N_k^p\right),\end{aligned} \tag{22}$$

where

$$\delta p_k = \frac{\mathrm{d}^+}{\mathrm{d}t} p_k(t)\big|_{t=0}. \tag{23}$$

Note that $\delta_{\mathrm{m}} p \in V_h$ but $\delta p \notin V_h$! That is, the material derivative is *conforming* to the finite element space, whereas the shape derivative is not. Also note that the material derivative is obtained by differentiating only the coefficients of p (and not the basis functions) with respect to the deformation.

Example 4. Consider the solution $u(t) \in H^1(\Omega(t))$ to the state equation (3). Sokolowski and Zolésio [14, §2.29] and Haslinger and Mäkinen [10, §2.5.2] discuss the existence of $\delta_{\mathrm{m}} p$ in similar situations, where they show that $\delta_{\mathrm{m}} u \in H^1(\Omega)$, provided that the domain deformations are sufficiently regular. As in Example 3, the material derivative is defined in the same space as the state, but since $\delta u = \delta_{\mathrm{m}} u - \delta\boldsymbol{\phi}\cdot\nabla u$, the shape derivative typically has less regularity.

5 Rules for the Material Derivative

It is immediate from Definition 1 that the product rule holds for the material derivatives of functions f, g on $\Omega(t) \times \mathbb{R}$:

$$\delta_{\mathrm{m}}(fg) = \delta_{\mathrm{m}} f \, g + f \, \delta_{\mathrm{m}} g, \tag{24}$$

where, for simplicity of notation, we have suppressed the evaluations at zero: the right side should really be $\delta_{\mathrm{m}} f \, g(0) + f(0) \, \delta_{\mathrm{m}} g$. The rest of the article adheres to the same convention: for a function f on $\Omega(t) \times \mathbb{R}$, the symbol "$f$" outside a material derivative will denote its restriction to $t = 0$.

The shape derivative commutes with the spatial gradient, that is, $\delta\nabla = \nabla\delta$, but the material derivative does not: $\delta_{\mathrm{m}}\nabla \neq \nabla\delta_{\mathrm{m}}$. However, it holds that

$$\delta_{\mathrm{m}}(\nabla p) = \nabla(\delta_{\mathrm{m}} p) - (\nabla \delta \boldsymbol{\phi})^{\mathrm{T}} \nabla p, \tag{25}$$

or, in Cartesian components,

$$\delta_{\mathrm{m}}\left(\frac{\partial p}{\partial x_i}\right) = \frac{\partial}{\partial x_i}\delta_{\mathrm{m}} p - \sum_{j=1}^{d} \frac{\partial p}{\partial x_j}\frac{\partial}{\partial x_i}\delta \phi_j, \quad i = 1, \ldots, d. \tag{26}$$

To prove the expression (25), consider a finite-difference approximation of the material derivative:

$$\mathrm{D}_m^+ p(t) \stackrel{\mathrm{def}}{=} \frac{p(t) \circ \boldsymbol{\tau}_t - p(0)}{t}. \tag{27}$$

Differentiating both sides of the expression (27) yields

$$\begin{aligned}
\nabla \mathrm{D}_m^+ p(t) &= \frac{\nabla(p(t) \circ \boldsymbol{\tau}_t) - \nabla p(0)}{t} \\
&= \frac{1}{t}\left[\left(\mathbf{I} + t(\nabla \delta \boldsymbol{\phi})^{\mathrm{T}}\right) \nabla p(t) \circ \boldsymbol{\tau}_t - \nabla p(0)\right] \\
&= \frac{\nabla p(t) \circ \boldsymbol{\tau}_t - \nabla p(0)}{t} + (\nabla \delta \boldsymbol{\phi})^{\mathrm{T}} \nabla p(t),
\end{aligned} \tag{28}$$

where the second equality follows from the chain rule applied on $\nabla(p(t) \circ \boldsymbol{\tau}_t)$ and from differentiation of $\boldsymbol{\tau}_t$ as defined in the expression (15). The expression (28) implies that

$$\nabla(\delta_{\mathrm{m}} p) = \lim_{t \to 0} \nabla \mathrm{D}_m^+ p(t) = \delta_{\mathrm{m}} \nabla p + (\nabla \delta \boldsymbol{\phi})^{\mathrm{T}} \nabla p, \tag{29}$$

which is the expression we wanted to show.

The product rule (24) and the expression (25) yields that

$$\delta_{\mathrm{m}}(\nabla q \cdot \nabla p) = \nabla \delta_{\mathrm{m}} q \cdot \nabla p + \nabla q \cdot \nabla \delta_{\mathrm{m}} p - \nabla q \cdot (\nabla_{\mathrm{S}}\, \delta \boldsymbol{\phi}) \nabla p, \tag{30}$$

where $\nabla_{\mathrm{S}}\, \delta \boldsymbol{\phi} = \nabla \delta \boldsymbol{\phi} + (\nabla \delta \boldsymbol{\phi})^{\mathrm{T}}$.

The rule for differentiating domain integrals that we will need in the following is [10, Lemma 3.3]

$$\delta\left(\int_{\Omega} f\right) = \frac{\mathrm{d}^+}{\mathrm{d}t}\left(\int_{\Omega(t)} f(t)\right)\Bigg|_{t=0} = \int_{\Omega} (\delta_{\mathrm{m}} f + f \nabla \cdot \delta \boldsymbol{\phi}). \tag{31}$$

The rules (25), (30), and (31) are the basic tools needed for a differentiation of the variational forms. Note that there are no direct counterparts to the expressions (25) and (30) for the shape derivative in the discrete case (when $p, q \in V_h$), and no shape-derivative counterpart to the expression (31) with $f = \nabla q \cdot \nabla p$, since such expressions would involve second derivatives of the finite-element functions, which are not functions.

6 Sensitivity Analysis Using Material Derivatives

Equipped with the tools of Sections 4 and 5, we now perform a derivation that simultaneously provides the directional derivatives (6) and (12).

Let $V(t) \subset H^1(\Omega(t))$ and define $V = V(0)$. In the case before discretization, $V(t) = H^1(\Omega(t))$, whereas $V(t) = V_h(t)$, a finite-element space, in the discrete case. The state equations (3) and (10) can then be written in the common form:

Let $u(t) \in V(t)$ such that

$$\int_{\Omega(t)} \nabla v(t) \cdot \nabla u(t) + \varepsilon \int_{\Omega(t)} v(t) u(t) = \int_{\Gamma_{\mathrm{io}}} v(t) g \quad \forall v(t) \in V(t), \tag{32}$$

and the objective functions (4) and (11) in the form

$$j(\delta\boldsymbol{\phi}; t) = \frac{1}{2} \int_{\Omega_{\mathrm{obs}}} |\nabla u(t) - \mathbf{u}_{\mathrm{obs}}|^2. \tag{33}$$

Differentiating the objective function (33) using the differentiation rule (31) and observing that $\delta\boldsymbol{\phi}|_{\Omega_{\mathrm{obs}}} \equiv 0$ yields

$$\delta j(\delta\boldsymbol{\phi}) = \int_{\Omega_{\mathrm{obs}}} \nabla \delta_{\mathrm{m}} u \cdot (\nabla u - \mathbf{u}_{\mathrm{obs}}). \tag{34}$$

Differentiating the state equation (32) at $t = 0$, using the rules (24), (30), and (31) yields that

$$\begin{aligned} 0 = &\int_{\Omega} (\nabla \delta_{\mathrm{m}} v \cdot \nabla u + \varepsilon\, (\delta_{\mathrm{m}} v)\, u) + \int_{\Omega} (\nabla v \cdot \nabla \delta_{\mathrm{m}} u + \varepsilon\, v\, \delta_{\mathrm{m}} u) \\ &+ \int_{\Omega} (\nabla v \cdot \nabla u\, \nabla \cdot \delta\boldsymbol{\phi} + v u\, \nabla \cdot \delta\boldsymbol{\phi} - \nabla v \cdot (\nabla_{\mathrm{S}} \delta\boldsymbol{\phi})\, \nabla u) \end{aligned} \tag{35}$$

for each $v \in V$. Since $\delta_{\mathrm{m}} v \in V$ (cf. Examples 3 and 4), the first integral in the expression (35) vanishes due to the state equation (32) evaluated at $t = 0$. Now let $u^* \in V$ satisfy the adjoint equation

$$\int_{\Omega} \nabla w \cdot \nabla u^* + \varepsilon \int_{\Omega} w u^* = \int_{\Omega_{\mathrm{obs}}} \nabla w \cdot (\nabla u - \mathbf{u}_{\mathrm{obs}}) \quad \forall w \in V. \tag{36}$$

By choosing $v = u^*$ in the expression (35) and making use of the equation (36) with $w = \delta_{\mathrm{m}} u$, the expression (35) reduces to

$$\begin{aligned} 0 = &\int_{\Omega_{\mathrm{obs}}} \nabla \delta_{\mathrm{m}} u \cdot (\nabla u - \mathbf{u}_{\mathrm{obs}}) \\ &+ \int_{\Omega} (\nabla u^* \cdot \nabla u\, \nabla \cdot \delta\boldsymbol{\phi} + u^* u\, \nabla \cdot \delta\boldsymbol{\phi} - \nabla u^* \cdot (\nabla_{\mathrm{S}} \delta\boldsymbol{\phi})\, \nabla u), \end{aligned} \tag{37}$$

from which we conclude that the expression (34) can be written

$$\delta j(\delta \boldsymbol{\phi}) = -\int_{\Omega} (\nabla u^* \cdot \nabla u \, \nabla \cdot \delta \boldsymbol{\phi} + \varepsilon u^* u \, \nabla \cdot \delta \boldsymbol{\phi} - \nabla u^* \cdot (\nabla_{\mathrm{S}} \delta \boldsymbol{\phi}) \, \nabla u). \quad (38)$$

Substituting $\delta \boldsymbol{\phi} = \delta \mathbf{x}_k \, N^1(\mathbf{x})$ into the expression (38) yields the discrete expression (12).

In order to proceed further, we need to integrate the expression (38) by parts in a way that respects the regularity properties of the involved functions. We will use a notation borrowed from the context of discontinuous Galerkin methods [1, §3]. Let $\mathcal{T}_h$ be the set of elements (triangles or tetrahedrons) in a triangulation of the domain Ω (that is, $\Omega(0)$). Note that the triangulation will be completely superficial, without any effect on the solution, in the case before discretization. Denote by $H^1(\mathcal{T}_h)$ the space of functions in $L^2(\Omega)$ whose restriction to each element $K \in \mathcal{T}_h$ is in $H^1(K)$ (functions in $H^1(\mathcal{T}_h)$ may, however, contain jump discontinuities between neighboring elements in the discrete case). Denote by Σ the union of the boundaries to all elements in the triangulation. Denote by $T(\Sigma)$ the space of traces of functions in $H^1(\mathcal{T}_h)$ on Σ; such traces are uniquely defined on the domain boundary $\partial\Omega$ but are, in general, double valued on the element boundaries $\Sigma_0 = \Sigma \setminus \partial\Omega$ interior to the domain. Consider two neighboring elements K_1 and K_2 that shares the surface (3D) or the edge (2D) $\sigma \in \Sigma_0$, and denote by $\mathbf{n}_1$ and $\mathbf{n}_2 = -\mathbf{n}_1$ the unit normals on σ that are outward directed with respect to K_1 and K_2, respectively. For $q \in H^1(\mathcal{T}_h)$, define jumps on σ by

$$[\![q]\!] = q\big|_{\partial K_1 \cap \sigma} \mathbf{n}_1 + q\big|_{\partial K_2 \cap \sigma} \mathbf{n}_2. \quad (39)$$

For $q \in H^1(\mathcal{T}_h)$ and $\boldsymbol{\psi} \in H^1(\Omega)^d$ hold the integration-by-parts formula

$$\int_{\Omega} \nabla \cdot \boldsymbol{\psi} \, q = -\sum_{K \in \mathcal{T}_h} \int_K \boldsymbol{\psi} \cdot \nabla q + \int_{\partial\Omega} \mathbf{n} \cdot \boldsymbol{\psi} \, q + \int_{\Sigma_0} \boldsymbol{\psi} \cdot [\![q]\!]. \quad (40)$$

We will now apply the formula (40) with $q = \nabla u^* \cdot \nabla u$ and $\boldsymbol{\psi} = \delta \boldsymbol{\phi}$. Note that $q \in H^1(\mathcal{T}_h)$ and $\boldsymbol{\psi} \in H^1(\Omega)^d$ hold for these choices: before discretization, $q|_K$ is smooth by internal regularity of the equations (32) and (36), and $\boldsymbol{\psi} \in H^1(\Omega)^d$ by assumption; after discretization, $q|_K$ is polynomial, and $\boldsymbol{\psi}$ is in an $H(\Omega)^d$ conforming finite-element space. Using the formula (40), the first term in the right side of the expression (38) can be written

$$\begin{aligned} -\int_{\Omega} \nabla u^* \cdot \nabla u \, \nabla \cdot \delta \boldsymbol{\phi} &= \sum_{K \in \mathcal{T}_h} \int_K \delta \boldsymbol{\phi} \cdot \nabla (\nabla u^* \cdot \nabla u) \\ &\quad - \int_{\Gamma_{\mathrm{d}}} \mathbf{n} \cdot \delta \boldsymbol{\phi} \, \nabla u^* \cdot \nabla u - \int_{\Sigma_0} \delta \boldsymbol{\phi} \cdot [\![\nabla u^* \cdot \nabla u]\!], \quad (41) \end{aligned}$$

where we have used that $\delta \boldsymbol{\phi}$ vanishes on $\partial\Omega \setminus \Gamma_{\mathrm{d}}$. Integration by parts on the second term in the right side of the expression (38) yields

$$-\varepsilon\int_{\Omega}u^*u\,\nabla\cdot\delta\boldsymbol{\phi}=\varepsilon\int_{\Omega}\delta\boldsymbol{\phi}\cdot\nabla(u^*u)-\varepsilon\int_{\Gamma_{\mathrm{d}}}\mathbf{n}\cdot\delta\boldsymbol{\phi}\,u^*u. \tag{42}$$

Substituting the expressions (41) and (42) into the expression (38) and collecting terms, we obtain

$$\begin{aligned}\delta j(\delta\boldsymbol{\phi})=&-\int_{\Gamma_{\mathrm{d}}}\mathbf{n}\cdot\delta\boldsymbol{\phi}\,(\nabla u^*\cdot\nabla u+\varepsilon\,u^*u)-\int_{\Sigma_0}\delta\boldsymbol{\phi}\cdot[\![\nabla u^*\cdot\nabla u]\!]\\&+\sum_{K\in\mathcal{T}_h}\int_K(\delta\boldsymbol{\phi}\cdot\nabla(\nabla u^*\cdot\nabla u)+\nabla u^*\cdot(\nabla_{\mathrm{S}}\delta\boldsymbol{\phi})\,\nabla u+\varepsilon\,\delta\boldsymbol{\phi}\cdot\nabla(u^*u)).\end{aligned} \tag{43}$$

The two first terms in the last integral in the expression (43) can be written

$$\delta\boldsymbol{\phi}\cdot\nabla(\nabla u^*\cdot\nabla u)+\nabla u^*\cdot(\nabla_{\mathrm{S}}\delta\boldsymbol{\phi})\,\nabla u=\nabla(\delta\boldsymbol{\phi}\cdot\nabla u^*)\cdot\nabla u+\nabla u^*\cdot\nabla(\delta\boldsymbol{\phi}\cdot\nabla u), \tag{44}$$

as shown by expanding in Cartesian components, for instance. Substituting the expression (44) into the expression (43) yields

$$\begin{aligned}\delta J=&-\int_{\Gamma_{\mathrm{d}}}\mathbf{n}\cdot\delta\boldsymbol{\phi}\,(\nabla u^*\cdot\nabla u+\varepsilon\,u^*u)-\int_{\Sigma_0}\delta\boldsymbol{\phi}\cdot[\![\nabla u^*\cdot\nabla u]\!]\\&+\sum_{K\in\mathcal{T}_h}\int_K(\nabla(\delta\boldsymbol{\phi}\cdot\nabla u^*)\cdot\nabla u+\varepsilon\,(\delta\boldsymbol{\phi}\cdot\nabla u^*)u)\\&+\sum_{K\in\mathcal{T}_h}\int_K(\nabla u^*\cdot\nabla(\delta\boldsymbol{\phi}\cdot\nabla u)+\varepsilon\,u^*\delta\boldsymbol{\phi}\cdot\nabla u).\end{aligned} \tag{45}$$

The expression (45) contains, as its first term, the "continuous" directional derivative expression (6), but also three "correction" terms. The first correction term involves jumps of $\nabla u^*\cdot\nabla u$ at inter-element boundaries, whereas the second and third terms contain some particular weighted element-wise residuals of the state and adjoint equations, respectively, for which $\delta\boldsymbol{\phi}\cdot\nabla u^*$ and $\delta\boldsymbol{\phi}\cdot\nabla u$ replace the test functions. Some or all of these "correction terms" may vanish, depending on the situation:

Case 1 (before discretization). When $V=H^1(\Omega)$ – the "continuous" case – the functions u and u^* are interior regular (and regular up to the boundary Γ_{d} when the boundary is smooth enough). In this case, the jump terms vanish due to the continuity of $\nabla u\cdot\nabla u^*$. Also, since $\delta\boldsymbol{\phi}\cdot\nabla u^*\in V$, $\delta\boldsymbol{\phi}\cdot\nabla u\in V$ in this case, the element residual terms will also vanish due to the state and adjoint equations (32), (36). Hence, in this case, the expression (45) reduces to the classic "continuous" expression (6).

Case 2 (lowest-order finite elements). If functions in V are linear on each element, the element residual terms vanish, since then $\nabla(\delta\boldsymbol{\phi}\cdot\nabla u^*)|_K=\nabla(\delta\boldsymbol{\phi}\cdot\nabla u)|_K\equiv 0$.

Case 3 (higher-order finite elements). Here, none of the terms vanishes, in general, and the expression (45) with $\delta\phi = \delta\mathbf{x}_k N^1(\mathbf{x})$ just provides a different way of evaluating the expression (12).

Case 4 (C^1 finite elements). When using the (rather unusual) class of C^1 finite elements (for instance, the Argyris element [3, §3.2.10]), the inter-element jump terms vanish since then $[\![\nabla u^* \cdot \nabla u]\!] \equiv 0$.

The expression (45) links together the "discrete" expression (12) and the "continuous" expression (6) and constitutes, therefore, hopefully a first step in a rigorous numerical analysis of finite-element shape optimization. For instance, the convergence rate of the discrete Frechet derivative could be estimated by estimates of the jumps and residual terms that the expression (45) exposes.

References

1. D. N. Arnold, F. Brezzi, B. Cockburn, and L. D. Marini. Unified analysis of discontinuous Galerkin methods for elliptic problems. *SIAM Journal on Numerical Analysis*, 39(5):1749–1779, 2002.
2. M. P. Bendsoe and O. Sigmund. *Topology optimization. Theory, methods, and applications.* Springer, 2003.
3. S. C. Brenner and L. R. Scott. *The mathematical theory of finite element methods.* Springer, New York, 2nd edition, 2002.
4. D. Bucur and G. Buttazzo. *Variational methods in shape optimization problems.* Birkhäuser, 2005.
5. N. Di Cesare, O. Pironneau, and E. Polak. Consistent approximations for an optimal design problem. Technical Report R 98005, Laboratoire d'Analyse Numérique, Université Pierre et Marie Curie, 1998.
6. M. C. Delfour and J.-P. Zolésio. *Shapes and geometries. Analysis, differential calculus, and optimization.* SIAM, Philadelphia, PA, 2001.
7. R. Glowinski and J. He. On shape optimization and related issues. In J. Borggaard, J. Burns, E. Cliff, and S. Schreck, editors, *Computational Methods for Optimal Design and Control, Proceedings of the AFOSR workshop on Optimal Design and Control (Arlington, VA, 1997)*, pages 151–179. Birkhäuser, 1998.
8. M. D. Gunzburger. *Perspectives in flow control and optimization.* SIAM, Philadelphia, PA, 2003.
9. M. E. Gurtin. *An introduction to continuum mechanics.* Academic Press, 2003.
10. J. Haslinger and R. A. E. Mäkinen. *Introduction to shape optimization. Theory, approximation, and computation.* SIAM, Philadelphia, 2003.
11. E. Laporte and P. Le Tallec. *Numerical methods in sensitivity analysis and shape optimization.* Birkhäuser, 2003.
12. B. Mohammadi and O. Pironneau. *Applied shape optimization for fluids.* Oxford University Press, 2001.
13. O. Pironneau. *Optimal shape design for elliptic systems.* Springer Series in Computational Physics. Springer, New York, 1984.
14. J. Sokolowski and J.-P. Zolésio. *Introduction to shape optimization. Shape sensitivity analysis.* Springer, Berlin, 1992.

A Novel Approach to Modeling Coronary Stents Using a Slender Curved Rod Model: A Comparison Between Fractured Xience-Like and Palmaz-Like Stents

Josip Tambača[1], Sunčica Čanić[2], and David Paniagua[3]

[1] Department of Mathematics, University of Zagreb, Bijenička 30, HR-10000 Zagreb, Croatia, tambaca@math.hr
[2] Department of Mathematics, University of Houston, 4800 Calhoun Road, Houston, TX 77204-3476, USA, canic@math.uh.edu
[3] Baylor College of Medicine, Texas Heart Institute at St Luke's Episcopal Hospital, P.O. Box 20345, Houston, TX 77225-0345, USA, dpaniag@pol.net

Summary. We present a novel mathematical model to study the mechanical properties of endovascular stents in their expanded state. The model is based on the theory of slender curved rods. Stent struts are modeled as linearly elastic curved rods that satisfy the kinematic and dynamic contact conditions at the vertices where the struts meet. A weak formulation for the stent problem is defined and a Finite Element Method for a numerical computation of its solution is used to study mechanical properties of two commonly used coronary stents (Palmaz-like and Xience-like stent) in their expanded, fractured state. A simple fracture (separation), corresponding to one stent strut being disconnected from one vertex in a stent, was considered. Our results show a drastic difference in the response of the two stents to the physiologically reasonable uniform compression and bending forces.

1 Motivation

Mathematical and computer modeling of endovascular stents is an efficient way to improve their design and performance [1, 5, 6, 8, 9, 12, 14–17, 22]. Currently available computational tools include "off the shelf", commercial software which is based on various three-dimensional Finite Element Method structure approximations of stent struts that form a three-dimensional stent mesh. Accurate, three-dimensional approximation of stents is often computationally very expensive in terms of time and memory requirements. This is why we developed a novel mathematical and computational algorithm which approximates three-dimensional stents as a *mesh of one-dimensional, elastic curved rods.*

W. Fitzgibbon et al. (eds.), *Applied and Numerical Partial Differential Equations*, Computational Methods in Applied Sciences 15, DOI 10.1007/978-90-481-3239-3_5,

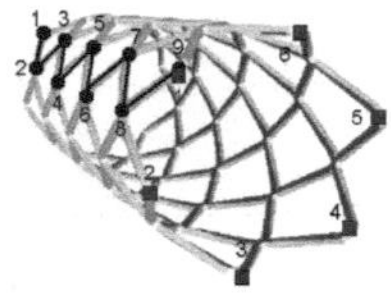

Fig. 1. A stent with $n_C = 6$ and $n_L = 9$.

Fig. 2. Deployment of a coronary stent.

Stent struts are modeled as linearly elastic, slender curved rods that satisfy the kinematic and dynamic contact conditions at the vertices where the struts meet. A weak formulation for the stent problem is defined and a Finite Element Method for a numerical computation of its solution was developed in [21]. The resulting FEM algorithm is incomparably simpler and faster than any corresponding three-dimensional solver, thereby enabling simulations of a large number of stent configurations in a short time.

Using this algorithm, we studied elastic deformation of stents *in their expanded state* (see Figure 1) exposed to physiologically reasonable pressure loads causing compression, expansion and bending. In particular, in this manuscript we compared the mechanical response to compression and bending of two commonly used coronary stents: a Palmaz-like stent and a Xience-like stent (see, e.g., Figure 2). Furthermore, a *fracture* (*separation*) was introduced prior to the computer simulations, corresponding to a separation of one stent strut from one vertex in the stent frame. Stent fractures and separation of coronary stent components are relatively rare (although fracture of stents used in larger arteries such as those of the legs, are more common) but they cause potentially serious complications of coronary artery stenting [13, 18]. Patients whose coronary stents suffer from stent fracture may present non-specific symptoms of angina as a result of restenosis (re-narrowing of a coronary artery) or in-stent thrombosis, or both [13, 18]. In order to insure proper recognition and treatment of this problem, physicians must be aware of its existence and of the stent behavior under these circumstances [3]. In this manuscript we present a few scenarios that shed light on the mechanical behavior of two commonly used coronary stents under the assumption of a disconnection of one of the struts in the stent frame. New insights related to the performance of such coronary stents are obtained.

2 The Model

We consider a stent to be a three-dimensional elastic body defined as a union of three-dimensional struts, see Figure 3 and Definition 1. The main novelty in this manuscript, over the standard approaches that can be found in literature [1, 5, 6, 9, 12, 14–17, 22], is the use of the one-dimensional curved rod model to approximate the slender three-dimensional stent struts, and a definition of a stent as a union of curved rods satisfying certain contact conditions. The one-dimensional approximation is given in terms of the arc-length of the middle curve of the rod as an unknown variable. The cross-section of a rod representing each stent strut is assumed to be rectangular, of width w and thickness t. The curved stent struts "lie" on a cylinder with reference (expanded) radius denoted by R, and reference (expanded) length denoted by L.

Struts themselves are assumed to be linearly elastic, with the elastic parameters given by the Lamé constants λ and μ, or, equivalently, by the Youngs modulus of elasticity E and the shear modulus μ.

2.1 Geometry: Parametrization of the Stent Frame

Without the loss of generality, we will be assuming that the stent struts form a uniform frame of diamonds with n_C vertices in the circumferential direction, and $n_L + 1$ vertices in the longitudinal direction, as shown in Figure 1. The assumption of uniform geometry is, however, not required for the implementation of the ideas described below, as they can be generalized to stents of arbitrary geometry with struts of different lengths. This will be utilized, for example, in Section 3.

Stent vertices will be denoted by $\mathbf{v}_{i,j}$, where $i = 1, \ldots, n_C$ and $j = 1, \ldots, n_L + 1$, see Figure 1. Vertices can be defined as

$$\mathbf{v}_{i,j} = \left(R\cos((i-1)\phi + (j-1)\phi/2), R\sin((i-1)\phi + (j-1)\phi/2), (j-1)\frac{L}{n_L} \right)^T,$$

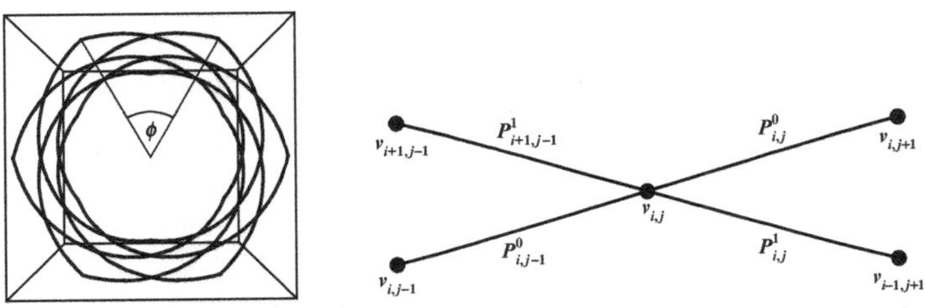

Fig. 3. Left: The figure shows the angle formed by a vertex of a stent, the center of the circular cross-section, and an adjacent vertex on the stent. The angle is denoted by $\phi = 2\pi/n_C$. Right: The geometry of an interior vertex $\mathbf{v}_{i,j}$ with incoming and outgoing struts.

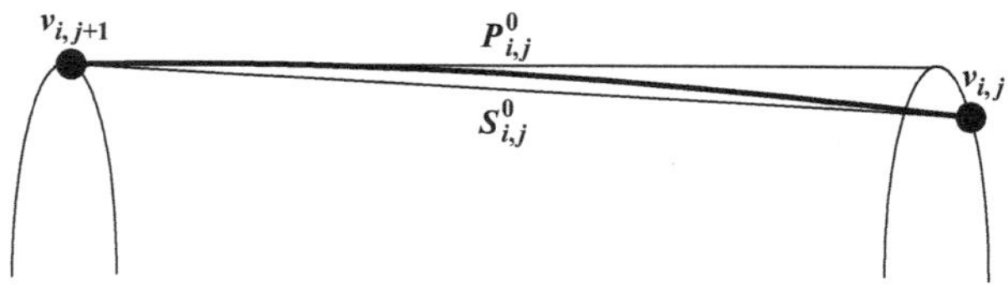

Fig. 4. Curved stent strut.

where $\phi = 2\pi/n_C$ is the angle formed by a vertex of a stent, the center of the circular cross-section of a stent, and an adjacent vertex on the same circumference of the stent, see Figure 3, left. The vertices on the adjacent circular cross-section are shifted by the angle $\phi/2$. Each interior vertex is characterized by two incoming and two outgoing struts. See Figure 3, right.

Struts of a high precision laser cut stainless steel stent are not straight, but curved and located on the cylinder of radius R. To write the equations for the curved stent struts we take a cord connecting the two vertices that define a strut, and then project the cord to the cylinder of radius R. See Figure 4. More precisely, denote by $R^k_{i,j}$, $k = 0, 1$, the two outgoing struts emerging from the vertex $\mathbf{v}_{i,j}$, and connecting to the vertices shifted by $\pm\phi/2$ at the level $j + 1$. Then the cords (straight lines) corresponding to the struts $R^k_{i,j}$, $k = 0, 1$, can be parameterized as

$$S^k_{i,j}(s) = s\mathbf{v}_{i,j} + (1 - s)\mathbf{v}_{((i-1-k) \bmod n_C)+1,j+1}, \quad s \in [0, 1],$$
$$i = 1, \ldots, n_C, \ j = 1, \ldots, n_L, \ k = 0, 1. \quad (1)$$

The *middle curve* of the curved stent struts $R^k_{i,j}$ can be expressed via the parameterization (see Figure 4)

$$P^k_{i,j} : [0, 1] \to \mathbb{R}^3,$$

where

$$P^k_{i,j}(s) = NS^k_{i,j}(s), \quad s \in [0, 1], \ i = 1, \ldots, n_C, \ j = 1, \ldots, n_L, \ k = 0, 1. \quad (2)$$

Here N is the operator that lifts the cord up to the cylinder of radius R:

$$N\mathbf{v} = P\mathbf{v} + R\frac{\mathbf{v} - P\mathbf{v}}{\|\mathbf{v} - P\mathbf{v}\|},$$

where P denotes the orthogonal projector on $\mathbf{e}_3$ in $\mathbb{R}^3$ with the standard scalar product, and $\{\mathbf{e}_1, \mathbf{e}_2, \mathbf{e}_3\}$ is the standard orthonormal basis of $\mathbb{R}^3$.

Using the parameterization $P^k_{i,j}$ of the middle curve of stent strut $R^k_{i,j}$, we can now introduce a parameterization of the *three-dimensional* stent strut $R^k_{i,j}$ as:

$$\mathbf{\Phi}^k_{i,j}(s_1, s_2, s_3) = P^k_{i,j}(s_1) + s_2\mathbf{n}^k_{i,j}(s) + s_3\mathbf{b}^k_{i,j}(s), \quad (3)$$

where $\mathbf{t}^k_{i,j}$, $\mathbf{n}^k_{i,j}$ and $\mathbf{b}^k_{i,j}(s)$ define a local basis at each point of the middle curve of stent strut $R^k_{i,j}$:

$$\mathbf{t}^k_{i,j}(s) = \frac{(P^k_{i,j})'(s)}{\|(P^k_{i,j})'(s)\|}, \quad \mathbf{n}^k_{i,j}(s) = \frac{(I-P)P^k_{i,j}(s)}{\|(I-P)P^k_{i,j}(s)\|}, \quad \mathbf{b}^k_{i,j}(s) = \mathbf{t}^k_{i,j}(s) \times \mathbf{n}^k_{i,j}(s),$$

for $s \in [0,1]$. The parameterization $\mathbf{\Phi}^k_{i,j}$ maps the set $[0,1] \times [-t/2, t/2] \times [-w/2, w/2]$ into $\mathbb{R}^3$.

Definition 1. *Three-dimensional stent Ω is a union of stent struts $R^k_{i,j}$ parameterized by $\mathbf{\Phi}^k_{i,j}$, given by (3):*

$$\Omega = \bigcup_{i=1}^{n_L} \bigcup_{j=1}^{n_C} \bigcup_{k=0}^{1} \mathbf{\Phi}^k_{i,j} \quad on\ [0,1] \times [-t/2, t/2] \times [-w/2, w/2]. \tag{4}$$

The interior *stent surface of a stent is defined by*

$$\Gamma_I = \bigcup_{i=1}^{n_L} \bigcup_{j=1}^{n_C} \bigcup_{k=0}^{1} \mathbf{\Phi}^k_{i,j} \quad on\ [0,1] \times \{-t/2\} \times [-w/2, w/2],$$

and the exterior *stent surface by*

$$\Gamma_E = \bigcup_{i=1}^{n_L} \bigcup_{j=1}^{n_C} \bigcup_{k=0}^{1} \mathbf{\Phi}^k_{i,j} \quad on\ [0,1] \times \{t/2\} \times [-w/2, w/2].$$

2.2 Mechanics: Stent as a Collection of Elastic Curved Rods

The curved rod model is a one-dimensional approximation of a three-dimensional rod-like structure to the ε^2 accuracy, where ε is the ratio between the largest dimension of the cross-section and the length of a rod. For a derivation and mathematical justification of the curved rod model see, e.g. [10, 11]. In general, the behavior of a three-dimensional rod-like elastic body is approximated by the behavior of its middle curve and of its cross-sections. In the curved rod model, the cross-sections behave approximately as infinitesimal rigid bodies that remain perpendicular to the deformed middle curve.

More precisely, let $P : [0, \ell] \to \mathbb{R}^3$ be the natural parameterization of the middle curve of the rod of length ℓ ($\|P'(s)\| = 1$, $s \in [0, \ell]$). Then the curved rod model can be formulated as a first-order system of differential equations for the following unknown functions:

$\tilde{\mathbf{u}} : [0, \ell] \to \mathbb{R}^3$, the displacement of the middle curve of the rod;

$\tilde{\boldsymbol{\omega}} : [0, \ell] \to \mathbb{R}^3$, the infinitesimal rotation of the cross-section of the rod;

$\tilde{\mathbf{q}} : [0, \ell] \to \mathbb{R}^3$, the contact moment; and

$\tilde{\mathbf{p}} : [0, \ell] \to \mathbb{R}^3$, the contact force.

(Here ℓ corresponds to the strut length, denoted by l_s.) For a given line force density $\tilde{\mathbf{f}}$, the equations of the curved rod model can be written as (see [19]):

$$\tilde{\mathbf{p}}' + \tilde{\mathbf{f}} = 0, \tag{5}$$

$$\tilde{\mathbf{q}}' + \mathbf{t} \times \tilde{\mathbf{p}} = 0, \tag{6}$$

describing the balance of contact force and contact moment, respectively, with

$$\tilde{\boldsymbol{\omega}}' - \mathbf{Q}\mathbf{H}^{-1}\tilde{\mathbf{Q}}^T\tilde{\mathbf{q}} = 0, \tag{7}$$

$$\tilde{\mathbf{u}}' + \mathbf{t} \times \tilde{\boldsymbol{\omega}} = 0, \tag{8}$$

describing the constitutive relations for a curved, linearly elastic rod. Here $\mathbf{t}$ is the unit tangent to the middle curve, $\mathbf{Q} = (\mathbf{t}, \mathbf{n}, \mathbf{b})$ is the orthogonal matrix containing the tangent vector $\mathbf{t}$ and vectors $\mathbf{n}$ and $\mathbf{b}$ that span the normal plane to the middle curve ($\mathbf{Q}$ describes the local basis at each point of the middle curve), and

$$\mathbf{H} = \begin{bmatrix} \mu K & 0 & 0 \\ 0 & EI_b & 0 \\ 0 & 0 & EI_n \end{bmatrix},$$

where $E = \mu\frac{3\lambda+2\mu}{\lambda+\mu}$ is the Young modulus of the material, I_n and I_b are moments of inertia of a cross-section and μK is the torsion rigidity of the cross-section. Therefore, $\mathbf{H}$ describes the elastic properties of the rod and the geometry of the cross-section.

The equation (8) is a condition that requires that the middle line is approximately inextensible and that allowable deformations of the cross-section are approximately orthogonal to the middle line. This condition has to be included in the solution space for the weak formulation of the problem (5)–(8) (pure traction problem for a single curved rod). Thus, introduce the space

$$V = \left\{(\tilde{\mathbf{v}}, \tilde{\mathbf{w}}) \in H^1(0,\ell)^6 : \tilde{\mathbf{v}}' + \mathbf{t} \times \tilde{\mathbf{w}} = 0\right\}. \tag{9}$$

Function $(\tilde{\mathbf{u}}, \tilde{\boldsymbol{\omega}}) \in V$ is called a weak solution of the problem (5)–(8) if

$$\int_0^\ell \mathbf{Q}\mathbf{H}\mathbf{Q}^T\tilde{\boldsymbol{\omega}}'\cdot\tilde{\mathbf{w}}'ds = \int_0^\ell \tilde{\mathbf{f}}\cdot\tilde{\mathbf{v}}ds + \tilde{\mathbf{q}}(\ell)\cdot\tilde{\mathbf{w}}(\ell) - \tilde{\mathbf{q}}(0)\cdot\tilde{\mathbf{w}}(0) + \tilde{\mathbf{p}}(\ell)\cdot\tilde{\mathbf{v}}(\ell) - \tilde{\mathbf{p}}(0)\cdot\tilde{\mathbf{v}}(0) \tag{10}$$

holds for all $(\tilde{\mathbf{v}}, \tilde{\mathbf{w}}) \in V$ (notice the difference in the notation between $\tilde{\boldsymbol{\omega}}$ and $\tilde{\mathbf{w}}$).

To model the mechanical behavior of a stent as a *collection of one-dimensional* linearly elastic, homogeneous, isotropic curved rods, we parameterize the struts using the one-dimensional parameterizations $P_{i,j}^k$ of the struts' middle curves, see (2). Now a stent can be defined as a union of *one-dimensional* parameterizations as follows:

$$\Omega_1 = \bigcup_{i=1}^{n_C} \bigcup_{j=1}^{n_L} \bigcup_{k=0}^{1} P_{i,j}^k([0,1]).$$

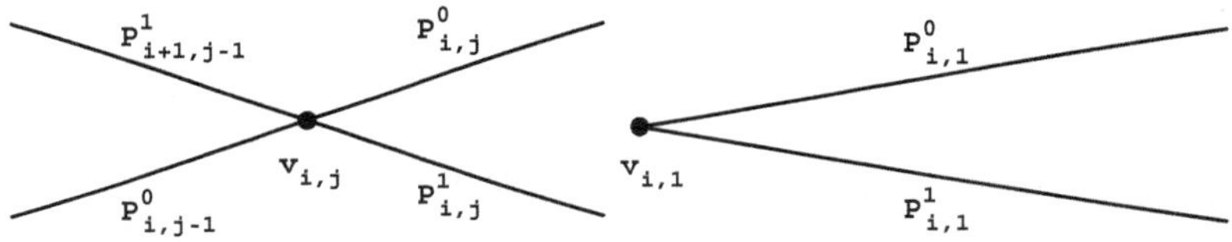

Fig. 5. Vertex $\mathbf{v}_{i,j}$.

Note that parameterizations $P^k_{i,j}$ are not arc-length parameterizations which is necessary for the formulation of the curved rod model (5)–(8). Nevertheless, they uniquely determine the middle curves of the stent struts and imply the existence of the arc-length parameterizations. Finding the arc-length parameterization in this case is a difficult task which is not necessary for the final formulation of the problem and the numerical method development.

Each of the curved rods approximating the stent struts $R^k_{i,j}$ satisfy a set of equations of the form (5)–(8). At the vertices where the curved rods meet, the kinematic and dynamic contact conditions determine the boundary condition for each curved rod in the stent frame structure. The kinematic contact condition describes the continuity of the kinematic quantities $\tilde{\mathbf{u}}^k_{i,j}$ and $\tilde{\boldsymbol{\omega}}^k_{i,j}$, stating that the displacement and the infinitesimal rotation for two struts meeting at a vertex, are the same. The dynamic contact condition is the equilibrium condition requiring that the sum of all contact forces at a vertex, and the sum of all contact moments at a vertex be equal zero. Thus, for each vertex $\mathbf{v}_{i,j}$ (see Figure 5) the kinematic contact conditions are given by

$$\tilde{\mathbf{u}}^0_{(i-1) \bmod n_C+1,j-1}(l_s) = \tilde{\mathbf{u}}^1_{i \bmod n_C+1,j-1}(l_s) = \tilde{\mathbf{u}}^0_{i,j}(0) = \tilde{\mathbf{u}}^1_{i,j}(0), \quad (11)$$

$$\tilde{\boldsymbol{\omega}}^0_{(i-1) \bmod n_C+1,j-1}(l_s) = \tilde{\boldsymbol{\omega}}^1_{i \bmod n_C+1,j-1}(l_s) = \tilde{\boldsymbol{\omega}}^0_{i,j}(0) = \tilde{\boldsymbol{\omega}}^1_{i,j}(0), \quad (12)$$

and the dynamic contact conditions are given by

$$\tilde{\mathbf{q}}^0_{(i-1) \bmod n_C+1,j-1}(l_s) + \tilde{\mathbf{q}}^1_{i \bmod n_C+1,j-1}(l_s) + \tilde{\mathbf{q}}^0_{i,j}(0) + \tilde{\mathbf{q}}^1_{i,j}(0) = 0, \quad (13)$$

$$\tilde{\mathbf{p}}^0_{(i-1) \bmod n_C+1,j-1}(l_s) + \tilde{\mathbf{p}}^1_{i \bmod n_C+1,j-1}(l_s) + \tilde{\mathbf{p}}^0_{i,j}(0) + \tilde{\mathbf{p}}^1_{i,j}(0) = 0, \quad (14)$$

for $i = 1, \ldots, n_C$, $j = 1, \ldots, n_L + 1$ with the convention that the quantity is removed for nonexistent indexes corresponding to the end vertices $\mathbf{v}_{i,1}$ and $\mathbf{v}_{i,n_L+1}$.

To define a weak formulation for the stent frame problem, introduce the following function space:

$$V_F = \left\{(\tilde{\mathbf{v}}^0_{1,1}, \tilde{\mathbf{w}}^0_{1,1}, \ldots, \tilde{\mathbf{v}}^1_{n_C,n_L}, \tilde{\mathbf{w}}^1_{n_C,n_L}) : (\tilde{\mathbf{v}}^k_{i,j}, \tilde{\mathbf{w}}^k_{i,j}) \in V^k_{i,j} \ \& \ (11), (12) \text{ hold}\right\},$$

where $V^k_{i,j}$ are the function spaces (9) corresponding to the struts $R^k_{i,j}$.

Now the *weak formulation for the stent frame structure consisting of curved rods* is given by the following:

Definition 2. *Function* $(\tilde{\mathbf{u}}_{1,1}^0, \tilde{\boldsymbol{\omega}}_{1,1}^0, \ldots, \tilde{\mathbf{u}}_{n_C,n_L}^1, \tilde{\boldsymbol{\omega}}_{n_C,n_L}^1) \in V_F$ *is a weak solution to the stent frame problem if*

$$\sum_{i=1}^{n_C}\sum_{j=1}^{n_L}\sum_{k=0,1}\int_0^{l_s} \mathbf{Q}_{i,j}^k \mathbf{H}(\mathbf{Q}_{i,j}^k)^T(\tilde{\boldsymbol{\omega}}_{i,j}^k)' \cdot (\tilde{\mathbf{w}}_{i,j}^k)' ds = \int_0^{l_s} \tilde{\mathbf{f}}_{i,j}^k \cdot \tilde{\mathbf{v}}_{i,j}^k ds \qquad (15)$$

holds for all $(\tilde{\mathbf{v}}_{1,1}^0, \tilde{\mathbf{w}}_{1,1}^0, \ldots, \tilde{\mathbf{v}}_{n_C,n_L}^1, \tilde{\mathbf{w}}_{n_C,n_L}^1) \in V_F$.

Notice again the difference in the notation for the infinitesimal rotation test functions $\tilde{\mathbf{w}}_{i,j}^k$ and the notation for the infinitesimal rotation solution functions $\tilde{\boldsymbol{\omega}}_{i,j}^k$. Also notice that all the intermediate boundary terms on the right-hand side of the equation (10) cancel out in the formulation (15) due to the kinematic and dynamics contact conditions.

Solution to the problem (15) is not unique. Namely, since only the derivative of $\tilde{\boldsymbol{\omega}}$ appears in the weak formulation, the solution will be determined up to a constant $\tilde{\boldsymbol{\omega}}_0$. Thus, if P is a point on the frame structure, then $\tilde{\boldsymbol{\omega}}(P) = \tilde{\boldsymbol{\omega}}_0$ is in the kernel of the problem. Furthermore, from the condition $\tilde{\mathbf{u}}' + \mathbf{t} \times \tilde{\boldsymbol{\omega}} = 0$, with $\tilde{\boldsymbol{\omega}}$ constant, one can solve the equation for $\tilde{\mathbf{u}}$ to obtain $\tilde{\mathbf{u}}(s) = \tilde{\mathbf{u}}_0 - P \times \tilde{\boldsymbol{\omega}}_0 = \tilde{\mathbf{u}}_0 + \tilde{\boldsymbol{\omega}}_0 \times P$. Thus, the infinitesimal rotation of the cross-section and displacement of P are unique up to the term

$$\begin{bmatrix} \tilde{\boldsymbol{\omega}}(P) \\ \tilde{\mathbf{u}}(P) \end{bmatrix} = \begin{bmatrix} \tilde{\boldsymbol{\omega}}_0 \\ \tilde{\mathbf{u}}_0 + \tilde{\boldsymbol{\omega}}_0 \times P \end{bmatrix},$$

for arbitrary vectors $\tilde{\mathbf{u}}_0, \tilde{\boldsymbol{\omega}}_0 \in \mathbb{R}^3$. This means that the solution is unique up to the translation and infinitesimal rotation of the frame structure. Thus we will be interested in the solution of (15) that satisfies an additional condition

$$\int_F \tilde{\mathbf{u}}(P) \cdot (\mathbf{a} + \mathbf{b} \times P) dP = 0, \quad \forall \mathbf{a}, \mathbf{b} \in \mathbb{R}^3. \qquad (16)$$

2.3 Numerical Implementation

The frame structure presented in this section is still extremely complex. The main obstacle for the numerical treatment of the problem of the form (15) is the implementation of the condition in the function spaces $V_{i,j}^k$ that should be satisfied by the test functions. For this reason, we made a further simplification that incorporates approximation of each curved rod by the piecewise straight rods. This approximation has been mathematically justified in [19, 20]. For details, please see [21]. A Finite Element Method was developed in [21] for a solution to this problem. Numerical results are presented next.

3 Numerical Results

The mechanical behavior of two types of stents is considered below: a Xience-like stent (nonuniform geometry) shown in Figure 6, and a Palmaz-like stent (uniform geometry), shown in Figure 14. Both stents are subject to two loading scenarios: uniform compression and bending.

Uniform Compression

A uniformly distributed force in the radial direction is applied to stents causing compression. Radial displacement from the expanded configuration is measured. The compression force corresponds to the pressure load of 0.5 atm. The force is calculated by considering the 0.5 atm pressure load of a *cylinder* (e.g., blood vessel) of length L acting on a *stent* of the same length L. This pressure load is physiologically reasonable. Namely, we can use the Law of Laplace to estimate exterior pressure loads to an inserted stent. Recall that the Law of Laplace relates the displacement u of the arterial wall with the transmurral pressure $p - p_0$ [7] via:

$$p - p_0 = \frac{Eh}{(1-\nu^2)R^2}u, \tag{17}$$

where E is the Young modulus of the vessel wall, h is the vessel wall thickness, R the vessel (reference) radius and ν the Poisson ratio. For incompressible materials such as arterial walls (nearly compressible), $\nu = 1/2$. The Young modulus of a coronary artery is between 10^5 and 10^6 Pa, see, e.g., [2] and the references therein. For our calculation let us take the intermediate value of $E = 5 \times 10^5$ Pa, and let us take the reference coronary artery radius to be around 1.3 mm with the vessel wall thickness $h = 1$ mm. Stents are typically oversized by 10% of the native vessel radius to provide reasonable fixation. Thus, 10% displacement of a coronary artery of radius 1.3 mm is 0.13 mm. This gives $u = 0.13$ mm. By plugging these values into the formula (17) one gets $p - p_0 \approx 5 \times 10^4$ Pa which equals 0.5 atm. Thus, a pressure load of 0.5 atm is necessary to expand a coronary artery by 10% of its reference radius. This force is applied to the stents studied below to capture the stent deformation under the coronary artery loading.

Bending

In the examples below we will be calculating stent deformation to forces causing bending. These forces will be applied pointwise to the center of a given

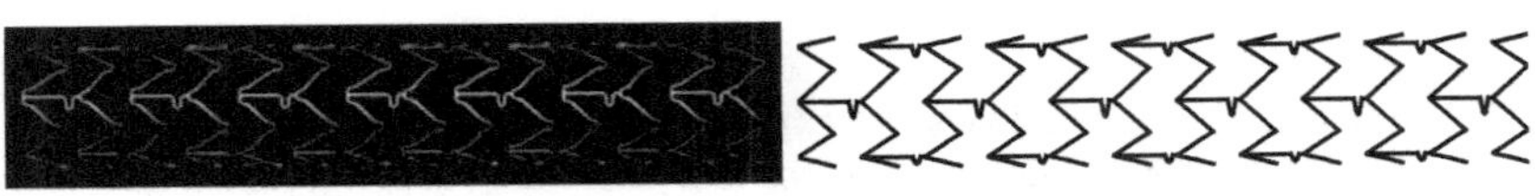

Fig. 6. Xience stent by Abbott (left); Computationally generated Xience-like stent (right) showing half of the mesh with $n_C = 6$ and $n_L = 24$.

stent (at 2–4 points in the center) and to the end points (at 1 point near each end of a stent). The force at the end points is applied in the opposite direction from the force applied to the center of the stent. The magnitude of the total applied force is calculated for each stents to be equal to the force that a curved vessel, with the radius of curvature $R_c = 2.5\,\text{cm}$, exerts on a straight stent that is inserted into the curved vessel. Stents with higher bending rigidity will deform less, while stents with low bending rigidity will deform more.

3.1 Xience-Like Stent (*Stent X*) (Non-Uniform Geometry)

The stent geometry is that of Multi-Link Mini Vision, resembling Xience stent by Abbott shown in Figure 6, left. Figure 6, right shows our computer-generated geometry of a Xience-like stent. The stent struts are made of Cobalt Chromium (CoCr) (L-605) (CoCr, Young's modulus $E = 2.43 \times 10^{11}\,\text{Pa}$) with thickness 0.08 mm. Stent struts are organized in zig-zag rings ("in-phase" rings) connected with horizontal struts which contain one wiggle near the protruding vertex of a zig-zag ring. Stent X has $n_C = 6$ vertices in the circumferential direction and $n_L = 24$ vertices in the longitudinal direction with reference radius $R = 1.5\,\text{mm}$.

In the examples below a fractured Xience-like stent will be considered, with a fracture corresponding to a disconnection of one strut from one vertex. In particular, a vertex in the middle of the stent is chosen to suffer component separation, see Figures 7 and 12. Namely, our simulations show that this vertex suffers from highest contact moments during bending (and compression). Denote this vertex by $\tilde{\mathbf{v}}$. There are three struts that meet at vertex $\tilde{\mathbf{v}}$: two symmetric, diagonally placed ones forming one zig-zag geometry in the zig-zag ring of stent struts, see Figure 7, bottom, and one horizontally placed strut

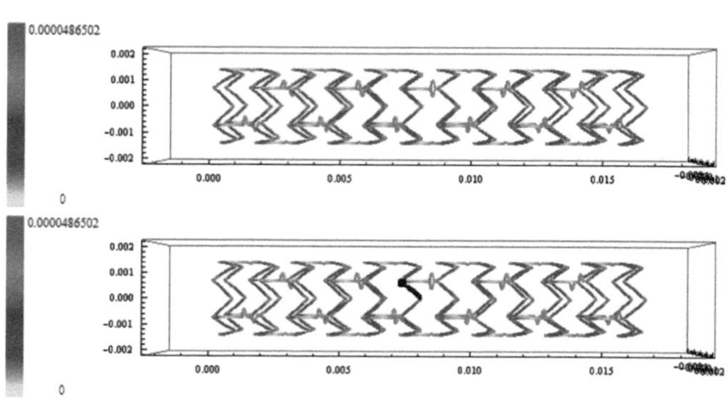

Fig. 7. Non-fractured Xience-like stent exposed to uniform compression. Stent struts are colored based on the magnitude of contact moment. The bottom figure shows the strut which will be disconnected from vertex $\tilde{\mathbf{v}}$, colored with a black dot.

connecting two different zig-zag rings, see Figure 12. We will consider below two examples: the first is an example of a Xience-like stent with a separated diagonally placed strut, and the second is an example of a Xience-like stent with a separated horizontally placed strut.

Example 1. Xience-like stent with a disconnected diagonally placed strut emerging from vertex $\tilde{\mathbf{v}}$ is exposed to uniform compression and bending. Figure 7 shows the bending moments for a non-fractured Xience-like stent, with a strut that is to be disconnected from vertex $\tilde{\mathbf{v}}$ shown in black. Figure 8 shows radial displacement under uniform compression of the fractured stent. The disconnected strut is shown in light blue (cyan). The two views show that the strut disconnected from vertex $\tilde{\mathbf{v}}$ protrudes into the lumen of the stented vessel by around 30% of the reference radius, causing potential for complications associated with in-stent thrombosis, as observed in clinical practice [13].

Figure 9 shows that the deformation of the disconnected strut causes higher contact moments. A comparison between the numbers on the scale shown on the left in Figures 7 and 9 indicate that the maximum bending moment for the deformed stent with a disconnected diagonally placed strut

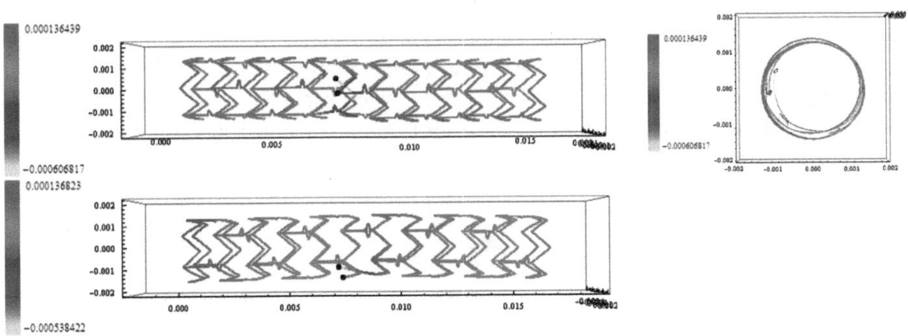

Fig. 8. Fractured Xience-like stent under uniform pressure load (three different views). The dislocated stent strut is shown in blue (cyan). The dots on the figure denote the points corresponding to the fractured vertex of a stent where the dislocated strut was broken away from the reference stent frame. The stent is colored based on the magnitude of the radial displacement.

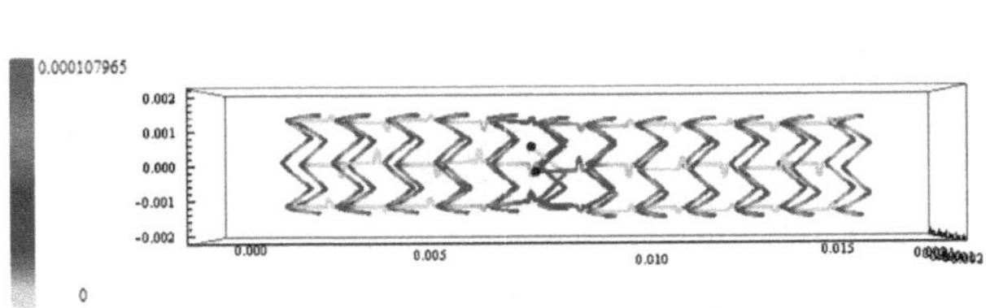

Fig. 9. Fractured Xience-like stent from Figure 8 under uniform pressure load. The stent is colored based on the magnitude of the contact moment.

is two times the contact moment of a non-fractured stent exposed to uniform compression. This is a precursor for possible further stent fractures that may be associated with this highly flexible and compliant stent.

Bending

Figure 10 shows contact moments for Xience-like stent exposed to bending forces. The bottom figure indicates the strut that is to be disconnected from vertex $\tilde{\mathbf{v}}$ (shown in black). The result of the bending load applied to the Xience-like stent with a disconnected diagonally placed strut is shown in Figure 11. The stent bends more than the non-fractured one. The calculated bending factors (reciprocal of the radius of curvature) for the non-fractured Xience-like stent (Stent X) and the fractured Xience-like stent (Stent X-frac1) are shown in Figure 19. Figure 11, top shows the stent from the side view, and the bottom figure shows the same stent from the bottom view. The two black dots denote the disconnected vertex, viewed from below of the curved stent.

Example 2. Xience-like stent with a disconnected horizontally placed strut emerging from vertex $\tilde{\mathbf{v}}$ is exposed to uniform compression and bending, see Figure 12. Figure 8 shows radial displacement under uniform compression of the fractured stent (the magnitude of the radial displacement is shown in the scale bar on the left of the figure). The disconnected strut is shown in light blue (cyan). The two views show that the disconnected strut protrudes into the lumen of the stented vessel causing potential for complications associated with thrombosis.

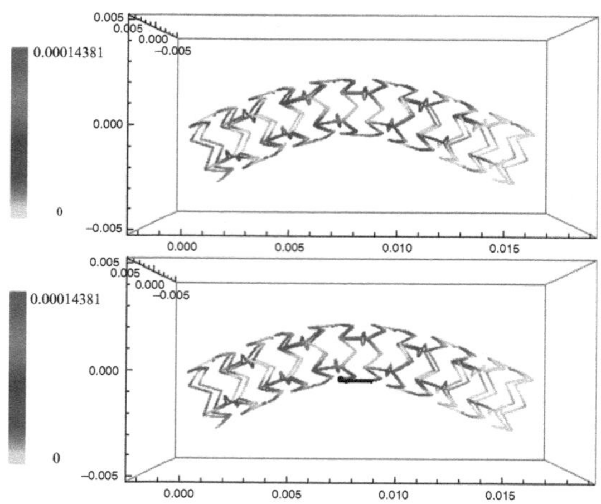

Fig. 10. Non-fractured Xience-like stent exposed to bending forces. Stent struts are colored based on the magnitude of the contact moment. The strut shown in black (right figure) denotes the strut that will be disconnected from the vertex denoted with a black dot.

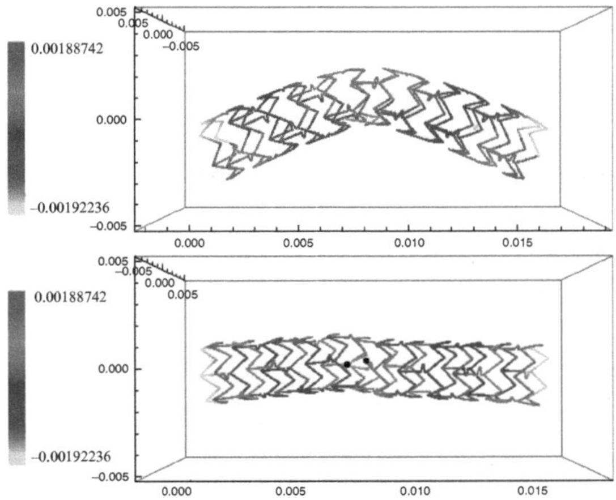

Fig. 11. Fractured Xience-like stent exposed to bending forces. Stent struts are colored based on the magnitude of the radial displacement. Two views are shown: a side view (top) and a view from the bottom of the deformed stent (bottom).

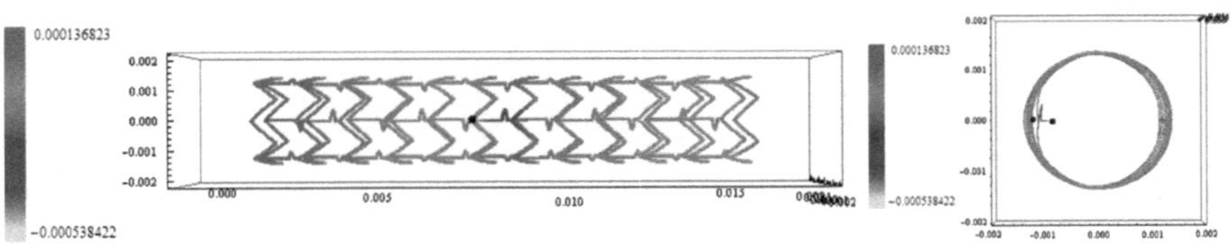

Fig. 12. Fractured Xience-like stent under uniform pressure load. Stent struts are colored based on the magnitude of the radial displacement. The circles on the figure denote the points corresponding to the fractured vertex of a stent. The disconnected strut, shown in blue, protrudes into the lumen with the largest radial displacement of all the struts.

Bending

Figure 13 shows a catastrophic deformation of a Xience-like stent with a disconnected horizontal strut under bending load. The disconnected strut is placed at the bottom, at the center of the bent stent. Figure 13 shows two views of the stent: the side view and the view from the bottom where the center of curvature of the bent stent lies. This deformation is too large for the model presented in this paper to be used to calculate accurate displacement and/or moments of the deformed stent. Our simulation, however, indicates that a disconnection of a central horizontal strut in a Xience-like stent will likely lead to unacceptable deformation under bending forces.

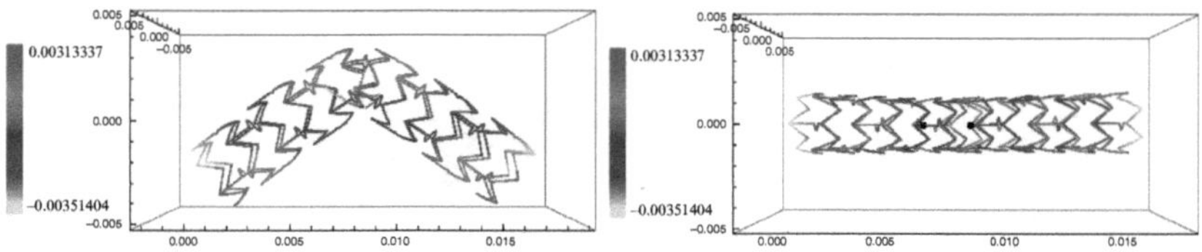

Fig. 13. Bending of a fractured Xience-like stent. Struts are colored based on the magnitude of the radial displacement. Two views are shown: a side view (left) and a view from the bottom of the stent (right). Catastophic deformation is observed.

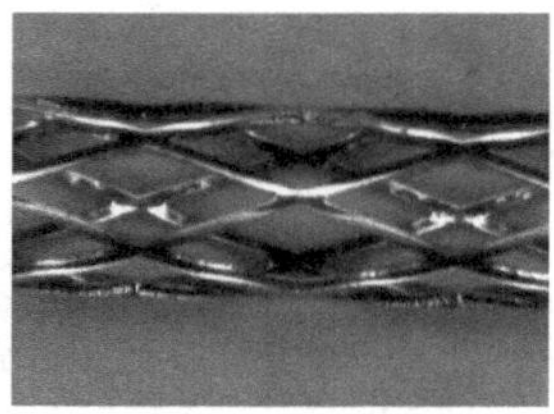

Fig. 14. A photograph of Palmaz stent by Cordis.

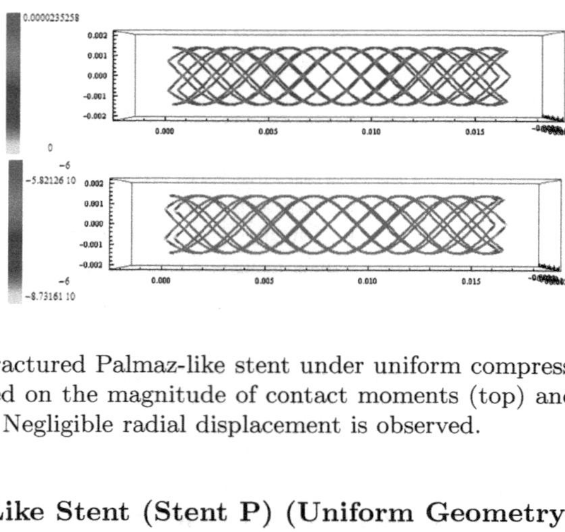

Fig. 15. Non-fractured Palmaz-like stent under uniform compression. Stent struts are colored based on the magnitude of contact moments (top) and radial displacement (bottom). Negligible radial displacement is observed.

3.2 Palmaz-Like Stent (Stent P) (Uniform Geometry)

A Palmaz-like stainless steel stent (316L) such as the one shown in Figure 14, with uniform geometry containing $n_C = 6$ vertices in the circumferential direction and $n_L = 24$ vertices in the longitudinal (axial) direction is considered. The stent has been expanded to the radius of 1.5 mm into its reference configuration.

Figure 15 shows contact moments and radial displacement of a Palmaz-like stent under uniform compression. This stent deforms more at the end

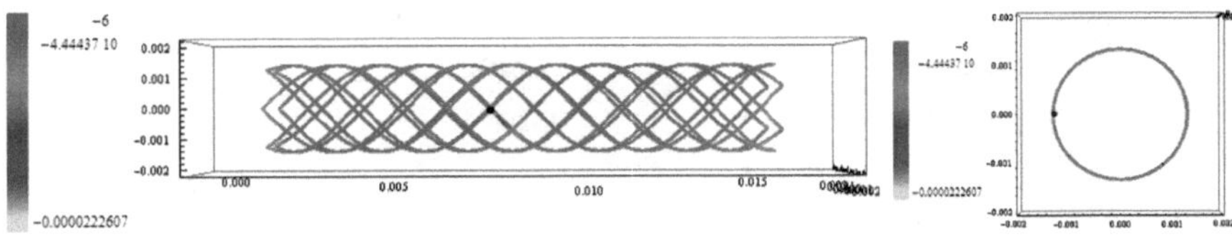

Fig. 16. Fractured Palmaz-like stent under uniform compression. Struts are colored based on the magnitude of radial displacement. The disconnected strut is shown in blue, with the black dot denoting the vertex from which the strut is disconnected. Two views are shown: a side view (left) and an axial view (right). The displacement of the disconnect strut is only 0.5% of the reference configuration.

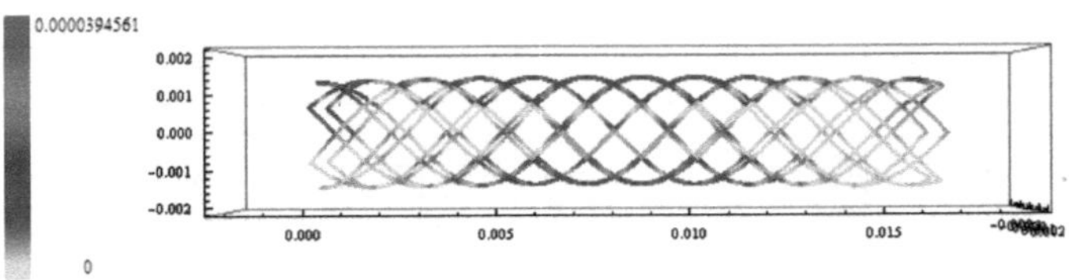

Fig. 17. Non-fractured Palmaz-like stent exposed to bending forces. Same bending forces are use as those corresponding to Figure 10. Stent struts are colored based on the magnitude of contact moment. Much smalled bending can be observed in comparison with the bending of a non-fractured Xience-like stent, shown in Figure 10.

points (radial displacement shown in light blue) than at the center (radial displacement shown in red). One of the diagonally placed struts was disconnected from a vertex $\tilde{\mathbf{v}}$ at the "center" of the stent, shown in Figure 16 with a black dot.

We see that, although the disconnected strut deforms more than the neighboring struts (shown in light blue versus red in Figure 16), the deformation is 25 times smaller (2×10^{-5} versus 5×10^{-4} m) than the deformation of the Xience-like stent with an "equivalent" disconnected strut shown in Figure 8. Thus, we conclude that a Palmaz-like stent with a disconnected strut in the center of a stent deforms less under uniform compression than a Xience-like stent with an equivalent disconnected strut (diagonally placed), see Figure 8.

Bending

In the remainder of this section we study the behavior of a fractured Palmaz-like stent under bending forces. Figure 17 shows the magnitude of the contact moment under the same bending forces as those that were applied to the Xience-like stent, shown in Figure 10. It is obvious that Palmaz-like stents have much higher bending rigidity than Xience-like stents. Figure 18 shows the magnitude of the contact moment for a Palmaz-like stent under bending forces with a disconnected strut from the vertex, shown in Figure 18 with a

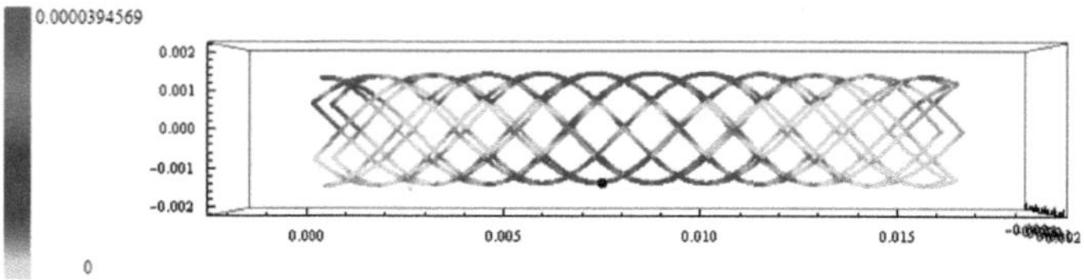

Fig. 18. Fractured Palmaz-like stent exposed to bending forces. Stent struts are colored based on the magnitude of contact moment. The black dot denotes the vertex from which a diagonally placed strut was disconnected. Much smaller bending can be observed in comparison with the bending of a fractured Xience-like stent shown in Figure 11.

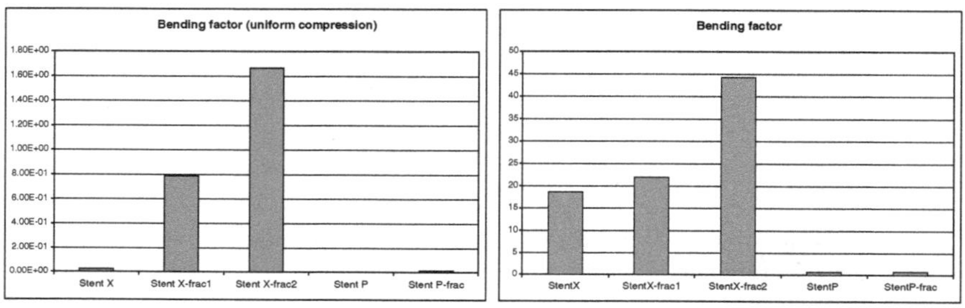

Fig. 19. Graphs showing the bending factor for the five stents: non-fractured Xience-like stent (stent X), fractured Xience-like stent from Example 1 (stent X-frac1), fractured Xience-like stent from Example 2 (stent X-frac2), non-fractured Palmaz-like stent (stent P) and fractured Palmaz-like stent (stent P-frac). Bending factor is calculated as the reciprocal of the radius of curvature for each deformed stent. Left: stents were exposed to the same uniform compression forces, as described at the beginning of Section 3. Right: stents were exposed to the same bending forces, as described at the beginning of Section 3.

black dot. Very small difference between the behavior of a non-fractured stent shown in Figure 17 and a fractured Palmaz-like stent shown in Figure 18 is observed.

4 Conclusions

Our model, based on the approximation of a three-dimensional stent strut mesh as a collection of slender curved rods, enables fast and accurate simulation of mechanical behavior of stents in their expanded state [21]. We used this model to study deformation of Palmaz-like stent and Xience-like stent with a fracture introduced prior to the simulation. The stent fractures considered in this manuscript correspond to a disconnection of one stent strut from one vertex. Drastic differences between the mechanical responses to uniform compression and bending of the Xience-like stent and of the Palmaz-like stent were detected. The following conclusions were obtained:

1. Palmaz-like stent is much stiffer than the Xience-like stent both under uniform compression and under bending force (compare Figures 7 and 15, and Figures 10 and 17). This, in turn, implies less deformation of a fractured Palmaz-like stent, see Figure 16, than the Xience-like stent, see Figure 8, and the overall smaller contact moments in the Palmaz-like stent introduced by a disconnection of a strut from the stent frame.
2. Disconnection of a horizontally placed strut in a Xience-like stent may lead to catastrophic deformation when such a stent is located in the tortuous (curved) geometry, which is the typical application of Xience-like stents, where the stent is naturally exposed to bending forces. See Figure 11.
3. Disconnection of any one strut in a Xience-like stent causes protrusion of a stent strut into the lumen of a stented artery by around 30% of its expanded radius, providing an environment that promotes coronary in-stent thrombosis and in-stent restenosis as clinicaly observed in [13, 18]. See Figure 8.
4. Disconnection of a diagonally-placed strut in a Xience-like stent causes visible bending of the stent even when the stent is exposed to uniform compression forces. See Figure 8 bottom and graphs in Figure 19 left.
5. Deformation of a fractured Xience-like stent (with one strut separated from one vertex) is significantly larger than the deformation of a fractured Palmaz-like stent when exposed to uniform compression during arterial pulsation and bending.

Acknowledgement. Research is supported in part by MZOS-Croatia under grant 037-0693014-2765 and by the NSF/NIH under grant DMS-0443826 and by the NSF under grant DMS-0806941, Texas Higher Education Board ARP 003652-0051-2006, NSF/NIH under grant DMS-0443826, by UH GEAR-2007 grant.

References

1. J. L. Berry, A. Santamarina, J. E. Moore, Jr., S. Roychowdhury, and W. D. Routh. Experimental and computational flow evaluation of coronary stents. *Ann. Biomed. Eng.*, 28(4):386–398, 2000.
2. S. Canic, C. J. Hartley, D. Rosenstrauch, J. Tambaca, G. Guidoboni, and A. Mikelic. Blood flow in compliant arteries: An effective viscoelastic reduced model, numerics and experimental validation. *Ann. Biomed. Eng.*, 34(4):575–592, 2006.
3. A. Carter. Stent strut fracture: Seeing is believing. *Catheter. Cardiovasc. Interv.*, 71(5):619–620, 2008.
4. P. G. Ciarlet. *Mathematical Elasticity. Volume I: Three-Dimensional Elasticity.* North-Holland, Amsterdam, 1988.

5. C. Dumoulin and B. Cochelin. Mechanical behavior modeling of balloon-expandable stents. *J. Biomech.*, 33(11):1461–1470, 2000.
6. A. O. Frank, P. W. Walsh, and J. E. Moore, Jr. Computational fluid dynamics and stent design. *Artificial Organs*, 26(7):614–621, 2002.
7. Y. C. Fung. *Biomechanics: Mechanical properties of living tissues.* Springer, second edition, 1993.
8. G. Hausdorf. Mechanical and biophysical aspects of stents. In P. Syamasundar Rao and Morton J. Kern, editors, *Catheter Based Devices for the Treatment of Non-coronary Cardiovascular Diseases in Adults and Children*, Philadelphia, PA, 2003. Lippincott Williams & Wilkins.
9. V. Hoang. Stent design and engineered coating over flow removal tool. Team #3 (Vimage), 10/29/04.
10. M. Jurak and J. Tambača. Derivation and justification of a curved rod model. *Math. Models Methods Appl. Sci.*, 9(7):991–1014, 1999.
11. M. Jurak and J. Tambača. Linear curved rod model. General curve. *Math. Models Methods Appl. Sci.*, 11(7):1237–1252, 2001.
12. K. W. Lau, A. Johan, U. Sigwart, and J. S. Hung. A stent is not just a stent: stent construction and design do matter in its clinical performance. *Singapore Med. J.*, 45(7):305–311, 2004.
13. A. N. Makaryus, L. Lefkowitz, and A. D. K. Lee. Coronary artery stent fracture. *Int. J. Cardiovasc. Imaging*, 23:305–309, 2007.
14. D. R. McClean and N. L. Eiger. Stent design: Implications for restenosis. *Rev. Cardiovasc. Med.*, 3(5):S16–22, 2002.
15. F. Migliavacca, L. Petrini, M. Colombo, F. Auricchio, and R. Pietrabissa. Mechanical behavior of coronary stents investigated through the finite element method. *J. Biomech.*, 35(6):803–811, 2002.
16. J. E. Moore Jr. and J. L. Berry. Fluid and solid mechanical implications of vascular stenting. *Ann. Biomed. Eng.*, 30(4):498–508, 2002.
17. A. C. Morton, D. Crossman, and J. Gunn. The influence of physical stent parameters upon restenosis. *Pathologie Biologie*, 52:196–205, 2004.
18. F. Shaikh, R. Maddikunta, M. Djelmami-Hani, J. Solis, S. Allaqaband, and T. Bajwa. Stent fracture, an incidental finding or a significant marker of clinical in-stent restenosis. *Catheter. Cardiovasc. Interv.*, 71(5):614–618, 2008.
19. J. Tambača. A model of irregular curved rods. In Z. Drmač, V. Hari, L. Sopta, Z. Tutek, and K. Veselić, editors, *Proceedings of the Conference on Applied Mathematics and Scientific Computing (Dubrovnik, 2001)*, pages 289–299. Kluwer, 2003.
20. J. Tambača. A numerical method for solving the curved rod model. *ZAMM Z. Angew. Math. Mech.*, 86(3):210–221, 2006.
21. J. Tambača, M. Kosor, S. Čanić, and D. Paniagua. Mathematical modeling of vascular stents. *SIAM J. Appl. Math.* Under revision.
22. L. H. Timmins, M. R. Moreno, C. A. Meyer, J. C. Criscione, A. Rachev, and J. E. Moore, Jr. Stented artery biomechanics and device design optimization. *Med. Bio. Eng. Comput.*, 45(5):505–513, 2007.

On the Stochastic Modelling of Interacting Populations. A Multiscale Approach Leading to Hybrid Models

Vincenzo Capasso and Daniela Morale

Department of Mathematics, University of Milan, IT-20133 Milan, Italy,
vincenzo.capasso, daniela.morale@unimi.it

Summary. In this paper a review by the research work of the authors on the stochastic modelling of interacting individuals is presented. Both cases of direct and indirect interaction (via underlying fields) are considered. Due to the strong coupling among individuals, the evolution of each individual is governed by a stochastic equation whose parameters are themselves stochastic; as a consequence we are dealing with a doubly stochastic system, and this is a source of complexity which may tremendously increase as the number of individuals becomes extremely large. A possible way to reduce complexity is to apply suitable laws of large numbers, at a mesoscale, in order to obtain a mean field governed now by deterministic PDEs. In this way we may obtain an approximation of the driving fields which are deterministic at the macroscale, thus driving, at the microscale, a simply stochastic evolution for the individuals. Such models are called hybrid models.

Key words: Stochastic differential equations, measure-valued processes, empirical measures, law of large numbers, invariant measures, ant colonies, tumour-induced angiogenesis, hybrid models, multiscales

1 Introduction

In biology and medicine it is possible to observe a wide spectrum of formation of patterns and clustering, usually due to self-organization phenomena. This may happen at any scale; from the cellular scale of embryonic tissue formation, wound healing or tumor growth, and angiogenesis, the microscopic scale of life cycles of bacteria or social amoebae, to the larger scale of animal grouping. Patterns are usually explained in terms of forces, external and/or internal, acting upon individuals. In this way formation of aggregating networks are shown as a consequence of collective behavior. Evidence of stochasticity are often shown. A fruitful approach to the mathematical description of such phenomena, suggested since long by various authors [10, 14, 19, 23, 28, 29], is based on the so called *individual based models*, i.e. the "movement" of each

W. Fitzgibbon et al. (eds.), *Applied and Numerical Partial Differential Equations*, Computational Methods in Applied Sciences 15, DOI 10.1007/978-90-481-3239-3_6,

individual embedded in the total population is described. This is known as *Lagrangian approach*, i.e. individuals are followed in their motion. Possible randomness may be included in the motion, so that the variation in time of the (random) location of the individuals in a group composed of $N(t)$ individuals at time $t \geq 0$, $X_N^k(t) \in \mathbb{R}^d$, $k = 1, \ldots, N(t)$, is described by a family of stochastic equations. On the other hand, particles are subject to specific forces of interaction which are responsible of the reaction term.

A classical widespread approach has been given in terms of PDEs [20, 24, 25]. This is due, above all, to the wider spread knowledge on nonlinear PDEs; so grouping behavior has been described by relevant quantities such as scalar or vector fields. Such kind of models are often called *Eulerian models*, since they describe the evolution of population densities; they are based on continuum equations, typically (deterministic) partial differential equations of the advection–reaction–diffusion type

$$\rho_t + \nabla \cdot (\mathbf{v}\rho) = \nabla \cdot (D\nabla\rho) + \nu(\rho), \tag{1}$$

where ρ is the population density and $\mathbf{v}$ is the velocity field, and $\nu(\rho)$ is a possible additive reaction term which may include birth and death processes. The advection term may describe the interaction mechanisms among individuals (via the velocity $\mathbf{v}$), while the non-convective (diffusive) flux takes into account the spatial spread of the population.

In conclusion, the two different approaches (Lagrangian and Eulerian) describe the system at different scales: the finer scale description is based on the (stochastic) behavior of individuals (microscale), and the larger scale description is based on the (continuum) behavior of population densities (macroscale). The central problem is to determine how information is transferred across scales; one of the aims of the modelling is to catch the main features of the interaction at the scale of single individuals that are responsible, at a larger scale, for a more complex behavior that leads to the formation of patterns [10]. Often a multiple scale approach is preferable: the global behavior of the population is described, at the macroscopic scale, by a continuum density whose evolution in terms of integro-differential equations is derived by a limiting process from the empirical distribution associated with a large number of particles. From the mathematical point of view this means to perform some kind of law of large numbers, in such a way that one may identify a possibly regular measure of the population distribution, having a density which satisfies a PDE similar to the equation (1).

This is a way to reduce the complexity of Lagrangian models. Indeed, the evolution equation of each individual is usually a stochastic equation whose parameters are themselves stochastic. This is a source of complexity which may tremendously increase as the number of individuals becomes extremely large, as it may happen in many cases of real interest. Applying suitable laws of large numbers at the mesoscale, we obtain an approximation of the driving fields which are deterministic at the macroscale. They drive, at the microscale, a simply stochastic evolution for the individuals. Here we consider a review of the investigation programme on the subject, that the authors have been carrying out during the last decade [1, 2, 5, 6, 21–23].

In Section 2 we discuss the mathematical modelling of the stochastic interacting population when the number of individuals is finite, both in the cases of direct and indirect interaction. We consider both Lagrangian and Eulerian (discrete) descriptions. In Sections 3 and 4, we look at two specific cases: a model for stochastic aggregating–repelling individuals (direct interaction), and a model for a branching and growth of vessels in tumor induced angiogenesis, an example of stochastic fiber processes, coupled with the continuum underlying field of a chemoattractor released by the tumor (indirect interaction). In Section 5 we study the derivation of the corresponding hybrid models, for the two working examples. In the first one we recall the mathematically rigorous derivation of the limit model as the number of individuals increases to infinity, via a law of large numbers; in the second example, we handle a heuristic derivation of an hybrid model. Finally in Section 6, we address the problem of the long time behavior of a stochastic interacting particle model, as the number of particle N is still finite. In particular, we consider the case of example one, discussed previously.

2 Individuals, Interactions and Evolution

We consider a population composed, at time $t \geq 0$, by a (possibly random) number $N(t)$ of individuals. Let the random variable $X_N^k(t)$ represent the random state in $\mathbb{R}^d$, e.g., the spatial location, of the kth individual, for $k = 1, \ldots, N(t)$. From a Lagrangian point of view, the state of the system of $N(t)$ particles may be described as a family of $N(t)$ stochastic processes $\{X_N^k(t)\}_{t \in \mathbb{R}_+}$, $k = 1, \ldots, N(t)$, defined on a common probability space $(\Omega, \mathcal{F}, P)$ and valued in $(\mathbb{R}^d, \mathcal{B}_{\mathbb{R}^d})$, where $\mathcal{B}_{\mathbb{R}^d}$ is the usual Borel σ-algebra generated by intervals. A convenient description of the state of the kth individual may achieved via a random Dirac-measure $\varepsilon_{X_N^k(t)}$, defined as follows:

$$\varepsilon_{X_N^k(t)}(B) = \begin{cases} 1 & \text{if } X_N^k(t) \in B \\ 0 & \text{if } X_N^k(t) \notin B \end{cases} \qquad \forall B \in \mathbb{R}^d. \tag{2}$$

It is a random element of $\mathcal{M}_P(\mathbb{R}^d)$, the space of probability measures on $\mathbb{R}^d$; for any sufficiently smooth function $f : \mathbb{R}^d \to \mathbb{R}$

$$\int_{\mathbb{R}^d} f(y) \varepsilon_{X_N^k(t)}(dy) = f\left(X_N^k(t)\right)$$

is a real valued random variable.

For any $t \geq 0$, given the particle locations $X_N^k(t)$, $k = 1, \ldots, N(t)$, an *Eulerian (discrete) description* of the system can be given in terms of the random probability measure on $\mathbb{R}^d$

$$X_N(t) = \frac{1}{N(t)} \sum_{k=1}^{N(t)} \varepsilon_{X_N^k(t)} \in \mathcal{M}_P(\mathbb{R}^d). \tag{3}$$

This measure may be regarded as the empirical distribution of the location of a single particle of the system in $\mathbb{R}^d$ at time $t \in \mathbb{R}_+$. Note that the number of particles may be either constant over time, say $N(t) = N$, for all $t \in \mathbb{R}_+$, or a dynamical variable itself, described, e.g., by a suitable birth and death process.

A key question concerns the modelling of the *interaction*; interaction among particles may be direct or indirect. In the first case individuals interact directly, i.e. the force exerted on each of them depends on the distribution of the individuals in the population. In the case of indirect interaction the force exerted on each particle depends on an underlying field whose evolution depends on the distribution of the entire population; as a consequence the dependence of the evolution of the spatial distribution of a single individual upon the spatial distribution of the whole population is mediated by the underlying field.

2.1 Direct Interaction and System Evolution

For sake of simplicity, let $N(t) = N$, independent of $t \in \mathbb{R}_+$. Generally speaking, in this first case we may describe the evolution of the system by a system of N random equations

$$dX_N^k(t) = h_N(X_N^1(t), \ldots, X_N^N(t), B_t, t)\, dt, \quad k = 1, 2, \ldots, N, \tag{4}$$

where $h_N : (\mathbb{R}^d)^n \times \mathbb{R}^d \times R_+ \to \mathbb{R}$ is a suitable function modelling the interaction. The random perturbing function B_t may model a random forcing factor.

If we consider pairwise interaction, the interaction between a couple of individuals is mathematically modelled by a reference potential K_1, depending on the distance between the two particles. In this way the range of the potential kernel represents the spatial region of influence of the interaction.

A good choice is $K_1 = W_1 * W_1$, a kernel given by the convolution of a sufficiently regular probability density W_1 with itself; we assume that the interaction of two particles, out of N, located in x and y, respectively, is modelled by

$$\frac{1}{N} K_N(x - y), \quad \text{where } K_N(z) = N^\beta K_1(N^{\beta/d} z), \tag{5}$$

which expresses the rescaling of K_1 with respect to the total member N of particles, in terms of a scaling coefficient $\beta \in [0, 1]$. Particles X_N^i and X_N^j

interact if the supports of the associated smoothed measures $W_N * \varepsilon_{X_N^l}$, $l = i, j$, overlap. As a consequence, if we denote by $W_N(z) = N^{\beta} W_1(N^{\beta/d} z)$, the interaction of the single k-particle, out of N, located at $X_N^k(t)$, with all the others in the population is given by

$$\begin{aligned} J\left(X_N^1(t), \dots, X_N^N(t)\right)\left(X_N^k(t)\right) &= \frac{1}{N} \sum_j \int_{\mathbb{R}^d} W_N(X_N^k - y) W_N(y - X_N^j) dy \\ &= (W_N * W_N * X_N(t))(X_N^k(t)) \\ &= \sum_{i=1}^N \frac{1}{N} K_N\left(X_N^i(t) - X_N^k(t)\right) \\ &= (K_N * X_N(t))(X_N^k(t)) \\ &=: I\left[X_N(t)\right](X_N^k(t)). \end{aligned} \tag{6}$$

In many cases a convenient way to model randomness is to consider an independent additive noise, acting on each particle; so that a possible model for (4) is

$$dX_N^k(t) = \left[f_N^k(t) + I\left[X_N(t)\right](X_N^k(t))\right] dt + \sigma dW^k(t), \quad k = 1, \dots, N; \tag{7}$$

the term given in (6) describes any interaction of the kth particle with other particles in the system, the function $f_N^k : \mathbb{R}_+ \to \mathbb{R}$ describes the individual dynamics which may depend only on time or on the state of the particle itself, and, finally, $\{W^k\}$, $k = 1, \dots N$ is a family of independent standard Wiener processes. In this review the diffusion coefficient σ is kept constant.

The system (7) offers a *Lagrangian description* of the stochastic model; from the fact that for any real function g on $\mathbb{R}^d \times \mathbb{R}_+$,

$$\int_{\mathbb{R}^d} g(x,t) X_N(t)(dx) = \frac{1}{N} \sum_{k=1}^N g(X_N^k(t), t).$$

Itô's formula leads to the *Eulerian (discrete) description* via an evolution equation for the empirical measure $X_N(t)$ [4, 6, 23]; indeed, for any $g \in C_b^{2,1}(\mathbb{R}^d \times \mathbb{R}_+)$,

$$\begin{aligned} \int_{\mathbb{R}^d} g(x,t) X_N(t)(dx) = \int_{\mathbb{R}^d} g(x,0) X_N(0)(dx) + \int_0^t Op_1\left(X_N(s), g(\cdot, s)\right) ds \\ + M_N\left[\underline{X}, \underline{W}\right](t), \end{aligned} \tag{8}$$

where

$$M_N\left[\underline{X}, \underline{W}\right](t) = \int_0^t \frac{\sigma}{2N} \sum_{k=1}^N \nabla g(X^k(s), s) dW^k(s) \tag{9}$$

is a zero mean martingale, so that, by the Doob inequality [4],

$$
\begin{aligned}
E\left[\sup_{t\leq T}|M_N\left[\underline{X},\underline{W}\right](t)|\right]^2 &\leq E\left[\sup_{t\leq T}|M_N\left[\underline{X},\underline{W}\right](t)|^2\right]\\
&\leq 4\frac{4\sigma^2}{N^2}\sum_{k=1}^{N}E\left[\int_0^T|\nabla g(X_N^k(s),s)|^2ds\right]\\
&\leq \frac{4\sigma^2\|\nabla g\|_\infty^2 T}{N}. \qquad (10)
\end{aligned}
$$

2.2 Indirect Interaction and System Evolution

As said above, in the case of indirect interaction the force exerted on each particle depends on an external field. As an example of self-organization mediated by a system of underlying fields, we may consider a process of individual organization that occurs at a microscopic scale, while diffusion of an underlying field occurs at a macroscopic scale. The dynamics of the field depends on the individuals themselves (for example, a degradation phenomenon may be due to an interaction with individuals at relevant spatial locations). Let $Z_N^k(t)$ be the state of the kth individual out of $N(t)$, at time t. Again, note that $N(t)$ may be itself a stochastic process. A general model might appear of the following form: for any $t\geq 0$

$$
dZ_N^k(t) = F\left[C(\cdot,t)\right](Z_N^k(t))dt + \sigma\, dW^k(t), \quad k=1,\ldots,N(t), \qquad (11)
$$

$$
\frac{\partial}{\partial t}C(x,t) = Op_2(C(\cdot,t))(x) + \widetilde{I}[Z_N(t)](x), \qquad x\in\mathbb{R}^d. \qquad (12)
$$

In this case the evolution of an individual state $Z_N^k(t)$ is driven by an underlying field $C(x,t)$, via the operator $F\left[C(\cdot,t)\right]$ depending on the field and acting on each individual; on the other hand, the evolution equation of the field $C(x,t)$ depends itself upon the structure of the system of individuals by means of $\widetilde{I}[Z_N(t)](x)$, an operator which depends on the empirical measure

$$
Z_N(t) = \frac{1}{N(t)}\sum_{k=1}^{N(t)}\varepsilon_{Z_N^k(t)}
$$

of individuals, acting at a spatial location x. For simplicity, also here we consider a diffusion coefficient σ in the SDEs (11) constant in time and space. Note that also the evolution of the stochastic process $\{N(t)\}_{t\in\mathbb{R}_+}$ may depend upon the underlying field $C(t,x)$.

Again, Itô's formula may lead to an *Eulerian (discrete) description* of the spatial structure of the population $Z_N(t)$ coupled with the equation (12) for $C(x,t)$, i.e. for any $g\in C_b^{2,1}(\mathbb{R}^d\times\mathbb{R}_+)$,

$$\int_{\mathbb{R}^d} g(x,t) Z_N(t)(dx) = \int_{\mathbb{R}^d} g(x,0) Z_N(0)(dx) + \int_0^t Op_3\left(Z_N(s), C(x,t), g(\cdot,s)\right) ds + M_N\left[\underline{Z}, \underline{W}\right](t). \quad (13)$$

In the next two sections we provide two examples of self organization phenomena, in which the dynamics depends upon direct interaction among individuals, in the first case, and upon indirect interaction in the second case.

3 Direct Interaction: an Aggregation–Repulsion Model

As an example of direct interaction we consider a stochastic system of $N(t) \equiv N$ individuals, subject to an advection term and a stochastic individual component. Here we specify the advection components on the basis of possible assumptions inducing self-organization of biological populations. "Social" forces are responsible for interaction of each individual with other individuals in the population within suitable neighborhoods. We consider both aggregating and repelling forces, which compete, but act at different scales. They are modelled by two regular kernels $G, K_N : \mathbb{R}^d \to \mathbb{R}$, with $G, K_N \in C_b^2(\mathbb{R}^d, \mathbb{R}_+)$, as given by (6).

In the case of aggregation the parameter β in (6) is equal to zero, so that the aggregating force exerted on the kth individual is given by $(\nabla G * X_N(t))X_N^k(t)$ (McKean–Vlasov interaction); in the case of repulsion, the repelling force is given by $(\nabla K_N * X_N(t))X_N^k(t)$, with $\beta \in (0,1)$, where K_N and the empirical measure X_N are given by (5) and (3) (moderate interaction) [23, 26, 27]. It is clear how the choice of β may determine the range and the strength of the influence of neighboring particles; indeed, any particle interacts (repelling) with $O(N^{1-\beta})$ other particles in a volume of order $O(N^{-\beta})$.

Additionally, the movement of each individual particle might be driven by an external information coming from the environment, expressed via a suitable potential $U : \mathbb{R}^d \to \mathbb{R}$. The potential

$$U \in C_b^2(\mathbb{R}^d, \mathbb{R}_+) \quad (14)$$

is taken as a smooth non-negative even function; we assume that it satisfies the following condition [33–35]: there exist constants $M_0 \geq 0$ and $r > 0$ such that

$$\left(\nabla U(x), \frac{x}{|x|}\right) \leq -\frac{r}{|x|}, \quad |x| \geq M_0, \quad (15)$$

where $(\cdot,\cdot)$ denotes the usual scalar product in $\mathbb{R}^d$.

Again the stochastic component is modelled by a family of independent standard Wiener processes $\{W^k,\ k = 1, \ldots\}$. These systems have been already discussed by the authors in several papers [1, 2, 6, 21–23].

Based on these modelling assumptions, we consider the following system of SDEs:

$$dX_N^k(t) = \left[\gamma_1 \nabla U(X_N^k(t)) + \gamma_2 \left(\nabla\left(G - K_N\right) * X_N\right)\left(X_N^k(t)\right)\right] dt + \sigma dW^k(t), \qquad k = 1, \ldots, N, \quad (16)$$

where $\gamma_1, \gamma_2, \sigma \in \mathbb{R}_+$. In the case $\gamma_1 = 0$, the system is a purely diffusive interacting particle system.

By standard arguments [4], we can prove that the system admits a unique solution $X(t) = (X_N^1(t), \cdots, X_N^N(t))$ for all $t \in [0, T]$, with almost surely continuous trajectories [6]. From the system (16), Itô's formula applied to a function $f \in C_b^{2,1}(\mathbb{R}^d \times \mathbb{R}_+)$ of $X_N^k(t)$, for any $k = 1, \ldots, N$, gives the evolution equation of the empirical measure (3) as follows:

$$\begin{aligned}\int_{\mathbb{R}^d} f(x,0) X_N(s)(dx) = \int_{\mathbb{R}^d} & f(x,0) X_N(0)(dx) \\ & \int_0^t \int_{\mathbb{R}^d} \left(\left[\gamma_1 \nabla U + \gamma_2 \left(\nabla\left(G - K_N\right) * X_N\right)\right](x)\right. \\ & \qquad\qquad \left.\nabla f(x,s)\right) X_N(s)(dx)ds \\ & + \int_0^t \int_{\mathbb{R}^d} \left(\frac{\sigma^2}{2} \Delta f(x,s) + \frac{\partial}{\partial s} f(x,s)\right) X_N(s)(dx)ds \\ & + \sigma \frac{1}{N} \int_0^t \sum_{k=1}^N \nabla f\left(X_N^k(s), s\right) dW^k(s), \qquad (17)\end{aligned}$$

where again the last term in (17) is a zero mean martingale with respect to the natural filtration of the process $\{X_N(t),\ t \in \mathbb{R}_+\}$.

In conclusion in the example presented here, the Lagrangian description of the system (7), discussed in the previous section, has the form of the system (16), while its Eulerian (discrete) description is given by the system (17). In Figure 1 simulation results for the same initial condition, and for different drifts, are shown. For more simulation results and comparison with experimental data, the interested reader may refer to [1, 22, 23].

4 Interaction via Underlying Fields: A Birth and Growth Model

An interesting example of formation of patterns may be found in the process of tumor growth and in particular in angiogenesis. Tumor-induced angiogenesis is believed to occur when normal tissue vasculature is no longer able to support growth of an avascular tumor. At this stage the tumor cells, lacking nutrients and oxygen, become hypoxic. This is assumed to trigger cellular release of

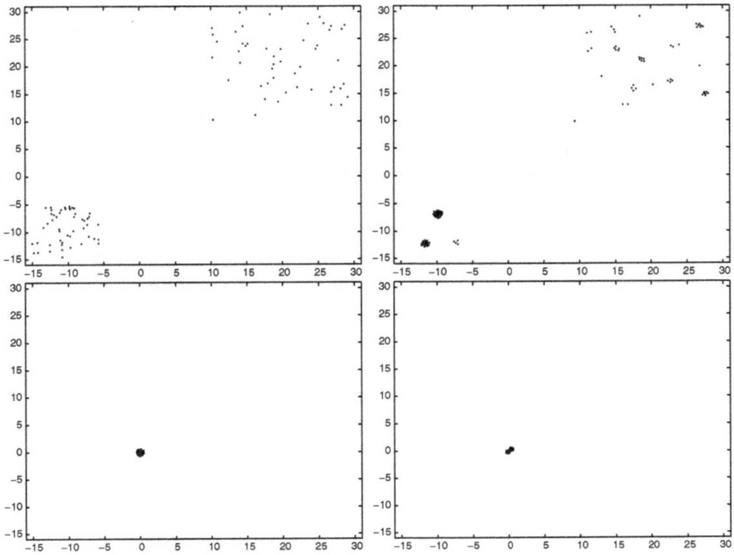

Fig. 1. Configuration of 100 particles for parameters values $\sigma = 0.02$, $\beta = 0.5$: (up left) $T = 0$, (up right) $T = 500$, $\gamma_1 = 0$, $\gamma_2 = 1$, (down left) $T = 1000$, $\gamma_1 = \gamma_2 = 1$, $\nabla U(x) = x/(1 + |x|)$, (down right) $T = 100$, $\gamma_1 = \gamma_2 = 1$, $\nabla U(x) = |x|^2$.

tumor angiogenic factors, TAF, which start to diffuse into the surrounding tissue and approach endothelial cells (ECs) of nearby blood vessels [13]. ECs subsequently respond to the TAF concentration gradients by forming sprouts, dividing and migrating towards the tumor. So, at an individual level, cells interact and perform a branching process coupled with elongation, under the stimulus of a chemical field produced by a tumor. In this way formation of aggregating networks (vessels) are shown as a consequence of collective behavior.

The initiation of sprouting from preexisting parental vessels is not considered here; in order to avoid further mathematical technicalities, we assume a given number N_0 of initial capillary sprouts; we refer to literature [16] for details on this topic. Let $N(t)$ be the number of tips at time t, and $X^i(t) \in \mathbb{R}^d$ the location of the tip of the ith vessel at time t. Furthermore, let us denote by T_i the branching time of the ith tip, i.e. the random time when the ith tip branches from an existing vessel. We model sprout extension by tracking the trajectory of individual capillary tips. The movement (extension) of the tips follows a Langevin model; at any $t > T^i$ and for any $k \in \{1, \ldots, N(t)\}$ we have

$$\begin{aligned} dX^i(t) &= v^i(t)(1 - \gamma \mathbb{I}_{X(t)}(X^i(t))dt, \\ dv^i(t) &= \left(-kv^i(t) + F\left(C(t, X^i(t))\right)\right)dt + \sigma dW^i(t), \end{aligned} \tag{18}$$

where $v^i(t)$ is the velocity of the ith tip at time t. According to a typical chemotaxis, velocity $v^i(t)$ is driven by a function F of the underlying field C. An example is $F(C(t, X^i(t))) = \nabla C(t, X^i(t)$, so that vessels follow the increasing density of the chemoattractor; the advection term includes the typical inertial component $-kv^i(t)$. A family of independent Wiener processes $W^i(t)$ model stochasticity. Finally, the network of endothelial cells is described by

$$X(t) = \bigcup_{i=1}^{N(t)} \{X^i(s), T_i \leq s \leq t\},$$

the union of the trajectories of the tips. In the equation (18) the parameter γ may assume only the 0 and 1 values; $\gamma = 0$ means that no impingement is considered; otherwise, for $\gamma = 1$ the phenomenon of anastomosis is taken into account (see [7] and references therein, for further information).

The branching process $\Phi_N(ds, dx)$ is modelled as a marked counting process with stochastic intensity

$$\alpha(t, x) = \alpha h(C(t, x)) \sum_{i=1}^{N(t^-)} \delta_{X^i(t)}(x), \tag{19}$$

where $h \in C_b(\mathbb{R}^d)$ is a non negative function. The equation (19) means that the probability that branching occurs exactly at the kth tip is given by

$$\text{prob}\left(\Phi(]t, t + dt] \times X^k(t)) \mid \mathcal{F}_{t^-}\right) = \frac{\alpha(t, X^k(t))}{\int_{\mathbb{R}^d} \alpha(t, x)dx} dt.$$

The counting process $N(t)$ is given by $N(t) = \Phi_N(]-\infty, t], \mathbb{R}^d)$, so that the probability of having a new tip during the time interval $]t, t + dt]$ is

$$\text{prob}\left(N(t + dt) - N(t) = 1 \mid \mathcal{F}_{t^-}\right) = \sum_{i=1}^{N(t^-)} \alpha(t, X^i(t))dt;$$

when a tip located in x branches, the initial value of the state of the new tip is taken as $(X^{N(t)+1}, v^{N(t)+1}) = (x, v_0)$, where v_0 is a non random velocity.

The chemotactic field $C(t, x)$ diffuses and degradates; the consumption is proportional to the extension velocities v^i, $i = 1, \ldots, N(t)$. So, for any $(t, x) \in \mathbb{R}_+ \times \mathbb{R}^d$,

$$\frac{\partial}{\partial t} C(t, x) = c_1 \delta_A(x) + d_1 \triangle C(t, x) - \eta C(t, x) \frac{1}{N} \sum_{i=1}^{N(t)} (v^i(t) \delta_{X^i(t)} * V_\varepsilon)(x). \tag{20}$$

We have considered a mollified version of the relevant random distributions, by means of a convolution with the kernel $V_\varepsilon(x)$, a smooth function with compact support of order ε. From a mathematical point of view, the use

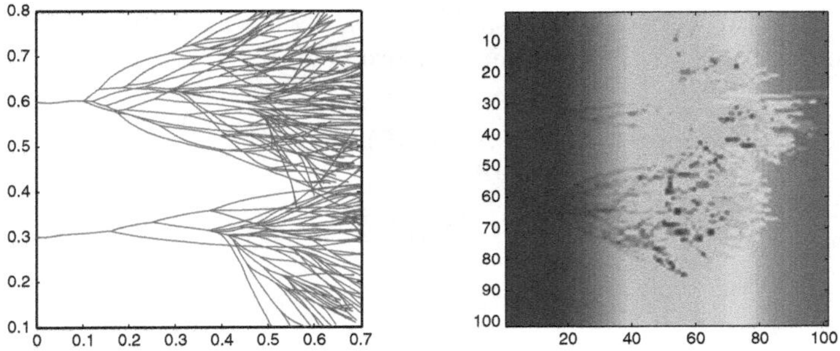

Fig. 2. A vessel network (on the left) interacting with a degrading TAF field (on the right) ($d_1 = 0, \gamma = 0$).

of mollifiers reduces analytical complexity; from a modelling point of view this might correspond to a nonlocal reaction with the relevant underlying fields. Parameters $c_1, d_1, \eta \in \mathbb{R}^+$ in the equation (20) represent the rate of production of a source located in a region $A \subset \mathbb{R}^d$, modelling, e.g., a tumor mass, the diffusivity and the rate of consumption, respectively. We have denoted by $\delta_{X^i(t)}(x)$ the random distribution (Dirac density) localized at the tip $X^i(t)$, for $i = 1, \ldots, N(t)$. Note that the equation (20) is a random partial differential equations, since the degradation term depends on the stochastic processes $\{(X^i(t), v^i(t))\}_t$, for any $i = 1, \ldots, N(t)$. The stochasticity of the underlying field leads to the stochasticity of the kinetic parameters of birth and growth of vessels. Figure 2 shows a simulation of the network coupled with a degradating field (for technical simplicity we have taken $d_1 = 0$, $\gamma = 0$).

To this process we may associate two fundamental random spatial measures, describing the network at time t; given a suitable scale parameter N, Q_N, the empirical measure associated with the processes $(X^k(t), v^k(t))$, $k = 1, \ldots, N(t)$, is given by

$$Q_N(t) = \frac{1}{N} \sum_{i=1}^{N(t)} \varepsilon_{(X^k(t), v^k(t))}, \tag{21}$$

while, $V_N(t)$, the empirical spatial distribution of velocities, is given by

$$V_N(t) = \frac{1}{N} \sum_{i=1}^{N(t)} v_k(t) \varepsilon_{X^k(t)} = \int_{\cdot \times \mathbb{R}^d} v \, Q_N(t)(d(x, v)).$$

We may write the equation (20) in the following form:

$$\frac{\partial}{\partial t} C(t, x) = c_1 \delta_A(x) + d_1 \triangle C(t, x) - \eta C(t, x)(V_N(t) * V_\varepsilon)(x). \tag{22}$$

Given a smooth function $g \in C_b(\mathbb{R}^d \times \mathbb{R}^d)$, by Itô's formula we obtain an evolution equation for the random measure Q_N [7]

$$\begin{aligned}\int_B g(x,v)Q_N(t)d(x,v) = \int_B g(x,v)Q_N(0)d(x,v)\\ + \int_0^t \int_B \Big[\nabla_x g(x,v)v + g(x,v)\alpha_1(s,x)\delta_{v_0}(v)\\ - \nabla_v g(x,v)\left[kv - F(C(t,x))\right]\\ + \frac{\sigma^2}{2}\Delta_v g(x,v)\Big] Q_N(t)(d(x,v))ds + \tilde{M}_N(t), \quad (23)\end{aligned}$$

where the last term

$$\begin{aligned}\tilde{M}_N(t) = \int_0^t \int_{\mathbb{R}^n} \left[\Phi_N(ds,dx) - N\alpha(s,x)Q_N(t)(dx \times \mathbb{R}^d)\,ds\right]\\ + \int_0^t \frac{\sigma}{2N}\sum_{k=1}^{N(t)} \nabla_v g((X^k(t), v^k(t)))dW^k(t)\end{aligned}$$

is a zero mean martingale, such that again by the Doob inequality, for N sufficiently large

$$E\left[\sup_{t\leq T}|\tilde{M}_N(t)|\right]^2 \leq C\frac{TN(t)}{N^2}(\|g\|_2^2 + \|\nabla g\|_2^2) < C\frac{T}{N}. \quad (24)$$

In conclusion in the example presented here, the Lagrangian description of the system (11)–(12), discussed in the previous section, has the form of the system (18), (19) and (20), while the Eulerian discrete description (12)–(13) is given by the system (22)–(23).

5 Hybrid Models: Large Population Behavior

Let us place our attention on the following facts. In the detailed models, in both examples, the evolution equation of each individual (either an individual in a population, or a tip in a vessel network) is a stochastic equation whose parameters are themselves stochastic; as a consequence we are dealing with a doubly stochastic system. A major difficulty, both analytical and computational, derives from the fact that, indeed, the parameters are $\{\mathcal{F}_{t^-}\}$-stochastic, i.e. their value at time $t > 0$ depends upon the actual *history* $\mathcal{F}_t$ of the whole system up to time t^-.

Let us remind the main features of the discrete systems, as already discussed in Section 2.

Direct Interaction

In this case each individual k, out of N, satisfies a system of SDEs of the form

$$dX_N^k(t) = Op\left[X_N(t)\right](X_N^k(t))dt + \sigma dW^k(t), \quad k = 1, \ldots, N, \tag{25}$$

where

$$X_N(t) = \frac{1}{N}\sum_{j=1}^{N} \varepsilon_{X_N^j(t)}$$

is the empirical measure at time t, and Op is a suitable operator which expresses the specific model of interaction.

Hence the analysis and the computation of the above system requires the knowledge of the evolution of all individuals up to time t; clearly $X_N(t)$ is an $\{\mathcal{F}_t\}$-stochastic quantity.

Indirect Interaction

In this case the individual dynamics is described by a system of the form

$$dZ_N^k(t) = Op\left[C(\cdot, t)\right](Z_N^k(t))dt + \sigma\, dW^k(t), \quad k = 1, \ldots, N(t), \tag{26}$$

whose kinetic parameters depend upon a biochemical underlying field $C(x,t)$ which obeys to a random evolution equation of the form

$$\frac{\partial}{\partial t}C(x,t) = Op_1[C(\cdot,t)](x) + Op_2\left[Z_N(t), C(\cdot,t)\right](x), \tag{27}$$

where $Z_N(t)$ is the empirical measure of the states $Z_N^k(t)$, and Op_1 and Op_2 are suitable operators which express the specific model of spatial spread and the interaction with the field produced by the whole system of individuals, respectively.

Once again, the analysis and the computation of the above system requires the knowledge of the evolution of all individuals up to time t; clearly $Z_N(t)$ is an $\{\mathcal{F}_{t-}\}$-stochastic quantity (in this case also the evolution of $N(t)$ is involved).

The strong coupling with the field (produced by the individuals themselves, in the first case, and external, in the second case) is a source of complexity which may tremendously increase as the number of individuals becomes extremely large, as it may happen in many cases of real interest. Under these circumstances, a possible way to reduce complexity, which has been suggested by the authors and by a large literature, is to apply suitable laws of large numbers at the mesoscale, i.e. in a suitable neighborhood of any relevant point $x \in \mathbb{R}^d$, such that, at that scale we may approximate, in the first case, $X_N(t)$ by a deterministic measure $X(t)$, possibly having a density $\rho(x,t)$ with respect to the usual Lebesgue measure; in the second case we may approximate $Z_N(t)$ by a deterministic measure, possibly having a density $w(x,t)$ with respect to the usual Lebesgue measure. The relevant densities $\rho(x,t)$, and $w(x,t)$ will satisfy suitable deterministic evolution equations. In this way we obtain an approximation of the driving fields which are deterministic at the macroscale, which now drive, at the microscale, a simply stochastic evolution for the individuals. More specifically, a typical individual k in the first model (25) will satisfy the following SDE:

$$dY^k(t) = Op\left[\rho(\cdot,t)\right](Y^k(t))dt + \sigma dW^k(t), \quad k = 1, \ldots, N, \tag{28}$$

coupled with a deterministic equation for $\rho(x,t)$. For the second model (26)–(27), a typical individual k will satisfy the following SDE:

$$dY^k(t) = Op\left[\widetilde{C}(\cdot,t)\right](Y^k(t))dt + \sigma\, dW^k(t), \quad k = 1, \ldots, N(t), \tag{29}$$

where the evolution equation for the underlying field has become

$$\frac{\partial}{\partial t}\widetilde{C}(x,t) = Op_1[\widetilde{C}(\cdot,t)](x) + Op_2\left[w(\cdot,t), \widetilde{C}(\cdot,t)\right](x), \tag{30}$$

coupled with a deterministic equation for $w(x,t)$.

A more detailed analysis follows for the two models described in Sections 3 and 4. Though, for the aggregation–repulsion model we have been able to carry out a detailed rigorous analysis, while for tumor-driven angiogenesis only an heuristic derivation has been obtained, which leads to a system of evolution equations which is compatible with existing deterministic models already available in literature [30–32].

We wish to stress that anyhow substituting mean densities of individuals in the first model, or mean densities of tips in the second model, to the corresponding stochastic quantities, leads to an acceptable coefficient of variation (percentage error) only when a law of large numbers can be applied, i.e. whenever the relevant numbers per unit volume are sufficiently large; otherwise stochasticity cannot be avoided, and, in addition, to mean values, the mathematical analysis and/or simulations should provide confidence bands for all quantities of interest. Indeed, numerical simulations carried out for the fully stochastic model show that local coefficients of variation are, indeed, much smaller in regions of largely crowded populations (either individuals or vessels) [3].

5.1 The Aggregation–Repulsion Model

Following [6], we show how to derive rigorously an hybrid model, as described at the beginning of this section in the case of the aggregation–repulsion model described in Section 3. For details the interested reader may refer to [6]. First note that from (17) we get an averaged equation

$$E\left[\langle X_N(t), f(\cdot,t)\rangle\right] = E\left[\langle X_N(0), f(\cdot,0)\rangle\right]$$
$$+ E\left[\int_0^t \langle X_N(s), [\gamma_1 \nabla U + \gamma_2 \left(\nabla\left(G - K_N\right) * X_N\right)](\cdot)\nabla f(\cdot,s)\rangle\, ds\right.$$
$$\left. + \int_0^t \left\langle X_N(s), \frac{\sigma^2}{2}\Delta f(\cdot,s) + \frac{\partial}{\partial s} f(\cdot,s)\right\rangle ds\right]. \quad (31)$$

Furthermore, thanks to the inequality (10), the quadratic variation of the martingale term vanishes, in a finite time interval $[0,T]$. So we might expect a deterministic behavior of the system in the limit.

Let us sketch the mathematically rigorous proof of this behavior in the case of large populations.

A Relative Compactness Result

We assume some regularity conditions for the initial empirical measure $X_N(0)$,

$$\sup_{N\in\mathbb{N}} E\left[\int_{\mathbb{R}^d} |x| X_N(0)(dx)\right] < \infty, \quad (32)$$

$$\sup_{N\in\mathbb{N}} E\left[\int_{\mathbb{R}^d} |h_N(x,0)|^2 dx\right] = \sup_{N\in\mathbb{N}} E\left[||h_N(\cdot,0)||_2^2\right] < \infty, \quad (33)$$

where

$$h_N(x,t) = (W_N * X_N(t))(x), \quad (34)$$

is a mollified measure.

Furthermore, let us impose the following restriction on β in the definition of the scaled kernel (5), $\beta \in (0, d/(d+2))$.

We have proven [6] the tightness and then the boundedness of small variations of the process X_N, in the bounded Lipschitz metric [6]. This leads, by means of the characterization of relative compactness by Ethier and Kurtz [11], to the following result on the sequence of laws $\mathcal{L}(X_N)$ of $X_N = \{X_N(t),\ t \in \mathbb{R}_+,\ N \in \mathbb{N}\}$:

Theorem 1 ([6]). *Under the hypotheses listed above and in Section 3, the sequence* $\{\mathcal{L}(X_N)\}_{N\in\mathbb{N}}$ *is relatively compact in the space* $\mathcal{M}_{\mathcal{P}}(C([0,T], \mathcal{M}_{\mathcal{P}}(\mathbb{R}^d)))$.

This is the main result needed for the asymptotics of the evolution equation of the measure-valued process $\{X_N(t), t \in \mathbb{R}_+\}$. Indeed, Theorem 1 implies the existence of a subsequence $N_k \subset \mathbb{N}$, $N_1 < N_2 < \ldots$, such that the sequence $\{\mathcal{L}(X_{N_k})\}_{k\in\mathbb{N}}$ converges in $\mathcal{M}_{\mathcal{P}}(C([0,T], \mathcal{M}_{\mathcal{P}}(\mathbb{R}^d)))$ to some limit $\mathcal{L}(X)$, which is the distribution of some process $X = \{X(t),\ t \in [0,T]\}$, with trajectories in $C([0,T], \mathcal{M}_{\mathcal{P}}(\mathbb{R}^d))$. We discuss the uniqueness of the limit later on. By now we assume uniqueness, so that we may take $\{N_k\} = \mathbb{N}$;

by Skorokhod theorem [4] we may assert that, corresponding to the possible unique limit law, we can also have an almost sure convergence, i.e.

$$\lim_{N\to\infty} \sup_{t\le T} d_{BL}(X_N(t), X(t)) = 0 \quad \mathbb{P} - a.s. \tag{35}$$

Regularity Properties of the Limit Measure

It is possible to show that there exists a positive (random) function h defined on $[0,T]\times\mathbb{R}^d$ such that

$$\lim_{N\to\infty} \mathbb{E}\left[\int_0^T \int_{\mathbb{R}^d} |h_N(x,t) - h(x,t)|^2 dxdt\right] = 0. \tag{36}$$

The equation (36) shows that the limit measure $X \in \mathcal{M}_P([0,T]\times\mathbb{R}^d)$ has P-a.s. a density

$$h \in L^2\left([0,T]\times\mathbb{R}^d\right) \tag{37}$$

with respect to the Lebesgue measure on $[0,T]\times\mathbb{R}^d$, i.e. for any $f \in C_b([0,T]\times\mathbb{R}^d)$

$$\int_0^T \int_{\mathbb{R}^d} f(t,x) X(dx,dt) = \int_0^T \int_{\mathbb{R}^d} f(t,x)\, h(t,x)(dx,dt). \tag{38}$$

By now, we do not know neither whether the measure $X(t)$ has a density for any fixed $t \in [0,T]$ nor that the density is deterministic. The next step is the identification of the limit by acquiring information on the *limit dynamics.* We have proven the following:

Proposition 1. *Let us suppose that a law of large numbers holds at initial time*

$$\lim_{N\to\infty} \mathcal{L}(X_N(0)) = \delta_{\mu_0} \quad in\ \mathcal{M}_{\mathcal{P}}(\mathcal{M}_{\mathcal{P}}(\mathbb{R}^d)), \tag{39}$$

where μ_0 has a density p_0 in $L^2(\mathbb{R}^d)$. Then, almost surely, for any $f \in C_b^{2,1}(\mathbb{R}^d, \mathbb{R}_+)$, $0 \le t \le T$,

$$\begin{aligned}\langle X(t), f(\cdot,t)\rangle = \langle \mu_0, f(\cdot,0)\rangle + \int_0^t \langle\, h(\cdot,s), \frac{1}{2}\sigma^2 \Delta f(\cdot,s) + \frac{\partial}{\partial s} f(\cdot,s) \\ + [(\nabla G_a * h(\cdot,s))(\cdot) + \nabla U(\cdot) - \nabla h(\cdot,s)]\cdot \nabla f(\cdot,s)\rangle ds. \end{aligned} \tag{40}$$

This means that any limit measure $X \in \mathcal{C}([0,T], \mathcal{M}_P(\mathbb{R}^d))$ is a solution of the equation (40), with $h \in L^2\left([0,T]\times\mathbb{R}^d\right)$, satisfying the relation (38).

So we have proven that for any $t \in [0,T]$, the measure $X(t)$ is absolutely continuous with respect to the Lebesgue measure, so that it admits a density for each t. We prove it by showing that the Fourier transform of the measure $X(t)$ is in L^2 for any $t \in [0,T]$, so that a density exists and the latter is also in $L^2(\mathbb{R}^d)$ and we prove that it is also L^2 uniformly bounded. So we have shown the following result:

Theorem 2. *Under the hypotheses of Theorem 1, let us suppose that a law of large numbers applies at initial time*

$$\lim_{N\to\infty} \mathcal{L}(X_N(0)) = \delta_{\mu_0} \quad \text{in } \mathcal{M}_{\mathcal{P}}(\mathcal{M}_{\mathcal{P}}(\mathbb{R}^d)), \tag{41}$$

where μ_0 has a density p_0 in $L^2(\mathbb{R}^d) \cap C_b^2(\mathbb{R}^d)$. Then, almost surely, the sequence X converges in law to a deterministic measure X. For any $t \in [0,T]$ the measure $X_N(t)$ has a density $h(\cdot,t)$ such that, for any $f \in C_b^{2,1}(\mathbb{R}^d, \mathbb{R}_+)$, $0 \le t \le T$,

$$\begin{aligned}\langle h(\cdot,t), f(\cdot,t)\rangle = \langle \mu_0, f(\cdot,0)\rangle + \int_0^t \langle h(\cdot,s), \frac{1}{2}\sigma^2 \Delta f(\cdot,s) + \frac{\partial}{\partial s} f(\cdot,s) \\ + [(\nabla G_a * h(\cdot,s))(\cdot) + \nabla U(\cdot) - \nabla h(\cdot,s)] \cdot \nabla f(\cdot,s)\rangle ds. \end{aligned}\tag{42}$$

One can easily see that the equation (42) is the weak form of the following partial differential equation:

$$\begin{aligned}\frac{\partial}{\partial t}\rho(x,t) &= \frac{\sigma^2}{2}\Delta\rho(x,t) + \nabla\cdot(\rho(x,t)\nabla U(x)) \\ &\quad + \nabla\cdot[\rho(x,t)\nabla(\rho(x,t) - G*\rho(\cdot,t))(x)], \quad x\in\mathbb{R}^d, t\ge 0, \\ \rho(x,0) &= p_0(x), \quad x\in\mathbb{R}^d.\end{aligned}\tag{43}$$

Regularity Properties of the Limit Measure

The uniqueness of the limit h derives from the uniqueness of the weak solution of the viscous equation (43), which can be achieved by classical arguments [12].

Hybrid Model

The equation (43) describes a mean field due to the large number of individuals. As far as the individual dynamics is concerned, for any k, we have that the typical particle $X^k(t) \sim Y^k(t)$, follows the SDE:

$$\begin{aligned}dY^k(t) = -\,&\big[\nabla U(Y^k(t)) + \nabla G_a * \rho(\cdot,t)(Y^k(t)) - \nabla\rho(Y^k(t)) \\ &-\nabla U(Y^k(t))\big]\, dt + \sigma dW^k(t),\end{aligned}$$

subject to the initial condition $Y^k(0) = X^k(0)$. While the Brownian stochasticity of the movement of each particle is preserved, the drift is now the same for each particle and depends on the mean field ρ in the equation (43).

5.2 The Branching and Growth Process

As discussed in [7], in the case of the branching and growth process described in Section 4, we may only give an heuristic convergence result. Starting from the system (23), if, formally, we take $Q_N(t)(d(x,v)) \to Q_\infty(t)(d(x,v)) = p(t,x,v)dxdv$, then

$$\begin{aligned}\int_B g(x,v)p(t,x,v)dxdv = \int_0^t \int_B p(s,x,v)ds\, dxdv \Big[\frac{\sigma^2}{2}\Delta_v g(x,v) \\ + \nabla_x g(x,v)v + g(x,v)\alpha_1(s,x)\delta_{\{v_0\}}(v) \\ - \nabla_v g(x,v)\left[kv - F\left(\widetilde{C}(t,x)\right)\right]\Big] \qquad (44)\end{aligned}$$

$$\frac{\partial}{\partial t}\widetilde{C}(t,x) = c_1\delta_A(x) + d_1\Delta\widetilde{C}(t,x) - \eta\widetilde{C}(t,x)\int_{\mathbb{R}^d} p(t,x,v)dv. \qquad (45)$$

The equation (44) may be seen as the weak form of the following partial differential equation for the density $p(t,x,v)$:

$$\begin{aligned}\frac{\partial}{\partial t}p(t,x,v) = -v\cdot\nabla_x p(t,x,v) + k\nabla_v\cdot(vp(t,x,v)) + \alpha_1(t,x)p(t,x,v_0) \\ - \nabla_v\cdot\left[F\left(\widetilde{C}(t,x)\right)p(t,x,v)\right] + \frac{\sigma^2}{2}\Delta_v p(t,x,v). \qquad (46)\end{aligned}$$

The individual processes $(Y^i(t), v^i(t))_t$ obey to the following stochastic system:

$$\begin{aligned}dY^i(t) &= v^i(t)dt, \\ dv^i(t) &= \left(-kv^i(t) + F(\widetilde{C}(t,Y^i(t)))\right)dt + \sigma dW^i(t),\end{aligned} \qquad (47)$$

coupled with a branching process with intensity

$$\alpha(t,x) = \alpha\, h(\widetilde{C}(t,x))\sum_{i=1}^{N(t^-)}\delta_{Y^i(t)}(x). \qquad (48)$$

Note that both (47) and (48) depend on the mean field $\widetilde{C}(t,x)$ in the equation (45).

6 Long Time Behavior

In this section we investigate the long time behavior of the particle system described in Section 3, for a fixed number N of particles.

6.1 Interacting–Diffusing Particles

First of all, let us consider the system (16) with $\gamma_1 = 0$, i.e. the case in which the advection is due only to interactions among particles. Following [17], from (16) it follows that the location of the center of mass $\bar{X}_N$ of the N particles,

$$\bar{X}_N(t) = \frac{1}{N}\sum_{k=1}^{N} X_N^k(t),$$

evolves according the following equation:

$$d\bar{X}_N(t) = -\frac{1}{N^2}\sum_{k,j=1}^{N} \nabla(K_N - G)(X_N^k(t) - X_N^j(t))dt + \sigma d\bar{W}(t), \qquad (49)$$

where $\bar{W}(t) = \frac{1}{N}\sum_{k=1}^{N} W^k(t)$ is still a Brownian motion; by the symmetry of the kernels K_1 and G, the first term on the right-hand side vanishes and we get

$$d\bar{X}_N(t) = \sigma d\bar{W}(t), \qquad (50)$$

i.e. the stochastic process $\bar{X}_N$ is a Wiener process. Hence, its law, conditional upon the initial state, is

$$\mathcal{L}\left(\bar{X}_N(t)|\bar{X}_N(0)\right) = \mathcal{L}\left(\bar{X}_N(0), \sigma^2\bar{W}(t)\right) = \mathcal{N}\left(\bar{X}_N(0), \frac{\sigma^2}{N}t\right);$$

with variance $\frac{\sigma^2}{N}t$, which, for any fixed N, increases as t tends to infinity. Consequently, we may claim that the probability law of the system does not converge to any non trivial probability law, since otherwise the same would happen for the law of the center of mass.

6.2 Complete System

Let us now consider the complete system of SDEs (16) with $\gamma_1 > 0$. This means that particles are also subject to a confining potential U. Equations of the type

$$dX_t = -\nabla P(X_t) + \sigma dW_t \qquad (51)$$

have been thoroughly analyzed in literature; under the sufficient condition of strict convexity of the symmetric potential U [8, 9, 17, 18], it has been shown that (51) does admit a nontrivial invariant distribution. From a biological point of view a strictly convex confining potential is difficult to explain; it would mean an infinite range of attraction of the force which becomes infinitely strong at infinity, with an at least constant drift even far from origin.

A weaker sufficient condition for the existence of a unique invariant measure has been more recently suggested by Veretennikov [34, 35], following Has'minski [15]. This condition states that there exist constants $M_0 \geq 0$ and $r > 0$ such that for $|x| \geq M_0$

$$\left(-\nabla P(\mu)(x), \frac{x}{|x|}\right) \leq -\frac{r}{|x|}. \qquad (52)$$

It is ease to prove that without any further condition on the interaction kernels K_N and G, by considering the condition (15) on U, we may apply the results by Veretennikov and prove the existence of an invariant measure for the joint law of the particles locations. The condition (15) means that ∇U may decay to zero as $|x|$ tends to infinity, provided that its tails are sufficiently "fat".

Proposition 2. *Under the hypotheses for the existence and uniqueness (hypotheses stated in Section 3) and the condition* (15)*, the system* (16) *admits a unique invariant measure.*

Let now $P_N^{x_0}(t)$ denote the joint distribution of the N particles at time t, conditional upon a non random initial condition x_0, and let P_S denote the invariant distribution. As far as the convergence of $P_N^{x_0}(t)$ is concerned, for t tending to infinity, as in [34], one can prove the following result.

Proposition 3. *Under the same assumptions of Proposition 2, for any k, $0 < k < \tilde{r} - \frac{Nd}{2} - 1$ with $m \in (2k+2, 2\tilde{r} - Nd)$ and $\tilde{r} = \gamma_1 N r$, there exists a positive constant c such that*

$$\left|P_N^{x_0}(t) - P_N^S\right| \leq c(1 + |x_0|^m)(1+t)^{-(k+1)},$$

where $\left|P_N^{x_0}(t) - P_N^S\right|$ denotes the total variation distance of the two measures, i.e.

$$\left|P_N^{x_0}(t) - P_N^S\right| = \sup_{A \in \mathcal{B}_{\mathbb{R}^d}} \left[P_N^{x_0}(t)(A) - P_N^S(A)\right],$$

and x_0 the initial data.

So Proposition 2 states a polynomial convergence rate to invariant measure. To improve the rate of convergence, one has to consider more restricted assumptions on U [35].

References

1. S. Boi, V. Capasso, and D. Morale. Modeling the aggregative behavior of ants of the species: Polyergus rufescens. *Nonlinear Anal. Real World Appl.*, 1(1):163–176, 2000.
2. M. Burger, V. Capasso, and D. Morale. On an aggregation model with long and short range interactions. *Nonlinear Anal. Real World Appl.*, 8(3):939–958, 2007.
3. M. Burger, V. Capasso, and L. Pizzocchero. Mesoscale averaging of nucleation and growth models. *Multiscale Model. Simul.*, 5(2):564–592, 2006.
4. V. Capasso and D. Bakstein. *An introduction to continuous-time stochastic processes. Theory, models, and applications to finance, biology and medicine.* Birkhäuser, Boston, MA, 2005.
5. V. Capasso and D. Morale. Rescaling stochastic processes: Asymptotics. In *Multiscale Problems in the Life Sciences from Microscopic to Macroscopic*, volume 1940 of *Lecture Notes in Mathematics/Fondazione C.I.M.E.*, pages 91–146. Springer, Heidelberg, 2008.

6. V. Capasso and D. Morale. Asymptotic behavior of a system of stochastic particles subject to nonlocal interactions. *Stoch. Anal. Appl.*, 27(3):574–603, 2009.
7. V. Capasso and D. Morale. Stochastic modelling of tumour-induced angiogenesis. *J. Math. Biol.*, 58(1–2):219–233, 2009.
8. J. Carrillo. Entropy solutions for nonlinear degenerate problems. *Arch. Rat. Mech. Anal.*, 147:269–361, 1999.
9. J. A. Carrillo, R. J. McCann, and C. Villani. Kinetic equilibration rates for granular media and related equations: entropy dissipation and mass transportation estimates. *Rev. Mat. Iberoamericana*, 19(3):971–1018, 2003.
10. R. Durrett and S. A. Levin. The importance of being discrete (and spatial). *Theor. Pop. Biol.*, 46:363–394, 1994.
11. S. N. Ethier and T. G. Kurtz. *Markov processes, characterization and convergence.* Wiley, New York, 1986.
12. L. C. Evans. *Partial differential equations.* AMS, Providence, 1998.
13. J. Folkman and M. Klagsbrun. Angiogenic factors. *Science*, 235:442–447, 1987.
14. S. Gueron and S. A. Levin. The dynamics of group formation. *Math. Biosci.*, 128:243–264, 1995.
15. R. Z. Has'minski. *Stochastic stability of differential equations.* Sijthoff & Noordhoff, Alphen aan den Rijn, 1980.
16. H. A. Levine, B. D. Sleeman, and M. Nilsen-Hamilton. Mathematical modelling of the onset of capillary formation initiating angiogenesis. *J. Math. Biol.*, 42(3):195–238, 2001.
17. F. Malrieu. Convergence to equilibrium for granular media equations and their Euler schemes. *Ann. Appl. Probab.*, 13(2):540–560, 2003.
18. P. A. Markowich and C. Villani. On the trend to equilibrium for the Fokker–Planck equation: an interplay between physics and functional analysis. *Mathematica Contemporanea*, 19:1–31, 2000.
19. S. Meleard and B. Fernandez. Asymptotic behaviour for interacting diffusion processes with space-time random birth. *Bernoulli*, 6:1–21, 2000.
20. A. Mogilner and L. Edelstein-Keshet. A non-local model for a swarm. *J. Math. Biol.*, 38(6):534–570, 1999.
21. D. Morale. Cellular automata and many-particles systems modeling aggregation behaviour among populations. *Int. J. Appl. Math. Comput. Sci.*, 10:157–173, 2000.
22. D. Morale. Modeling and simulating animal grouping: individual-based models. *Future Generation Computer Systems*, 17(7):883–891, 2001.
23. D. Morale, V. Capasso, and K. Oelschläger. An interacting particle system modelling aggregation behavior: from individuals to populations. *J. Math. Biol.*, 50(1):49–66, 2005.
24. T. Nagai and M. Mimura. Asymptotic behaviour for a nonlinear degenerate diffusion equation in population dynamics. *SIAM J. Appl. Math.*, 43:449–464, 1983.
25. T. Nagai and M. Mimura. Some nonlinear degenerate diffusion equations related to population dynamics. *J. Math. Soc. Japan*, 35:539–561, 1983.
26. K. Oelschläger. A law of large numbers for moderately interacting diffusion processes. *Z. Wahrscheinlichkeitstheorie verw. Gebiete*, 69:279–322, 1985.
27. K. Oelschläger. On the derivation of reaction-diffusion equations as lilit dynamics of systems of moderately interacting stochastic processes. *Probab. Theory Relat. Fields*, 82:565–586, 1989.

28. A. Okubo. Dynamical aspects of animal grouping: swarms, school, flocks and herds. *Adv. BioPhys.*, 22:1–94, 1986.
29. A. Okubo and S. Levin. *Diffusion and ecological problems: Modern perspectives.* Springer, Heidelberg, 2002.
30. M. J. Plank and B. D. Sleeman. A reinforced random walk model of tumour angiogenesis and anti-angiogenic strategies. *IMA J. Math. Med. Biol.*, 20:135–181, 2003.
31. M. J. Plank and B. D. Sleeman. Lattice and non-lattice models of tumour angiogenesis. *Bull. Math. Biol.*, 66(6):1785–1819, 2004.
32. S. Sun, M. F. Wheeler, M. Obeyesekere, and C. W. Patrick Jr. A deterministic model of growth factor-induced angiogenesis. *Bull. Math. Biol.*, 67(2):313–337, 2005.
33. A. Y. Veretennikov. On polynomial mixing bounds for stochastic differential equations. *Stochastic Process. Appl.*, 70:115–127, 1997.
34. A. Y. Veretennikov. On polynomial mixing and convergence rate for stochastic differential equations. *Theory Probab. Appl.*, 44:361–374, 1999.
35. A. Y. Veretennikov. On subexponential mixing rate for Markov processes. *Theory Probab. Appl.*, 49:110–122, 2005.

Remarks on the Controllability of Some Parabolic Equations and Systems

Enrique Fernández-Cara

University of Sevilla, Dpto. E.D.A.N., Aptdo. 1160, 41080 Sevilla, Spain,
cara@us.es

Summary. This paper is devoted to present a review of recent results concerning the controllability of some (linear and nonlinear) parabolic systems. Among others, we will consider the classical heat equation, the Burgers, Navier–Stokes and Boussinesq equations, etc.

1 Introduction: Controllability and Observability

Let us first recall some general ideas. Suppose that we are considering an abstract *state equation* of the form

$$\begin{cases} y_t - A(y) = Bv, \quad t \in (0,T), \\ y(0) = y^0, \end{cases} \tag{1}$$

which governs the behavior of a physical system. It is assumed that

- $y : [0,T] \mapsto H$ is the *state*, i.e. the variable that serves to identify the physical properties of the system.
- $v : [0,T] \mapsto U$ is the *control*, i.e. the variable we can choose (for simplicity, we assume that U and H are Hilbert spaces).
- $A : D(A) \subset H \mapsto H$ is a (generally nonlinear) operator with $A(0) = 0$, $B \in \mathcal{L}(U;H)$ and $y^0 \in H$.

Suppose that (1) is well-posed in the sense that, for each $y^0 \in H$ and each $v \in L^2(0,T;U)$, it possesses exactly one solution. Then the *null controllability* problem for (1) can be stated as follows:

> *For each $y^0 \in H$, find $v \in L^2(0,T;U)$ such that the corresponding solution of (1) satisfies $y(T) = 0$.*

W. Fitzgibbon et al. (eds.), *Applied and Numerical Partial Differential Equations*, Computational Methods in Applied Sciences 15, DOI 10.1007/978-90-481-3239-3_7,

More generally, the *exact controllability to the trajectories* problem for (1) is the following:

> *For each free trajectory* $\overline{y} : [0,T] \mapsto H$ *and each* $y^0 \in H$, *find* $v \in L^2(0,T;U)$ *such that the corresponding solution of (1) satisfies* $y(T) = \overline{y}(T)$.

Here, by a *free* or *uncontrolled* trajectory we mean any (sufficiently regular) function $\overline{y} : [0,T] \mapsto H$ satisfying $\overline{y}(t) \in D(A)$ for all t and

$$\overline{y}_t - A(\overline{y}) = 0, \quad t \in (0,T).$$

Notice that exact controllability to the trajectories is a very useful property from the viewpoint of applications: if we can find such a control, then after time T we can switch off the control and the system will follow the "ideal" trajectory $\overline{y}$.

For each system of the form (1), these problems lead to several interesting questions. Among them, let us mention the following:

- First, are there controls v such that $y(T) = 0$ and/or $y(T) = \overline{y}(T)$?
- Then, if this is the case, which is the *cost* we have to pay to drive y to zero and/or $\overline{y}(T)$? In other words, which is the minimal norm of a control $v \in L^2(0,T;U)$ satisfying these properties?
- How can these controls be computed?

The controllability of differential systems is a very relevant area of research and has been the subject of many papers the last years. In particular, in the context of partial differential equations, the null controllability problem was first analyzed in [26, 29–31, 33, 34]. For semilinear systems of this kind, the first contributions have been given in [9, 19, 35].

In this paper, we will be mainly concerned with the case of parabolic partial differential systems. The typical situation corresponds to the classical heat equation in a bounded N-dimensional domain, complemented with appropriate initial and boundary-value conditions; see Section 2.

The paper is organized as follows. In Section 2, we consider the heat equation and some linear variants. We explain the role of observability and Carleman estimates in control theory, we recall the main results in this framework and we mention some open problems. Section 3 deals with the viscous Burgers equation. We show that, for this equation, the null controllability problem (with distributed and locally supported control) is well understood.[1] In Sections 4 and 5, we consider the Navier–Stokes and Boussinesq equations and some other systems from mechanics. We recall several results concerning the local exact controllability to the trajectories and we explain how to deal with a reduced number of controls. Several open problems are also indicated.

---

[1] More precisely, if we denote by $T^*(r)$ the minimal time needed to drive any initial state with L^2 norm $\le r$ to zero, we show that $T^*(r) > 0$, with explicit sharp estimates from above and from below.

2 The Classical Heat Equation: Observability and Carleman Estimates

Let us consider the following control system for the heat equation:

$$\begin{cases} y_t - \Delta y = v 1_\omega, & (x,t) \in \Omega \times (0,T), \\ y(x,t) = 0, & (x,t) \in \partial\Omega \times (0,T), \\ y(x,0) = y^0(x), & x \in \Omega. \end{cases} \tag{2}$$

Here (and also in the following sections), $\Omega \subset \mathbb{R}^N$ is a nonempty regular and bounded domain, $\omega \subset\subset \Omega$ is a (small) nonempty open subset (1_ω is the characteristic function of ω) and $y^0 \in L^2(\Omega)$.

It is well known that, for every $y^0 \in L^2(\Omega)$ and every $v \in L^2(\omega \times (0,T))$, there exists a unique solution y to (2), with $y \in L^2(0,T;H_0^1(\Omega)) \cap C^0([0,T];L^2(\Omega))$.

In this context, the null controllability problem reads:

For each $y^0 \in L^2(\Omega)$, find $v \in L^2(\omega \times (0,T))$ such that the associated solution of (2) satisfies $y(x,T) = 0$ in Ω.

Since the state equation (2) is linear, null controllability is equivalent in this case to *exact controllability to the trajectories*. This means that, for any uncontrolled solution $\overline{y}$ and any $y^0 \in L^2(\Omega)$, there exists $v \in L^2(\omega \times (0,T))$ such that the associated state y satisfies

$$y(x,T) = \overline{y}(x,T) \quad \text{in } \Omega.$$

A related notion is *approximate controllability*. It is said that (2) is approximately controllable in $L^2(\Omega)$ at time T if, for any $y^0, y^1 \in L^2(\Omega)$ and any $\varepsilon > 0$, there exist controls $v \in L^2(\omega \times (0,T))$ such that the solutions to (2) associated to these v and the initial state y^0 satisfy

$$\|y(\cdot\,,T) - y^1\|_{L^2} \le \varepsilon. \tag{3}$$

It is not difficult to prove that this is weaker notion: the null controllability of (2) at any time T implies the approximate controllability of (2) in $L^2(\Omega)$ at any T. On the other hand, since $\omega \subset\subset \Omega$, in view of the regularizing effect of the heat equation, *exact controllability*, i.e. approximate controllability with $\varepsilon = 0$, does not hold.

Together with (2), for each $\varphi^1 \in L^2(\Omega)$, we can introduce the associated adjoint system

$$\begin{cases} -\varphi_t - \Delta \varphi = 0, & (x,t) \in \Omega \times (0,T), \\ \varphi(x,t) = 0, & (x,t) \in \partial\Omega \times (0,T), \\ \varphi(x,T) = \varphi^1(x), & x \in \Omega. \end{cases} \tag{4}$$

Then, it is well known that the null controllability of (2) is equivalent to the following property:

There exists $C > 0$ such that

$$\|\varphi(\cdot,0)\|_{L^2}^2 \le C \iint_{\omega\times(0,T)} |\varphi|^2 \, dx \, dt \qquad \forall \varphi^1 \in L^2(\Omega). \tag{5}$$

This is called an observability estimate for the solutions of (4). We thus find that, in order to solve the null controllability problem for (2), it suffices to prove (5).

The estimate (5) is implied by the so called global Carleman inequalities. These have been introduced in the context of the controllability of PDEs by Fursikov and Imanuvilov, see [19, 26]. When they are applied to the solutions of the adjoint system (4), they take the form

$$\iint_{\Omega\times(0,T)} \rho^2 \, |\varphi|^2 \, dx \, dt \le K \iint_{\omega\times(0,T)} \rho^2 \, |\varphi|^2 \, dx \, dt \quad \forall \varphi^1 \in L^2(\Omega), \tag{6}$$

where $\rho = \rho(x,t)$ is an appropriate weight depending on Ω, ω and T and the constant K only depends on Ω and ω.[2]

Combining (6) and the dissipativity of the backwards heat equation (4), it is not difficult to deduce (5) for some C only depending on Ω, ω and T.

As a consequence, we have:

Theorem 1. *The linear system (2) is null controllable. In other words, for each $y^0 \in L^2(\Omega)$, there exists $v \in L^2(\omega \times (0,T))$ such that the corresponding solution of (2) satisfies*

$$y(x,T) = 0 \quad in\ \Omega. \tag{7}$$

Remark 1. Notice that Theorem 1 ensures the null controllability of (2) for *any* ω and T. This is a consequence of the fact that, in a parabolic equation, the transmission of information is instantaneous. For instance, this is not the case for the transport equation. Thus, let us consider the control system

$$\begin{cases} y_t + y_x = v 1_\omega, & (x,t) \in (0,L) \times (0,T), \\ y(0,t) = 0, & t \in (0,T), \\ y(x,0) = y^0(x), & x \in (0,L), \end{cases} \tag{8}$$

with $\omega = (a,b) \subset\subset (0,L)$. Then, if $0 < T < a$, null controllability does not hold, since the solution always satisfies

$$y(x,T) = y^0(x-T) \quad \forall x \in (T,a),$$

independently of the choice of v; see [7] for more details and similar results concerning other control systems for the wave, Schrödinger and Korteweg–De Vries equations. ■

[2] In order to prove (6), we have to use a weight ρ decreasing to zero, as $t \to 0$ and also as $t \to T$, for instance, exponentially.

There are many generalizations and variants of Theorem 1 that provide the null controllability of other similar linear (parabolic) state equations:

- Time–space dependent (and sufficiently regular) coefficients can appear in the equation, other boundary conditions can be used, boundary control (instead of distributed control) can be imposed, etc.; see [19]. For a review of recent applications of Carleman inequalities to the controllability of parabolic systems, see [11].
- The null controllability of Stokes-like systems can also be analyzed with these techniques. This includes systems of the form

$$y_t - \Delta y + (a\cdot\nabla)y + (y\cdot\nabla)b + \nabla p = v1_\omega, \quad \nabla\cdot y = 0, \tag{9}$$

  where a and b are regular enough. See, for instance, [14]; see also [8] for other controllability properties.
- Other linear parabolic (non-scalar) systems can also be considered, etc.

However, there are several interesting problems related to the controllability of linear parabolic systems that still remain open. Let us mention two of them.

First, let us consider the controlled system

$$\begin{cases} y_t - \nabla\cdot(a(x)\nabla y) = v1_\omega, & (x,t)\in\Omega\times(0,T),\\ y(x,t) = 0, & (x,t)\in\partial\Omega\times(0,T),\\ y(x,0) = y^0(x), & x\in\Omega, \end{cases} \tag{10}$$

where y^0 and v are as before and the coefficient a is assumed to satisfy

$$a\in L^\infty(\Omega), \quad 0 < a_0 \le a(x) \le a_1 < +\infty \quad \text{a.e.} \tag{11}$$

It is natural to consider the null controllability problem for (10). Of course, this is equivalent to the observability of the associated adjoint system

$$\begin{cases} -\varphi_t - \nabla\cdot(a(x)\nabla\varphi) = 0, & (x,t)\in\Omega\times(0,T),\\ \varphi(x,t) = 0, & (x,t)\in\partial\Omega\times(0,T),\\ y\varphi(x,T) = \varphi^1(x), & x\in\Omega, \end{cases} \tag{12}$$

that is to say, to the fact that an inequality like (5) holds for the solutions to (12).

To our knowledge, it is at present unknown whether (10) is null controllable. In fact, it is also unknown whether approximate controllability holds.

Remark 2. Recently, some partial results have been obtained in this context. Thus, when $N = 1$, the null controllability of (10) has been established in [1]. When $N \ge 2$, the best known result up to now is that this property holds under the following assumption:

$$\exists \textit{ smooth open set } \Omega_0 \subset\subset \Omega \textit{ such that } a \textit{ is } C^1 \textit{ in } \overline{\Omega_0} \textit{ and } \overline{\Omega\setminus\Omega_0}. \tag{13}$$

This has been proved in [28]. In both cases, the proofs use that a is independent of t in an essential way. In fact, it is an open question whether a Carleman estimate like (6) holds for the solutions to (12) even if $N = 1$ or (13) holds. ∎

Our second open problem concerns the system

$$\begin{cases} y_t - D\Delta y = Ay + Bv1_\omega, & (x,t) \in \Omega \times (0,T), \\ y(x,t) = 0, & (x,t) \in \partial\Omega \times (0,T), \\ y(x,0) = y^0(x), & x \in \Omega, \end{cases} \tag{14}$$

where $y = (y_1, \ldots, y_n)$ is the state, $v = (v_1, \ldots, v_m)$ is the control and D, A and B are constant matrices, with $D, A \in \mathcal{L}(\mathbb{R}^n; \mathbb{R}^n)$ and $B \in \mathcal{L}(\mathbb{R}^m; \mathbb{R}^n)$. It is assumed that D is definite positive, that is,

$$D\xi \cdot \xi \geq d_0 |\xi|^2 \quad \forall \xi \in \mathbb{R}^n, \ d_0 > 0. \tag{15}$$

When D is diagonal (or similar to a diagonal matrix), the null controllability problem for (14) is well understood. In view of the results in [2], (14) is null controllable if and only if

$$\text{rank}[(-\lambda_i D + A); B] = n \quad \forall i \geq 1, \tag{16}$$

where the λ_i are the eigenvalues of the Dirichlet–Laplace operator and, for any matrix $H \in \mathcal{L}(\mathbb{R}^n; \mathbb{R}^n)$, $[H; B]$ stands for the $n \times nm$ matrix

$$[H; B] := [B|HB| \cdots |H^{n-1}B].$$

Therefore, it is natural to search for (algebraic) conditions on D, A and B that ensure the null controllability of (14) in the general case. But, to our knowledge, this is unknown.

Remark 3. The results in [2] have been extended recently to the case of any D having no eigenvalue of geometric multiplicity > 3; see [10]. ∎

Remark 4. As we have said, global Carleman estimates are the main tool we can use to establish the observability property (5). These two open questions can be viewed as consequences of the limitations of Carleman estimates: first, they need regular coefficients; then, they are, in fact, a tool proper of *scalar* equations. ∎

As mentioned above, an interesting question related to Theorem 1 concerns the cost of null controllability. One has the following result from [16]:

Theorem 2. *For each $y^0 \in L^2(\Omega)$, let us denote by $C(y^0)$ the minimal norm in $L^2(\omega \times (0,T))$ of a control v such that the associated solution of (2) satisfies (7). Then, for some C only depending on Ω and ω, the following estimate holds:*

$$C(y^0) \leq \exp\left[C\left(1 + \frac{1}{T}\right)\right] \|y^0\|_{L^2}. \tag{17}$$

Remark 5. We can be more explicit on the way C depends on Ω and ω: there exist "universal" constants $C_0 > 0$ and $m \geq 1$ such that C can be taken of the form

$$C = \exp\left(C_0 \|\psi\|_{C^2}^m\right),$$

where $\psi \in C^2(\overline{\Omega})$ is any function satisfying $\psi > 0$ in Ω, $\psi = 0$ on $\partial\Omega$ and $\nabla\psi \neq 0$ in $\overline{\Omega} \setminus \omega$. All this is a consequence of the particular form that must have ρ in order to ensure (6); see [16] for more details. ∎

3 Positive and Negative Controllability Results for the One-Dimensional Burgers Equation

In this section, we will be concerned with the null controllability of the following system for the viscous Burgers equation:

$$\begin{cases} y_t - y_{xx} + yy_x = v1_\omega, & (x,t) \in (0,1) \times (0,T), \\ y(0,t) = y(1,t) = 0, & t \in (0,T), \\ y(x,0) = y^0(x), & x \in (0,1). \end{cases} \tag{18}$$

Recall that some controllability properties of (18) have been studied in [19, Chapter 1, Theorems 6.3 and 6.4]. There, it is shown that, in general, a stationary solution of (18) with large L^2-norm cannot be reached (not even approximately) at any time T. In other words, with the help of one control, the solutions of the Burgers equation cannot go anywhere at any time.

For each $y^0 \in L^2(0,1)$, let us introduce

$$T(y^0) = \inf\{T > 0 :\ (18) \text{ is null controllable at time } T\}.$$

Then, for each $r > 0$, let us define the quantity

$$T^*(r) = \sup\{T(y^0) : \|y^0\|_{L^2} \leq r\}.$$

Our main purpose is to show that $T^*(r) > 0$, with explicit sharp estimates from above and from below. In particular, this will imply that (global) null controllability at any positive time does not hold for (18).

More precisely, let us set $\phi(r) = (\log \frac{1}{r})^{-1}$. We have the following result from [13]:

Theorem 3. *One has*

$$C_0\phi(r) \leq T^*(r) \leq C_1\phi(r) \quad \textit{as } r \to 0, \tag{19}$$

for some positive constants C_0 and C_1 not depending of r.

Remark 6. The same estimates hold when the control v acts on system (18) through the boundary *only* at $x = 1$ (or only at $x = 0$). Indeed, it is easy to transform the boundary controlled system

$$\begin{cases} y_t - y_{xx} + yy_x = 0, & (x,t) \in (0,1) \times (0,T), \\ y(0,t) = 0, \quad y(1,t) = w(t), & t \in (0,T), \\ y(x,0) = y^0(x), & x \in (0,1) \end{cases} \tag{20}$$

into a system of the kind (18). The boundary controllability of the Burgers equation with *two* controls (at $x = 0$ and $x = 1$) has been analyzed in [23]. There, it is shown that even in this more favorable situation null controllability does not hold for small time. It is also proved in that paper that exact controllability does not hold for large time.[3] ■

The proof of the estimate from above in (19) can be obtained by solving the null controllability problem for (18) via a (more or less) standard fixed point argument, using global Carleman inequalities to estimate the control and energy inequalities to estimate the state and being very careful with the role of T in these inequalities.

The proof of the estimate from below is inspired by the arguments in [3] and is implied by the following property: there exist positive constants C_0 and C_0' such that, for any sufficiently small $r > 0$, we can find initial data y^0 and associated states y satisfying $\|y^0\|_{L^2} \le r$ and

$$|y(x,t)| \ge C_0' r \quad \text{for some } x \in (0,1) \text{ and any } t : 0 < t < C_0 \phi(r).$$

For more details, see [13].

4 The Navier–Stokes and Boussinesq Systems

There are a lot of more realistic nonlinear equations and systems from mechanics that can also be considered in this context. First, we have the well known Navier–Stokes equations:

$$\begin{cases} y_t + (y \cdot \nabla)y - \Delta y + \nabla p = v 1_\omega, \quad \nabla \cdot y = 0, & (x,t) \in Q, \\ y = 0, & (x,t) \in \Sigma, \\ y(x,0) = y^0(x), & x \in \Omega. \end{cases} \tag{21}$$

Here and below, Q and Σ respectively stand for the sets $Q = \Omega \times (0,T)$ and $\Sigma = \partial\Omega \times (0,T)$, where $\Omega \subset \mathbb{R}^N$ is a nonempty regular and bounded domain, $N = 2$ or $N = 3$ and (again) $\omega \subset\subset \Omega$ is a nonempty open set.

[3] Let us remark that the results in [23] do not allow to estimate $T(r)$; in fact, the proofs are based in contradiction arguments.

In (21), (y, p) is the state (the velocity field and the pressure distribution) and v is the control (a field of external forces applied to the fluid particles located at ω). To our knowledge, the best results concerning the controllability of this system have been given in [14, 15].[4] Essentially, these results establish the local exact controllability of the solutions of (21) to uncontrolled trajectories.

In order to be more specific, let us recall the definition of some usual spaces in the context of Navier–Stokes equations:

$$V = \{y \in H_0^1(\Omega)^N : \nabla \cdot y = 0 \text{ in } \Omega\}$$

and

$$H = \{y \in L^2(\Omega)^N : \nabla \cdot y = 0 \text{ in } \Omega,\ y \cdot n = 0 \text{ on } \partial\Omega\}.$$

Of course, it will be said that (21) is *exactly controllable to the trajectories* if, for any trajectory $(\overline{y}, \overline{p})$, i.e. any solution of the uncontrolled Navier–Stokes system

$$\begin{cases} \overline{y}_t + (\overline{y} \cdot \nabla)\overline{y} - \Delta\overline{y} + \nabla\overline{p} = 0, \quad \nabla \cdot \overline{y} = 0, & (x,t) \in Q, \\ \overline{y} = 0, & (x,t) \in \Sigma \end{cases} \tag{22}$$

and any $y^0 \in H$, there exist controls $v \in L^2(\omega \times (0,T))^N$ and associated solutions (y, p) such that

$$y(x,T) = \overline{y}(x,T) \quad \text{in } \Omega. \tag{23}$$

At present, we do not know any global result concerning exact controllability to the trajectories for (21). However, the following local result holds:

Theorem 4. *Let $(\overline{y}, \overline{p})$ be a strong solution of (22), with*

$$\overline{y} \in L^\infty(Q)^N, \quad \overline{y}(\cdot\,,0) \in V. \tag{24}$$

Then, there exists $\delta > 0$ such that, for any $y^0 \in H \cap L^{2N-2}(\Omega)^N$ satisfying $\|y^0 - \overline{y}^0\|_{L^{2N-2}} \le \delta$, we can find a control $v \in L^2(\omega \times (0,T))^N$ and an associated solution (y,p) to (21) such that (23) holds.

In other words, the local exact controllability to the trajectories holds for (21) in the space $X = L^{2N-2}(\Omega)^N \cap H$; see [14] for a slightly stronger result. Similar questions were addressed (and solved) in [17, 18]. The fact that we consider here Dirichlet boundary conditions and locally supported distributed control increases a lot the mathematical difficulty of the control problem.

Remark 7. It is clear that we cannot expect exact controllability for the Navier–Stokes equations with an arbitrary target function, because of the dissipative and non reversible properties of the system. On the other hand,

[4] The main ideas come from [20,27]; some additional results will appear soon in [21].

approximate controllability is still an open question for this system. Some results in this direction have been obtained in [6] for different boundary conditions (Navier slip boundary conditions) and in [8] for a different nonlinearity. However, the notion of approximate controllability does not appear to be optimal from a practical viewpoint. Indeed, even if we could reach an arbitrary neighborhood of a given target y^1 at time T by the action of a control, the question of what to do after time T to stay in the same neighbourhood would remain open. ■

The proof of Theorem 4 can be obtained as an application of *Liusternik's inverse mapping theorem* in an appropriate framework.

A key point in the proof is a related null controllability result for the linearized Navier–Stokes system at $(\overline{y}, \overline{p})$, that is to say

$$\begin{cases} y_t + (\overline{y} \cdot \nabla) y + (y \cdot \nabla)\overline{y} - \Delta y + \nabla p = v 1_\omega, & (x,t) \in Q, \\ \nabla \cdot y = 0, & (x,t) \in Q, \\ y = 0, & (x,t) \in \Sigma, \\ y(x,0) = y^0(x), & x \in \Omega. \end{cases} \tag{25}$$

This control result is a consequence of a global Carleman inequality of the kind (6) that can be established for the solutions to the adjoint of (25), which is the following:

$$\begin{cases} -\varphi_t - (\nabla \varphi + \nabla \varphi^t)\,\overline{y} - \Delta \varphi + \nabla \pi = g, & (x,t) \in Q, \\ \nabla \cdot \varphi = 0, & (x,t) \in Q, \\ \varphi = 0, & (x,t) \in \Sigma, \\ \varphi(T) = \varphi^0, & x \in \Omega. \end{cases} \tag{26}$$

The details can be found in [14].

Similar results have been given in [22] for the Boussinesq equations

$$\begin{cases} y_t + (y \cdot \nabla) y - \Delta y + \nabla p = v 1_\omega + \theta\, e_N, \quad \nabla \cdot y = 0 & (x,t) \in Q, \\ \theta_t + y \cdot \nabla \theta - \Delta \theta = h 1_\omega, & (x,t) \in Q, \\ y = 0, \quad \theta = 0, & (x,t) \in \Sigma, \\ y(x,0) = y^0(x), \quad \theta(x,0) = \theta^0(x), & x \in \Omega. \end{cases} \tag{27}$$

Here, the state is the triplet (y, p, θ) (θ is interpreted as a temperature distribution) and the control is (v, h) (as before, v is a field of external forces; h is an external heat source).

An interesting question concerning both (21) and (27) is whether we can still get local exact controllability to the trajectories with a reduced number of scalar controls. This is partially answered in [15], where the following results are proved:

Theorem 5. *Assume that the following property is satisfied:*

$$\exists x^0 \in \partial\Omega,\ \exists \varepsilon > 0 \ \text{ such that } \overline{\omega} \cap \partial\Omega \supset B(x^0;\varepsilon) \cap \partial\Omega. \tag{28}$$

Here, $B(x^0;\varepsilon)$ is the ball centered at x^0 of radius ε. Then, for any $T > 0$, (21) is locally exactly controllable at time T to the trajectories satisfying (24) with controls $v \in L^2(\omega \times (0,T))^N$ having one component identically zero.

Theorem 6. *Assume that ω satisfies (28) with $n_k(x^0) \neq 0$ for some $k < N$. Then, for any $T > 0$, (27) is locally exactly controllable at time T to the trajectories $(\overline{y},\overline{p},\overline{\theta})$ satisfying (24) and*

$$\overline{\theta} \in L^\infty(Q), \quad \overline{\theta}(\cdot\,,0) \in H_0^1(\Omega). \tag{29}$$

with controls $v \in L^2(\omega \times (0,T))^N$ and $h \in L^2(\omega \times (0,T))$ such that $v_k \equiv v_N \equiv 0$. In particular, if $N = 2$, we have local exact controllability to these trajectories with controls $v \equiv 0$ and $h \in L^2(\omega \times (0,T))$.

The proofs of Theorems 5 and 6 are similar to the proof of Theorem 4. We have again to rewrite the controllability property as a nonlinear equation in a Hilbert space. Then, we have to check that the hypotheses of Liusternik's theorem are fulfilled.

Again, a crucial point is to prove the null controllability of certain linearized systems, this time with *reduced* controls. For instance, when dealing with (21), the task is reduced to prove that, for some $\rho = \rho(x,t)$ and $K > 0$, the solutions to (25) satisfy the following Carleman-like estimates:

$$\iint_{\Omega\times(0,T)} \rho^2|\varphi|^2\,dx\,dt \le K \iint_{\omega\times(0,T)} \rho^2(\varphi_1^2+\varphi_2^2)\,dx\,dt \quad \forall \varphi^1 \in L^2(\Omega). \tag{30}$$

This inequality can be proved using the assumption (28) and the incompressibility identity $\nabla\cdot\varphi = 0$; see [15].

5 Some Other Nonlinear Systems from Mechanics

The previous arguments can be applied to other similar partial differential systems arising in mechanics. For instance, this is made in [12] in the context of micro-polar fluids.

To fix ideas, let us assume that $N = 3$. The behavior of a micro-polar three-dimensional fluid is governed by the following system:

$$\begin{cases} y_t - \Delta y + (y\cdot\nabla)y + \nabla p = \nabla\times w + v1_\omega, \quad \nabla\cdot y = 0, & (x,t)\in Q,\\ w_t + (y\cdot\nabla)w - \Delta w - \nabla(\nabla\cdot w) = \nabla\times y + u1_\omega, & (x,t)\in Q,\\ y = 0, \quad w = 0 & (x,t)\in \Sigma,\\ y(x,0) = y^0(x), \quad w(x,0) = w^0(x) & x\in\Omega. \end{cases} \tag{31}$$

Here, the state is (y, p, w) and the control is (v, u). As usual, y and p stand for the velocity field and pressure and w is the microscopic velocity of rotation of the fluid particles. Then, the following result holds:

Theorem 7. *Let $(\overline{y}, \overline{p}, \overline{w})$ be such that*

$$\overline{y},\, \overline{w} \in L^\infty(Q) \cap L^2(0,T; H^2(\Omega)), \quad \overline{y}_t,\, \overline{w}_t \in L^2(Q) \tag{32}$$

and

$$\begin{cases} \overline{y}_t - \Delta \overline{y} + (\overline{y} \cdot \nabla)\overline{y} + \nabla \overline{p} = \nabla \times \overline{w}, \quad \nabla \cdot \overline{y} = 0, & (x,t) \in Q, \\ \overline{w}_t + (\overline{y} \cdot \nabla)\overline{w} - \Delta \overline{w} - \nabla(\nabla \cdot \overline{w}) = \nabla \times \overline{y}, & (x,t) \in Q, \\ \overline{y} = 0, \quad \overline{w} = 0 & (x,t) \in \Sigma. \end{cases} \tag{33}$$

Then, for each $T > 0$, (31) is locally exactly controllable to $(\overline{y}, \overline{p}, \overline{w})$ at time T. In other words, there exists $\delta > 0$ such that, for any initial data $(y^0, w^0) \in (H^2(\Omega) \cap V) \times H_0^1(\Omega)$ satisfying

$$\|(y^0, w^0) - (\overline{y}(\cdot\,,0), \overline{w}(\cdot\,,0))\|_{H^2 \times H_0^1} \le \delta, \tag{34}$$

there exist L^2 controls u and v and associated solutions (y, p, w) satisfying

$$y(x,T) = \overline{y}(x,T), \quad w(x,T) = \overline{w}(x,T) \quad \text{in } \Omega. \tag{35}$$

Notice that this case involves a nontrivial difficulty. Indeed, w is a non-scalar variable and the equations satisfied by its components w_i are coupled through the second-order terms $\partial_i(\nabla \cdot w)$. This is a serious inconvenient. An appropriate strategy has to be applied in order to deduce the required Carleman estimates.

Let us also mention [4, 24, 25], where the controllability of the MHD and other related equations has been analyzed.

For all these systems, the proof of the controllability can be achieved arguing as in the first part of the proof of Theorem 4. This is the general structure of the argument:

- First, rewrite the original controllability problem as a nonlinear equation in a space of admissible "state-control" pairs.
- Then, prove an appropriate global Carleman inequality and a regularity result and deduce that the linearized equation possesses at least one solution. This provides a controllability result for a related linear problem.
- Check that the hypotheses of a suitable implicit function theorem are satisfied and deduce a local result.

Remark 8. Recall that an alternative strategy was introduced in [35] in the context of the semilinear wave equation: first, consider a linearized similar problem and rewrite the original controllability problem in terms of a fixed point equation; then, prove a global Carleman inequality and deduce an observability estimate for the adjoint system and a controllability result for the linearized problem; finally, prove appropriate estimates for the control and the state (this usually needs some kind of *smallness* of the data), prove an appropriate compactness property of the state and deduce that there exists at least one fixed point. This method has been used in [21] to prove a result similar to Theorem 4.

Remark 9. Observe that all these results are positive, in the sense that they provide local controllability properties. At present, no negative result is known to hold for these nonlinear systems (except for the already considered one-dimensional Burgers equation).

To end this section, let us mention another system from fluid mechanics, apparently not much more complex than (21), for which local exact controllability (and even local null controllability) is an open question:

$$\begin{cases} y_t + (y \cdot \nabla)y - \nabla \cdot (\nu(|Dy|)Dy) + \nabla p = v1_\omega, & (x,t) \in Q, \\ \nabla \cdot y = 0, & (x,t) \in Q, \\ y = 0, & (x,t) \in \Sigma, \\ y(x,0) = y^0(x), & x \in \Omega. \end{cases} \tag{36}$$

Here, $Dy = \frac{1}{2}(\nabla y + \nabla y^t)$ and $\nu : \mathbb{R}_+ \mapsto \mathbb{R}_+$ is a regular function (for example, we can take $\nu(s) \equiv a + bs^{r-1}$ for some $a, b, r > 0$).

This system models the behavior of a *quasi-Newtonian* fluid; for a mathematical analysis, see [5, 32]. In view of the new nonlinear diffusion term $\nabla \cdot (\nu(|Dy|)Dy)$, its control properties are much more difficult to analyze than for (21).

Acknowledgement. The author has been partially supported by D.G.I. (Spain), Grants BFM2003–06446 and MTM2006–07932.

References

1. G. Alessandrini and L. Escauriaza. Null-controllability of one-dimensional parabolic equations. *ESAIM Control Optim. Calc. Var.*, 14(2):284–293, 2007.
2. F. Ammar-Khodja, A. Benabdallah, C. Dupaix, and M. González-Burgos. A Kalman rank condition for the localized distributed controllability of a class of linear parabolic systems. *J. Evol. Equ.*, 2009. DOI 10.1007/s00028-009-0008-8.
3. S. Anita and D. Tataru. Null controllability for the dissipative semilinear heat equation. *Appl. Math. Optim.*, 46:97–105, 2002.
4. V. Barbu, T. Havarneanu, C. Popa, and S. S. Sritharan. Exact controllability for the magnetohydrodynamic equations. *Comm. Pure Appl. Math.*, 56:732–783, 2003.
5. H. Bellout, F. Bloom, and J. Nečas. Young measure-valued solutions for non-Newtonian incompressible fluids. *Comm. Partial Differential Equations*, 19 (11–12):1763–1803, 1994.

6. J. M. Coron. On the controllability of the 2-D incompressible Navier–Stokes equations with the Navier slip boundary conditions. *ESAIM Control Optim. Calc. Var.*, 1:35–75, 1995/96.
7. J. M. Coron. *Control and nonlinearity*, volume 136 of *Mathematical surveys and monographs*. American Mathematical Society, Providence, RI, 2007.
8. C. Fabre. Uniqueness results for Stokes equations and their consequences in linear and nonlinear control problems. *ESAIM Contrôle Optim. Calc. Var.*, 1:267–302, 1995/96.
9. C. Fabre, J.-P. Puel, and E. Zuazua. Approximate controllability of the semilinear heat equation. *Proc. Roy. Soc. Edinburgh Sect. A*, 125(1):31–61, 1995.
10. E. Fernández-Cara, M. González-Burgos, and L. De Teresa. 2009. In preparation.
11. E. Fernández-Cara and S. Guerrero. Global Carleman inequalities for parabolic systems and applications to controllability. *SIAM J. Control Optim.*, 45(4):1399–1446, 2006.
12. E. Fernández-Cara and S. Guerrero. Local exact controllability of micropolar fluids. *J. Math. Fluid Mech.*, 9(3):419–453, 2007.
13. E. Fernández-Cara and S. Guerrero. Null controllability of the Burgers system with distributed controls. *Systems Control Lett.*, 56(5):366–372, 2007.
14. E. Fernández-Cara, S. Guerrero, O. Yu. Imanuvilov, and J.-P. Puel. Local exact controllability of the Navier–Stokes system. *J. Math. Pures Appl. (9)*, 83(12):1501–1542, 2004.
15. E. Fernández-Cara, S. Guerrero, O. Yu. Imanuvilov, and J.-P. Puel. Some controllability results for the n-dimensional Navier–Stokes and Boussinesq systems with $n-1$ scalar controls. *SIAM J. Control and Optim.*, 45(1):146–173, 2006.
16. E. Fernández-Cara and E. Zuazua. The cost of approximate controllability for heat equations: The linear case. *Adv. Differential Equations*, 5(4–6):465–514, 2000.
17. A. V. Fursikov. Exact boundary zero-controllability of three-dimensional Navier–Stokes equations. *J. Dynam. Control Systems*, 1(3):325–350, 1995.
18. A. V. Fursikov and O. Yu. Imanuvilov. On exact boundary zero-controllability of two-dimensional Navier–Stokes equations. *Acta Appl. Math.*, 37(1–2):67–76, 1994.
19. A. V. Fursikov and O. Yu. Imanuvilov. Controllability of evolution equations. Lecture Notes Series 34, Seoul National University, Research Institute of Mathematics, Global Analysis Research Center, 1996.
20. A. V. Fursikov and O. Yu. Imanuvilov. Exact controllability of the Navier–Stokes and Boussinesq equations. *Uspekhi Mat. Nauk*, 54(3(327)):93–146, 1999. In Russian, translation in Russian Math. Surveys 54(3):565–618, 1999.
21. M. González-Burgos, S. Guerrero, and J. P. Puel. On the exact controllability of the Navier–Stokes and Boussinesq equations. 2009. To appear.
22. S. Guerrero. Local exact controllability to the trajectories of the Boussinesq system. *Ann. Inst. H. Poincaré Anal. Non Linéaire*, 23(1):29–61, 2006.
23. S. Guerrero and O. Yu. Imanuvilov. Remarks on global controllability for the Burgers equation with two control forces. *Ann. Inst. H. Poincaré Anal. Non Linéaire*, 24(6):897–906, 2007.
24. T. Havarneanu, C. Popa, and S. S. Sritharan. Exact internal controllability for the magnetohydrodynamic equations in multi-connected domains. *Adv. Differential Equations*, 11(8):893–929, 2006.

25. T. Havarneanu, C. Popa, and S. S. Sritharan. Exact internal controllability for the two-dimensional magnetohydrodynamic equations. *SIAM J. Control Optim.*, 46(5):1802–1830, 2007.
26. O. Yu. Imanuvilov. Boundary controllability of parabolic equations. *Russian Acad. Sci. Sb. Math.*, 186:109–132, 1995. (In Russian).
27. O. Yu. Imanuvilov. Remarks on exact controllability for the Navier–Stokes equations. *ESAIM Control Optim. Calc. Var.*, 6:39–72, 2001.
28. J. Le Rousseau and L. Robbiano. Carleman estimate for elliptic operators with coefficients with jumps at an interface in arbitrary dimension and application to the null controllability of linear parabolic equations. *Arch. Rational Mech. Anal.*, 2009. To appear.
29. G. Lebeau and L. Robbiano. Contrôle exact de l'équation de la chaleur. *Comm. Partial Differential Equations*, 20:335–356, 1995.
30. J.-L. Lions. *Contrôlabilité exacte, perturbations et stabilisation de systèmes distribués, Tomes 1 & 2.* Recherches en Mathématiques Appliquées, 8 & 9. Masson, Paris, 1988.
31. J.-L. Lions. Exact controllability, stabilizability and perturbations for distributed systems. *SIAM Review*, 30:1–68, 1988.
32. J. Málek, J. Nečas, M. Rokyta, and M. Ružička. *Weak and measure-valued solutions to evolutionary PDEs*, volume 13 of *Applied Mathematics and Mathematical Computation.* Chapman & Hall, London, 1996.
33. D. L. Russell. A unified boundary controllability theory for hyperbolic and parabolic partial differential equations. *Studies in Appl. Math.*, 52:189–211, 1973.
34. D. L. Russell. Controllability and stabilizability theory for linear partial differential equations. Recent progress and open questions. *SIAM Review*, 20:639–739, 1978.
35. E. Zuazua. Exact boundary controllability for the semilinear wave equation. In H. Brezis and J.-L. Lions, editors, *Nonlinear Partial Differential Equations and their Applications, Vol. X*, pages 357–391. Pitman, New York, 1991.

Goal Oriented Mesh Adaptivity for Mixed Control-State Constrained Elliptic Optimal Control Problems

Michael Hintermüller[1] and Ronald H.W. Hoppe[2]

[1] Institute of Mathematics, Humboldt University, DE-10117 Berlin, Germany, and Department of Mathematics and Scientific Computing, University of Graz, AT-8010 Graz, Austria, hint@mathematik.hu-berlin.de, michael.hintermueller@uni-graz.at

[2] Department of Mathematics, University of Houston, Houston, TX 77204-3008, USA, and Institute of Mathematics, University of Augsburg, DE-86159 Augsburg, Germany, rohop@math.uh.edu, hoppe@math.uni-augsburg.de

1 Introduction

Adaptive finite element methods for the numerical solution of partial differential equations consist of successive cycles of the loop

$$\mathsf{SOLVE} \implies \mathsf{ESTIMATE} \implies \mathsf{MARK} \implies \mathsf{REFINE}.$$

Here, SOLVE stands for the finite element solution of the problem with respect to a given triangulation of the computational domain. The following step ESTIMATE is devoted to the estimation of the global discretization error in some appropriate norm or a user specified quantity of interest by a cheaply computable a posteriori error estimator. The estimator is assumed to consist of local contributions whose actual magnitude is then used in the step MARK to specify elements of the triangulation for refinement. The final step REFINE deals with the generation of a new triangulation based on the refinement of the elements selected in the previous step according to specific refinement rules. Adaptive finite elements are by now well established. There are various approaches such as residual-type a posteriori error estimators which rely on the proper evaluation of the residuals with respect to a computed approximation in the norm of the dual space and hierarchical type estimators where the equation satisfied by the error is suitably localized along with a solution of the local problems by higher order finite elements (cf., e.g. [1, 3, 35]). Averaging-type estimators typically use some sort of gradient recovery on element-related patches (cf., e.g. [1, 35]), whereas the theory of guaranteed error majorants provides reliable upper bounds for the error (see [31]). Finally, the goal oriented weighted dual approach extracts information on the error via the dual problem (cf. [4, 12]).

W. Fitzgibbon et al. (eds.), *Applied and Numerical Partial Differential Equations*, Computational Methods in Applied Sciences 15, DOI 10.1007/978-90-481-3239-3_8,

As far as the optimal control of PDEs are concerned, the goal oriented dual weighted approach has been applied to unconstrained problems in [4,5], to control constrained ones in [17,36] and to state constrained problems in [16,19]. Residual-type a posteriori error estimators for control constrained problems have been developed and analyzed in [13,14,18,20,23,26,27]. State constrained optimal control problems are more difficult to handle than control constrained ones, since the Lagrange multiplier for the state constraints typically lives in a measure space. An appropriate way to cope with this problem is to use a regularization of the state constrained problems by means of mixed control-state constraints (Lavrentiev regularization). With regard to numerical solution techniques the regularized problems can be formally treated as in the case of control constraints (cf., e.g. [2,9,29,32–34]).

In this paper, we will develop, analyze and implement the goal oriented weighted dual approach to mixed control-state constrained distributed optimal control problems for linear second order elliptic boundary value problems. The paper is organized as follows: In Section 2, we consider a model distributed optimal control problem for a two-dimensional, second order elliptic PDE with a quadratic objective functional and mixed unilateral constraints on the state and on the control. The finite element discretization is based on standard P1 conforming finite elements with respect to simplicial triangulations of the computational domain and gives rise to a finite dimensional constrained minimization problem. In both the continuous and discrete regime, the optimality conditions are stated in terms of the associated Lagrangians. Section 3 is devoted to a representation of the error in the quantity of interest which is chosen as the objective functional. The error representation involves primal–dual residuals, a primal–dual mismatch in complementarity due to a possible mismatch between the continuous and discrete active and non-active sets, and data oscillation terms. In Section 4, we derive the goal oriented a posteriori error estimator based on appropriate upper bounds both for the primal–dual residuals and the primal–dual mismatch in complementarity. The final section, Section 5 contains a brief description of the marking and refinement strategy as well as numerical results for an example illustrating the performance of the error estimator.

2 The Mixed Control-State Elliptic Optimal Control Problem and Its Finite Element Approximation

We assume Ω to be a bounded domain in $\mathbb{R}^2$ with boundary $\Gamma := \overline{\Gamma}_D \cup \overline{\Gamma}_N$, $\Gamma_D \cap \Gamma_N = \emptyset$. We use standard notation from Lebesgue and Sobolev space theory. In particular, we refer to $L^2(\Omega)$ as the Hilbert space with inner product $(\cdot,\cdot)_{0,\Omega}$ and norm $\|\cdot\|_{0,\Omega}$ and to $H^k(\Omega)$, $k \in \mathbb{N}$, as the Sobolev space with norm $\|\cdot\|_{k,\Omega}$. The set $L^2_+(\Omega)$ stands for the positive cone in $L^2(\Omega)$ with respect to the canonical ordering.

Given a desired state $y^d \in L^2(\Omega)$, a shift control $u^d \in L^2(\Omega)$, regularization parameters $\alpha > 0$, $\varepsilon > 0$, and a function $\psi \in L^\infty(\Omega)$, we consider the mixed control-state constrained distributed optimal control problem:
Find $(y, u) \in V \times L^2(\Omega)$, where $V := \{v \in H^1(\Omega) \mid v|_{\Gamma_D} = 0\}$, such that

$$\inf_{y,u} J(y,u) := \frac{1}{2}\|y - y^d\|^2_{0,\Omega} + \frac{\alpha}{2}\|u - u^d\|^2_{0,\Omega}, \tag{1a}$$

$$\text{subject to} \quad a(y,v) = (u,v)_{0,\Omega}, \quad v \in V, \tag{1b}$$

$$\varepsilon u + y \in K := \{v \in L^2(\Omega) \mid v(x) \le \psi(x) \text{ f.a.a. } x \in \Omega\}. \tag{1c}$$

Here, $a(\cdot,\cdot) : V \times V \to \mathbb{R}$ stands for the bounded, V-elliptic bilinear form

$$a(u,v) := \int_\Omega (\nabla u \cdot \nabla v + cuv)\, dx, \quad c \in \mathbb{R}_+.$$

Denoting by $A : V \to V^*$ the operator associated with $a(\cdot,\cdot)$, we introduce the Lagrangian $\mathcal{L} : V \times L^2(\Omega) \times V \times L^2_+(\Omega) \to \mathbb{R}$ according to

$$\mathcal{L}(y,u,p,\sigma) := J(y,u) + \langle Ay - u, p\rangle + (\varepsilon u + y - \psi, \sigma)_{0,\Omega}, \tag{2}$$

where $\langle\cdot,\cdot\rangle$ denotes the dual pairing between V^* and V. Then, the minimization problem (1a)–(1c) can be equivalently stated as the saddle point problem

$$\inf_{y,u} \sup_{p,\sigma} \mathcal{L}(y,u,p,\sigma). \tag{3}$$

Setting $x := (y,u,p) \in X := V \times L^2(\Omega) \times V$, the optimality conditions read as follows:

$$\nabla_x \mathcal{L}(x,\sigma) = 0, \tag{4a}$$

$$\nabla_\sigma \mathcal{L}(x,\sigma)(\mu - \sigma) \le 0, \quad \mu \in L^2_+(\Omega), \tag{4b}$$

where $\nabla_x \mathcal{L}(x,\sigma)$ and $\nabla_\sigma \mathcal{L}(x,\sigma)$ stand for the derivatives of $\mathcal{L}$ with respect to x and σ in (x,σ). The multiplier p is referred to as the adjoint state. We note that (4a) gives rise to the state equation (1b), the adjoint state equation

$$a(p,v) = (y^d - y - \sigma, v)_{0,\Omega}, \quad v \in V, \tag{5}$$

and the equation

$$p = \alpha(u - u^d) + \varepsilon\sigma, \tag{6}$$

whereas the variational inequality (4b) can be equivalently written in terms of the complementarity conditions

$$\sigma \in L^2_+(\Omega), \quad \psi - (\varepsilon u + y) \in L^2_+(\Omega), \quad (\varepsilon u + y - \psi, \sigma)_{0,\Omega} = 0. \tag{7}$$

We define the active set $\mathcal{A}$ as the maximal open set $A \subset \Omega$ such that $\varepsilon u(x) + y(x) = \psi(x)$ f.a.a. $x \in A$ and the inactive set $\mathcal{I}$ according to $\mathcal{I} := \bigcup_{\kappa>0} B_\kappa$, where B_κ is the maximal open set $B \subset \Omega$ such that $\varepsilon u(x) + y(x) \leq \psi(x) - \kappa$ for almost all $x \in B$.

For the finite element discretization of (1a)–(1c) we consider a family $\{\mathcal{T}_\ell(\Omega)\}$ of shape-regular simplicial triangulations of Ω which align with Γ_D, Γ_N on Γ. We denote by $\mathcal{N}_\ell(D)$ and $\mathcal{E}_\ell(D)$, $D \subseteq \overline{\Omega}$, the sets of vertices and edges of $\mathcal{T}_\ell(\Omega)$ in $D \subseteq \overline{\Omega}$, and we refer to h_T and $|T|$ as the diameter and the area of an element $T \in \mathcal{T}_\ell(\Omega)$, whereas h_E stands for the length of an edge $E \in \mathcal{E}_\ell(D)$. For $E \in \mathcal{E}_\ell(\Omega)$ such that $E = T_+ \cap T_-$, $T_\pm \in \mathcal{T}_\ell(\Omega)$, we define $\omega_E := T_+ \cup T_-$. Further, we denote by $S_\ell := \{v_\ell \in C_0(\Omega) \mid v_\ell|_T \in P_1(T),\ T \in \mathcal{T}_\ell(\Omega)\}$ the finite element space of continuous, piecewise linear finite elements and we refer to V_ℓ as its subspace $V_\ell := \{v_\ell \in S_\ell \mid v_\ell|_{\Gamma_D} = 0\}$. We will also use the following notation: If A and B are two quantities, then $A \preceq B$ means that there exists a positive constant C such that $A \leq CB$, where C only depends on the shape regularity of the triangulations, but not on their granularities.

Then, given approximations $y_\ell^d \in S_\ell$, $u_\ell^d \in S_\ell$ and $\psi_\ell \in S_\ell$ of y^d, u^d and ψ, the finite element approximation of (1a)–(1c) is given by
Find $(y_\ell, u_\ell) \in V_\ell \times S_\ell$ such that

$$\inf_{y_\ell, u_\ell} J_\ell(y_\ell, u_\ell) := \frac{1}{2}\|y_\ell - y_\ell^d\|_{0,\Omega}^2 + \frac{\alpha}{2}\|u_\ell - u_\ell^d\|_{0,\Omega}^2, \tag{8a}$$

$$\text{subject to} \quad a(y_\ell, v_\ell) = (u_\ell, v_\ell)_{0,\Omega}, \quad v_\ell \in V_\ell, \tag{8b}$$

$$\varepsilon u_\ell + y_\ell \in K_\ell := \{v_\ell \in S_\ell \mid v_\ell \leq \psi_\ell \quad \text{in } \Omega\}. \tag{8c}$$

We proceed as in the continuous regime and introduce the Lagrangian $\mathcal{L}_\ell$: $V_\ell \times S_\ell \times V_\ell \times (S_\ell \cap L_+^2(\Omega))$ by

$$\mathcal{L}_\ell(y_\ell, u_\ell, p_\ell, \sigma_\ell) := J_\ell(y_\ell, u_\ell) + \langle Ay_\ell - u_\ell, p_\ell\rangle + (\varepsilon u_\ell + y_\ell - \psi_\ell, \sigma_\ell)_{0,\Omega} \tag{9}$$

such that (8a)–(8c) is equivalent to the saddle point problem

$$\inf_{y_\ell, u_\ell} \sup_{p_\ell, \sigma_\ell} \mathcal{L}_\ell(y_\ell, u_\ell, p_\ell, \sigma_\ell). \tag{10}$$

The optimality conditions turn out to be

$$\nabla_x \mathcal{L}_\ell(x_\ell, \sigma_\ell) = 0, \tag{11a}$$

$$\nabla_\sigma \mathcal{L}_\ell(x_\ell, \sigma_\ell)(\mu_\ell - \sigma_\ell) \leq 0, \quad \mu_\ell \in S_\ell \cap L_+^2(\Omega), \tag{11b}$$

where $x_\ell := (y_\ell, u_\ell, p_\ell) \in X_\ell := V_\ell \times S_\ell \times V_\ell$. Again, (11a) comprises the discrete state equation (8b), the discrete adjoint state equation

$$a(p_\ell, v_\ell) = (y_\ell^d - y_\ell - \sigma_\ell, v_\ell)_{0,\Omega}, \quad v_\ell \in V_\ell, \tag{12}$$

and the equation

$$p_\ell = \alpha(u_\ell - u_\ell^d) + \varepsilon\sigma_\ell. \tag{13}$$

On the other hand, (11b) represents the discrete complementarity conditions

$$\sigma_\ell \in S_\ell \cap L^2_+(\Omega), \quad \psi_\ell - (\varepsilon u_\ell + y_\ell) \in S_\ell \cap L^2_+(\Omega), \quad (\varepsilon u_\ell + y_\ell - \psi_\ell, \sigma_\ell)_{0,\Omega} = 0. \tag{14}$$

We define the discrete active set $\mathcal{A}_\ell$ according to $\mathcal{A}_\ell := \{x \in \overline{\Omega} \mid \varepsilon u_\ell(x) + y_\ell(x) = \psi_\ell(x)\}$ and refer to $\mathcal{I}_\ell := \overline{\Omega} \setminus \mathcal{A}_\ell$ as the discrete inactive set.

3 Error Representation in the Quantity of Interest

We derive an error representation in the quantity of interest which involves the second derivative of the Lagrangian $\mathcal{L}$ with respect to x. Since this second derivative does depend neither on x nor on σ, we simply write $\nabla_{xx}\mathcal{L}(z,z')$, $z, z' \in X$, instead of $\nabla_{xx}\mathcal{L}(x,\sigma)(z,z')$. We will use the same simplifying notation for the second derivative of $\mathcal{L}_h$.

Theorem 1. *Let $(x,\sigma) \in X \times L^2_+(\Omega)$ and $(x_\ell, \sigma_\ell) \in X_\ell \times (S_\ell \cap L^2_+(\Omega))$ be the solutions of* (3) *and* (10), *respectively. Then there holds*

$$J(y,u) - J_\ell(y_\ell, u_\ell) = -\frac{1}{2}\nabla_{xx}\mathcal{L}_\ell(x_\ell - x, x_\ell - x) + (\varepsilon u_\ell + y_\ell - \psi, \sigma)_{0,\Omega} + \mathrm{osc}_\ell^{(1)}, \tag{15}$$

where $\mathrm{osc}_\ell^{(1)}$ stands for the data oscillations

$$\mathrm{osc}_\ell^{(1)} := \sum_{T \in \mathcal{T}_\ell(\Omega)} \mathrm{osc}_T^{(1)}, \tag{16}$$

$$\mathrm{osc}_T^{(1)} := (y_\ell - y_\ell^d, y_\ell^d - y^d)_{0,T} + \alpha(u_\ell - u_\ell^d, u_\ell^d - u^d)_{0,T} + \frac{1}{2}\|y^d - y_\ell^d\|_{0,T}^2 + \frac{\alpha}{2}\|u^d - u_\ell^d\|_{0,T}^2.$$

Proof. We note that for $z_\ell = (\delta y_\ell, \delta u_\ell, \delta p_\ell) \in X_\ell$ there holds

$$\mathcal{L}(x,\sigma_\ell) = \mathcal{L}(x,\sigma) + (\varepsilon u + y - \psi, \sigma_\ell - \sigma)_{0,\Omega}, \tag{17a}$$

$$\nabla_x\mathcal{L}(x_\ell,\sigma_\ell)(z_\ell) = \nabla_x\mathcal{L}(x_\ell,\sigma_\ell)(z_\ell) + (\varepsilon \delta u_\ell + \delta y_\ell, \sigma_\ell - \sigma)_{0,\Omega}. \tag{17b}$$

Using the optimality conditions (4a), (4b) and (11a), (11b) as well as (17a), (17b), Taylor expansion yields

$$\begin{aligned}
&J(y,u) - J_\ell(y_\ell,u_\ell)\mathcal{L}(x,\sigma) - \mathcal{L}_\ell(x_\ell,\sigma_\ell) \\
&= \mathcal{L}(x,\sigma) - \mathcal{L}_\ell(x,\sigma_\ell) - \nabla_x\mathcal{L}_\ell(x,\sigma_\ell)(x_\ell - x) - \frac{1}{2}\nabla_{xx}\mathcal{L}_\ell(x_\ell - x, x_\ell - x) \\
&= J(y,u) - J_\ell(y,u) - (\varepsilon u + y - \psi_\ell, \sigma_\ell)_{0,\Omega} \\
&\quad - \nabla_x\mathcal{L}_\ell(x,\sigma_\ell)(x_\ell - x) - \frac{1}{2}\nabla_{xx}\mathcal{L}_\ell(x_\ell - x, x_\ell - x)
\end{aligned}$$

$$= -\nabla_x \mathcal{L}(x,\sigma_\ell)(x_\ell - x) - \frac{1}{2}\nabla_{xx}\mathcal{L}_\ell(x_\ell - x, x_\ell - x)$$
$$- (\varepsilon u + y - \psi_\ell, \sigma_\ell)_{0,\Omega} + \mathrm{osc}_\ell^{(1)}$$
$$= -\frac{1}{2}\nabla_{xx}\mathcal{L}_\ell(x_\ell - x, x_\ell - x) - (\varepsilon u + y - (\varepsilon u_\ell + y_\ell), \sigma_\ell)_{0,\Omega}$$
$$+ (\varepsilon u_\ell + y_\ell - (\varepsilon u + y), \sigma - \sigma_\ell)_{0,\Omega} + \mathrm{osc}_\ell^{(1)}$$
$$= -\frac{1}{2}\nabla_{xx}\mathcal{L}_\ell(x_\ell - x, x_\ell - x) + (\varepsilon u_\ell + y_\ell - \psi, \sigma)_{0,\Omega} + \mathrm{osc}_\ell^{(1)},$$

from which we conclude. □

Remark 1. We note that the error representation (15) reduces to the result from [5] in the unconstrained case, i.e. when $\sigma = \sigma_\ell = 0$.

For a further evaluation of the error, we introduce interpolation operators

$$i_\ell^y : V \to V_\ell, \quad i_\ell^p : V \to V_\ell, \quad i_\ell^u : L^2(\Omega) \to S_\ell, \quad i_\ell^\sigma : L^2(\Omega) \to S_\ell, \tag{18}$$

such that for all $y, p \in V$ and $u \in L^2(\Omega)$ there holds

$$\|i_\ell^y y - y\|_{0,T}^2 + h_T^{1/2}\|i_\ell^y y - y\|_{0,\partial T}^2 \preceq h_T\|y\|_{1,D_T},$$
$$\|i_\ell^p p - p\|_{0,T}^2 + h_T^{1/2}\|i_\ell^p p - p\|_{0,\partial T}^2 \preceq h_T\|p\|_{1,D_T},$$
$$\|i_\ell^u u - u\|_{0,T}, \quad \|i_\ell^\sigma \sigma - \sigma\|_{0,T} \to 0 \quad \text{as } h_T \to 0.$$

where $D_T := \{T' \in \mathcal{T}_\ell(\Omega) \mid \mathcal{N}_\ell(T') \cap \mathcal{N}_\ell(T) \neq \emptyset\}$. We may choose, for instance, Clément-type quasi-interpolation operators (cf., e.g. [35]) or the Scott–Zhang interpolation operators (cf., e.g. [8]).

Theorem 2. *In addition to the assumptions of Theorem 1, let $i_\ell^x = (i_\ell^y, i_\ell^u, i_\ell^p)$ be the interpolation operators as given by* (18). *Then there holds*

$$J(y,u) - J_\ell(y_\ell, u_\ell) = -r(i_\ell^y y - y) - r(i_\ell^p p - p) + \mu_\ell(x,\sigma) + \mathrm{osc}_\ell^{(1)} + \mathrm{osc}_\ell^{(2)}, \tag{19}$$

where $r(i_\ell^y y - y)$ and $r(i_\ell^p p - p)$ stand for the primal–dual residuals

$$r(i_\ell^y y - y) := \frac{1}{2}\left((y_\ell - y_\ell^d + \sigma_\ell, i_\ell^y y - y)_{0,\Omega} + (\nabla p_\ell, \nabla(i_\ell^y y - y))_{0,\Omega}\right), \tag{20a}$$
$$r(i_\ell^p p - p) := \frac{1}{2}\left((\nabla y_\ell, \nabla(i_\ell^p p - p))_{0,\Omega} - (u_\ell, i_\ell^p p - p)_{0,\Omega}\right), \tag{20b}$$

Moreover, $\mu_\ell(x,\sigma)$ is the primal–dual mismatch in complementarity and $\mathrm{osc}_\ell^{(2)}$ a further data oscillation term given by

$$\mu_\ell(x,\sigma) := \frac{1}{2}\left((\varepsilon u_\ell + y_\ell - \psi, \sigma)_{0,\Omega} + (\psi_\ell - (\varepsilon u + y), \sigma_\ell)_{0,\Omega}\right), \tag{21a}$$
$$\mathrm{osc}_\ell^{(2)} := \frac{1}{2}(y^d - y_\ell^d, y_\ell - i_\ell^y y)_{0,\Omega} + \frac{1}{2}(y^d - y_\ell^d, i_\ell^y y - y)_{0,\Omega}$$
$$+ \frac{\alpha}{2}(u^d - u_\ell^d, u_\ell - i_\ell^u u)_{0,\Omega} + \frac{\alpha}{2}(u^d - u_\ell^d, i_\ell^u u - u)_{0,\Omega}. \tag{21b}$$

Proof. Using (11a) and (17b), for $z_\ell = (\delta y_\ell, \delta u_\ell, \delta p_\ell) \in X_\ell$ we find

$$\begin{aligned}
0 &= \nabla_x \mathcal{L}(x,\sigma)(z_\ell) \\
&= \nabla_x \mathcal{L}(x_\ell,\sigma_\ell)(z_\ell) + \nabla_{xx}\mathcal{L}(x - x_\ell, z_\ell) + (\varepsilon\delta u_\ell + \delta y_\ell, \sigma - \sigma_\ell)_{0,\Omega} \\
&= \nabla_{xx}\mathcal{L}(x - x_\ell, z_\ell) + (\varepsilon\delta u_\ell + \delta y_\ell, \sigma - \sigma_\ell)_{0,\Omega} + (y_\ell^d - y^d, \delta y_\ell)_{0,\Omega} \\
&\quad + \alpha(u_\ell^d - u^d, \delta u_\ell)_{0,\Omega},
\end{aligned}$$

from which we deduce

$$\nabla_x \mathcal{L}(x_\ell,\sigma)(x - x_\ell - z_\ell) = \nabla_{xx}\mathcal{L}(x_\ell - x, x - x_\ell - z_\ell), \tag{22a}$$

$$\begin{aligned}
\nabla_{xx}\mathcal{L}(x_\ell - x, x_\ell - x) &= \nabla_{xx}\mathcal{L}(x_\ell - x, x_\ell - x + z_\ell) \\
&\quad - (\varepsilon\delta u_\ell + \delta y_\ell, \sigma - \sigma_\ell)_{0,\Omega}.
\end{aligned} \tag{22b}$$

Taking advantage of (22a),(22b) in (15), it follows that

$$\begin{aligned}
&J(y,u) - J_\ell(y_\ell,u_\ell) \\
&= \frac{1}{2}\nabla_{xx}\mathcal{L}(x,\sigma_\ell)(x - x_\ell, x_\ell - x + z_\ell) \\
&\quad + \frac{1}{2}(\varepsilon\delta u_\ell + \delta y_\ell, \sigma - \sigma_\ell)_{0,\Omega} + \frac{1}{2}(y_\ell^d - y^d, \delta y_\ell)_{0,\Omega} \\
&\quad + \frac{\alpha}{2}(u_\ell^d - u^d, \delta u_\ell)_{0,\Omega} + (\varepsilon\delta u_\ell + y_\ell - \psi, \sigma)_{0,\Omega} + \mathrm{osc}_\ell^{(1)} \\
&= -\frac{1}{2}\nabla_x \mathcal{L}(x_\ell,\sigma_\ell)(x_\ell - x + z_\ell) + \frac{1}{2}(\varepsilon u_\ell + y_\ell - (\varepsilon u + y), \sigma_\ell + \sigma)_{0,\Omega} \\
&\quad + \frac{1}{2}(y^d - y_\ell^d, y_\ell - y)_{0,\Omega} + \frac{\alpha}{2}(u_\ell^d - u^d, \delta u_\ell)_{0,\Omega} + \mathrm{osc}_\ell^{(1)}.
\end{aligned}$$

We conclude by choosing $z_\ell = (i_\ell^y y - y_\ell, i_\ell^p p - p_\ell, i_\ell^u - u_\ell)$ and observing (7) and (14). □

Remark 2. The primal–dual residuals $r(i_\ell^y y - y)$ and $r(i_\ell^p p - p)$ will be further estimated in the following section and will be made fully a posteriori in a standard way (cf., e.g. [4]). The term $\mu_\ell(x,\sigma)$ as given by (21a) represents the primal–dual mismatch in complementarity due to a possible mismatch in the approximation of the active and inactive sets $\mathcal{A}$ and $\mathcal{I}$ by their discrete counterparts $\mathcal{A}_\ell$ and $\mathcal{I}_\ell$. In its present form it is not yet a posteriori. In the subsequent section, we will show how $\mu_\ell(x,\sigma)$ can be made fully a posteriori and thus be included in the refinement strategy. A similar remark applies to the term $\mathrm{osc}_\ell^{(2)}$ which is essentially a data oscillation term, but as given by (21b) not a posteriori due to the occurrence of y. It will be made fully a posteriori as well.

4 Weighted Primal–Dual A Posteriori Error Estimator

By straightforward estimation of the right-hand sides in the representations (20a), (20b) of the primal–dual residuals the following result can be easily established.

Theorem 3. *The primal–dual residuals can be estimated according to*

$$|r(i_\ell^y y - y)| \preceq \sum_{T\in\mathcal{T}_\ell(\Omega)} \omega_T^y \rho_T^y, \tag{23a}$$

$$|r(i_\ell^p p - p)| \preceq \sum_{T\in\mathcal{T}_\ell(\Omega)} \omega_T^p \rho_T^p. \tag{23b}$$

Here, ρ_T^y and ρ_T^p are the L^2-norms of the residuals associated with the state and the adjoint state equation

$$\rho_T^y := \left(\|u_\ell\|_{0,T}^2 + h_T^{-1} \|\frac{1}{2}\nu \cdot [\nabla y_\ell]\|_{0,\partial T}^2 \right)^{1/2}, \tag{24a}$$

$$\rho_T^p := \left(\|y_\ell - y_\ell^d - \sigma_\ell\|_{0,T}^2 + h_T^{-1} \|\frac{1}{2}\nu \cdot [\nabla p_\ell]\|_{0,\partial T}^2 \right)^{1/2}. \tag{24b}$$

The corresponding dual weights ω_T^y and ω_T^p are given by

$$\omega_T^y := \left(\|i_\ell^p p - p\|_{0,T}^2 + h_T \|i_\ell^p p - p\|_{0,\partial T}^2 \right)^{1/2}, \tag{25a}$$

$$\omega_T^p := \left(\|i_\ell^y y - y\|_{0,T}^2 + h_T \|i_\ell^y y - y\|_{0,\partial T}^2 \right)^{1/2}. \tag{25b}$$

Remark 3. If the state y of the purely state constrained problem (i.e. $\varepsilon = 0$) is in $W^{1,r}(\Omega)$ for some $r > 2$ and hence represents a continuous function, the adjoint state p lives in $W^{1,s}(\Omega)$ with s being conjugate to r. The multiplier σ turns out to be a bounded Borel measure, and the discrete multipliers σ_ℓ are chosen as a linear combination of Dirac delta functionals associated with the nodal points of the triangulation. In this case, the primal–dual residuals have to be estimated in the respective L^r- and L^s-norms and the multipliers have to be treated separately (cf. [19]).

There are several ways to provide approximations of the weights ω_T^y and ω_T^p, $T \in \mathcal{T}_\ell(\Omega)$. We refer to [4] for a detailed discussion. Here, we use piecewise quadratic interpolations $i_{\ell,2}^y y_\ell$ and $i_{\ell,2}^p p_\ell$ of the computed P1 approximations y_ℓ and p_ℓ of the state y and the adjoint state p with respect to the coarser triangulation $\mathcal{T}_{\ell-1}(\Omega)$. This results in the computable weights

$$\hat{\omega}_T^y := \left(\|i_{\ell,2}^p p_\ell - p_\ell\|_{0,T}^2 + h_T \|i_{\ell,2}^p p_\ell - p_\ell\|_{0,\partial T}^2 \right)^{1/2}, \tag{26a}$$

$$\hat{\omega}_T^p := \left(\|i_{\ell,2}^y y_\ell - y_\ell\|_{0,T}^2 + h_T \|i_{\ell,2}^y y_\ell - y_\ell\|_{0,\partial T}^2 \right)^{1/2}. \tag{26b}$$

We now concentrate on the primal–dual mismatch in complementarity $\mu_\ell(x, \sigma)$ where for notational simplicity we drop the argument (x, σ). Taking the complementarity conditions (7) and (14) into account, we find

$$\mu_\ell|_{\mathcal{I}\cap\mathcal{I}_\ell} = 0, \tag{27a}$$

$$\begin{aligned}\mu_\ell|_{\mathcal{A}\cap\mathcal{I}_\ell} &= \frac{1}{2}\big((\varepsilon u_\ell + y_\ell - \psi, i_\ell^\sigma \sigma)_{0,\mathcal{A}\cap\mathcal{I}_\ell} \\ &\quad + (\varepsilon u_\ell + y_\ell - \psi, \sigma - i_\ell^\sigma \sigma)_{0,\mathcal{A}\cap\mathcal{I}_\ell}\big),\end{aligned} \tag{27b}$$

$$\begin{aligned}\mu_\ell|_{\mathcal{I}\cap\mathcal{A}_\ell} &= \frac{1}{2}(\psi_\ell - (\varepsilon u + y), \sigma_\ell)_{0,\Omega} \\ &= \frac{1}{2}\big((\varepsilon(u_\ell - i_\ell^u u) + y_\ell - i_\ell^y y, \sigma_\ell)_{0,\mathcal{I}\cap\mathcal{A}_\ell} \\ &\quad + (\varepsilon(i_\ell^u u - u) + i_\ell^y y - y, \sigma_\ell)_{0,\mathcal{I}\cap\mathcal{A}_\ell}\big),\end{aligned} \tag{27c}$$

$$\begin{aligned}\mu_\ell|_{\mathcal{A}\cap\mathcal{A}_\ell} &= \frac{1}{2}\left((\varepsilon u_\ell + y_\ell - \psi, \sigma)_{0,\mathcal{A}\cap\mathcal{A}_\ell} + (\psi_\ell - (\varepsilon u + y), \sigma_\ell)_{0,\mathcal{A}\cap\mathcal{A}_\ell}\right) \\ &= \frac{1}{2}\big((\psi_\ell - \psi, i_\ell^\sigma \sigma + \sigma_\ell)_{0,\mathcal{A}\cap\mathcal{A}_\ell} + (\psi_\ell - \psi, \sigma - i_\ell^\sigma \sigma)_{0,\mathcal{A}\cap\mathcal{A}_\ell}\big).\end{aligned} \tag{27d}$$

We further need to provide computable approximations of the sets $\mathcal{A}$ and $\mathcal{I}$. We use a modification of the approximation of the indicator function $\chi(\mathcal{A})$ of the continuous coincidence set $\mathcal{A}$ from [26] (cf. also [17]) according to

$$\chi_\ell^{\mathcal{A}} := 1 - \frac{\psi - (\varepsilon i_{\ell,2}^u u_\ell + i_{\ell,2}^y y_\ell)}{\gamma h_\ell^r + \psi - (\varepsilon i_{\ell,2}^u u_\ell + i_{\ell,2}^y y_\ell)}, \tag{28}$$

where $0 < \gamma \le 1$ and $r > 0$ are fixed and $i_{\ell,2}^u u_\ell$ is defined in the same way as $i_{\ell,2}^y y_\ell$. Indeed, for $T \subset \mathcal{A}$ we find

$$\|\chi(\mathcal{A}) - \chi_\ell^{\mathcal{A}}\|_{0,T} \le \min(|T|^{1/2}, \gamma^{-1} h_\ell^{-r} \|\varepsilon u + y - (\varepsilon i_{\ell,2}^u u_\ell + i_{\ell,2}^y y_\ell)\|_{0,T})$$

which converges to zero whenever $\|\varepsilon u + y - (\varepsilon i_{\ell,2}^u u_\ell + i_{\ell,2}^y y_\ell)\|_{0,T} = O(h_\ell^q)$, $q > r$. By the same arguments, for $T \subset \mathcal{I}$ one can show as well that $\|\chi(\mathcal{A}) - \chi_\ell^{\mathcal{A}}\|_{0,T} \to 0$ as $h_\ell \to 0$. Now, for fixed $0 < \kappa \le 1$ and $0 < s \le r$ we provide approximations $\hat{\mathcal{A}}_\ell$ of $\mathcal{A}$ and $\hat{\mathcal{I}}_\ell$ of $\mathcal{I}$ according to

$$\hat{\mathcal{A}}_\ell := \bigcup\{T \in \mathcal{T}_\ell(\Omega) \mid \chi_\ell^{\mathcal{A}}(x) \ge 1 - \kappa h_\ell^s \text{ for all } x \in T\}, \tag{29a}$$

$$\hat{\mathcal{I}}_\ell := \bigcup\{T \in \mathcal{T}_\ell(\Omega) \mid \chi_\ell^{\mathcal{A}}(x) < 1 - \kappa h_\ell^s \text{ for some } x \in T\}. \tag{29b}$$

We define approximations $\mathcal{T}_{\mathcal{A}\cap\mathcal{A}_\ell}$, $\mathcal{T}_{\mathcal{I}\cap\mathcal{A}_\ell}$ and $\mathcal{T}_{\mathcal{A}\cap\mathcal{I}_\ell}$ of $\mathcal{A}\cap\mathcal{A}_\ell$, $\mathcal{I}\cap\mathcal{A}_\ell$ and $\mathcal{A}\cap\mathcal{I}_\ell$ by means of

$$\mathcal{T}_{\mathcal{A}\cap\mathcal{A}_\ell} := \hat{\mathcal{A}}_\ell \cap \mathcal{A}_\ell, \quad \mathcal{T}_{\mathcal{I}\cap\mathcal{A}_\ell} := \hat{\mathcal{I}}_\ell \cap \mathcal{A}_\ell, \quad \mathcal{T}_{\mathcal{A}\cap\mathcal{I}_\ell} := \hat{\mathcal{A}}_\ell \cap \mathcal{I}_\ell.$$

We further define

$$\begin{aligned}\tilde{\omega}_T^y &:= \|i_{\ell,2}^y y_\ell - y_\ell\|_{0,T}, \\ \tilde{\omega}_T^u &:= \|i_{\ell,2}^u u_\ell - u_\ell\|_{0,T}, \\ \tilde{\omega}_T^\sigma &:= \|i_{\ell,2}^\sigma \sigma_\ell - \sigma_\ell\|_{0,T},\end{aligned}$$

where $i_{\ell,2}^{\sigma}\sigma_\ell$ is also given by piecewise quadratic interpolation. Then, we can estimate the contributions to the primal–dual mismatch in complementarity in (27b)–(27d) according to

$$|\mu_\ell|_{\mathcal{A}\cap\mathcal{I}_\ell}| \le \sum_{T\in\mathcal{T}_{\mathcal{A}\cap\mathcal{I}_\ell}} \bar{\mu}_T^{(1)}, \tag{30a}$$

$$\bar{\mu}_T^{(1)} := \frac{1}{2}\|\varepsilon u_\ell + y_\ell - \psi\|_{0,T}(\|i_{\ell,2}^{\sigma}\sigma_\ell\|_{0,T} + \tilde{\omega}_T^{\sigma}),$$

$$|\mu_\ell|_{\mathcal{I}\cap\mathcal{A}_\ell}| \le \sum_{T\in\mathcal{T}_{\mathcal{I}\cap\mathcal{A}_\ell}} \bar{\mu}_T^{(2)}, \tag{30b}$$

$$\bar{\mu}_T^{(2)} := \|\sigma_\ell\|_{0,T}(\varepsilon\tilde{\omega}_T^{u} + \tilde{\omega}_T^{y}),$$

$$|\mu_\ell|_{\mathcal{A}\cap\mathcal{A}_\ell}| \le \sum_{T\in\mathcal{T}_{\mathcal{A}\cap\mathcal{A}_\ell}} \bar{\mu}_T^{(3)}, \tag{30c}$$

$$\bar{\mu}_T^{(3)} := \frac{1}{2}\|\psi_\ell - \psi\|_{0,T}(\|i_{\ell,2}^{\sigma}\sigma_\ell + \sigma_\ell\|_{0,T} + \tilde{\omega}_T^{\sigma}).$$

This leads to the following upper bound for the primal–dual mismatch in complementarity:

$$|\mu_\ell(x,\sigma)| \le \sum_{T\in\mathcal{T}_\ell(\Omega)} \bar{\mu}_T, \tag{31}$$

where

$$\bar{\mu}_T := \begin{cases} 0, & T\in\mathcal{T}_{\mathcal{I}\cap\mathcal{I}_\ell},\\ \bar{\mu}_T^{(1)}, & T\in\mathcal{T}_{\mathcal{A}\cap\mathcal{I}_\ell},\\ \bar{\mu}_T^{(2)}, & T\in\mathcal{T}_{\mathcal{I}\cap\mathcal{A}_\ell},\\ \bar{\mu}_T^{(3)}, & T\in\mathcal{T}_{\mathcal{A}\cap\mathcal{A}_\ell}. \end{cases}$$

The oscillation term $\mathrm{osc}_\ell^{(2)}$ as given by (21b) is treated analogously which results in

$$|\mathrm{osc}_\ell^{(1)} + \mathrm{osc}_\ell^{(2)}| \le \sum_{T\in\mathcal{T}_\ell(\Omega)} \mathrm{osc}_T, \quad \mathrm{osc}_T := \mathrm{osc}_T^{(1)} + \mathrm{osc}_T^{(2)}, \tag{32}$$

where $\mathrm{osc}_T^{(1)}$ is given by (16) and $\mathrm{osc}_T^{(2)}$ by

$$\mathrm{osc}_T^{(2)} := \tilde{\omega}_T^{y}\|y^d - y_\ell^d\|_{0,T} + \tilde{\omega}_T^{u}\|u^d - u_\ell^d\|_{0,T}.$$

Hence, we end up with the computable upper bound

$$|J(y,u) - J_\ell(y_\ell,u_\ell)| \preceq \sum_{T\in\mathcal{T}_\ell(\Omega)} (\hat{\omega}_T^{y}\rho_T^{y} + \hat{\omega}_T^{p}\rho_T^{p} + \bar{\mu}_T + \mathrm{osc}_T). \tag{33}$$

5 Numerical Results

The marking strategy for selection of elements of the triangulation for refinement is based on a bulk criterion (cf. [11,30]) where we select a set $\mathcal{M}_\ell \subset \mathcal{T}_\ell(\Omega)$ of elements such that with respect to a given constant $0 < \Theta < 1$ there holds

$$\Theta \sum_{T\in\mathcal{T}_\ell(\Omega)} (\hat{\omega}_T^y \rho_T^y + \hat{\omega}_T^p \rho_T^p + \bar{\mu}_T + \mathrm{osc}_T) \le \sum_{T\in\mathcal{M}} (\hat{\omega}_T^y \rho_T^y + \hat{\omega}_T^p \rho_T^p + \bar{\mu}_T + \mathrm{osc}_T)\,.$$

The bulk criterion is realized by a greedy algorithm (cf., e.g. [23]). The refinement is realized by newest vertex bisection.

We conclude this section with the results for an example which was chosen as a test case in [28]. The data of the problem are as follows:

$$\begin{aligned} \Omega &:= B(0,1), \quad \Gamma_D = \emptyset, \quad \alpha := 1.0, \quad c = 1.0,\\ y^d(r) &:= 4 + \frac{1}{\pi} - \frac{1}{4\pi} r^2 + \frac{1}{2\pi}\ln(r),\\ u^d(r) &:= 4 + \frac{1}{4\pi} r^2 - \frac{1}{2\pi}\ln(r), \quad \psi(r) := r + 4. \end{aligned}$$

The optimal solution in the pure state constrained case is given by

$$\begin{aligned} y(r) &\equiv 4, \qquad p(r) = \frac{1}{4\pi} r^2 - \frac{1}{2\pi}\ln(r),\\ u(r) &\equiv 4, \qquad \sigma = \delta_0. \end{aligned}$$

As regularization parameter ε for the Lavrentiev regularization we have chosen $\varepsilon = 10^{-4}$. The finite element discretized optimal control problem has been solved by the Moreau–Yosida based active set strategy from [6]. Moreover, $\Theta = 0.4$ has been used for the bulk criterion in the step MARK of the adaptive loop.

Figure 1 shows the computed optimal state (left) and optimal control (right). We note that the peaks at the origin are numerical artefacts due

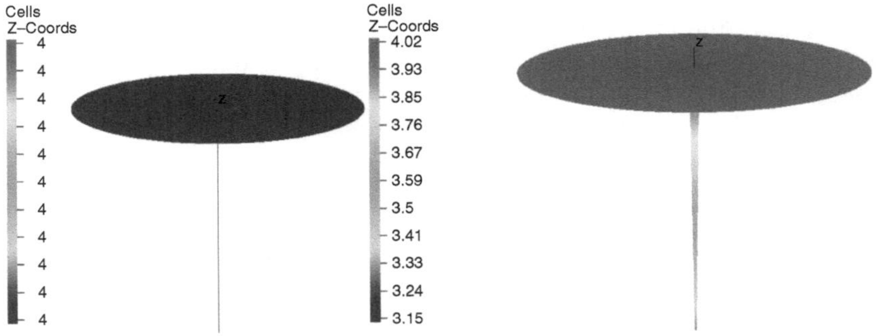

Fig. 1. Optimal state (left) and optimal control (right).

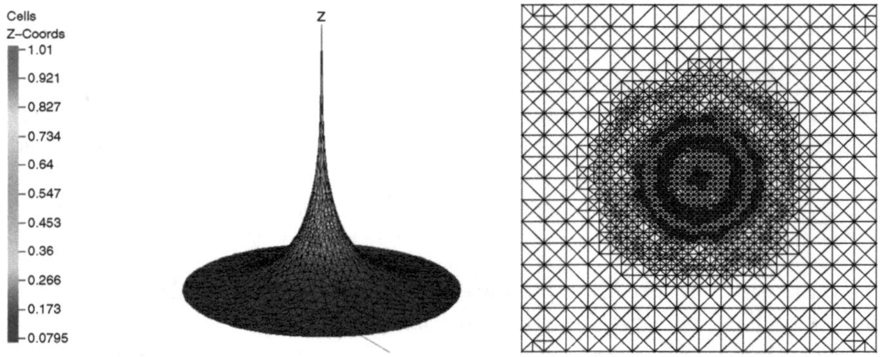

Fig. 2. Optimal adjoint state (left) and adaptively refined triangulation after 14 refinement steps of the adaptive loop (right).

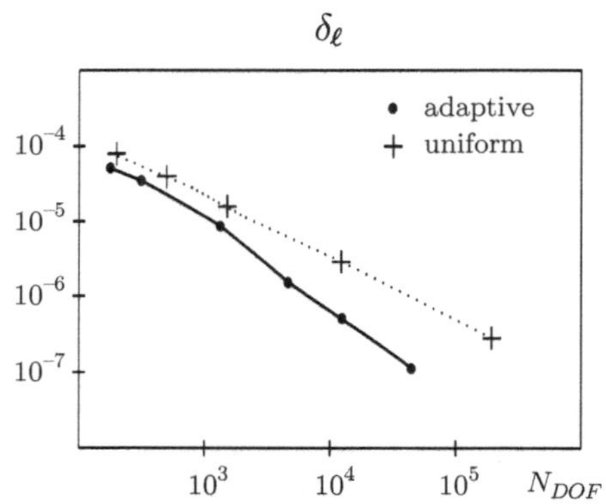

Fig. 3. Decrease of the quantity of interest $\delta_\ell := |J(y,u) - J_\ell(y_\ell, u_\ell)|$ as a function of the total number of degrees of freedom for adaptive and uniform refinement.

to the singularity of the optimal adjoint state in the origin (see Figure 2 (left)). Figure 2 (right) displays the computed adaptively refined mesh after 14 refinement steps of the adaptive loop. Finally, Figure 3 shows the decrease of the error $\delta_\ell := |J(y,u) - J_\ell(y_\ell, u_\ell)|$ measured in the quantity of interest as a function of the total number of degrees of freedom on a logarithmic scale both for adaptive and uniform refinement.

Acknowledgement. The first author acknowledges support by the Austrian Science Foundation (FWF) within the START-program Y305 'Interfaces and Free Boundaries'. The work of the second author has been supported by the NSF under Grant No. DMS-0511611, DMS-0707602 as well as by the DFG within the Priority Program SPP 1253 'PDE Constrained Optimization'.

References

1. M. Ainsworth and J. T. Oden. *A posteriori error estimation in finite element analysis*. Wiley, Chichester, 2000.
2. N. Arada and J.-P. Raymond. Optimal control problems with mixed control-state constraints. *SIAM J. Control Optim.*, 39(5):1391–1407, 2000.
3. I. Babuska and T. Strouboulis. *The finite element method and its reliability*. Clarendon Press, Oxford, 2001.
4. W. Bangerth and R. Rannacher. *Adaptive finite element methods for differential equations*. Lectures in Mathematics, ETH Zürich. Birkhäuser, Basel, 2003.
5. R. Becker, H. Kapp, and R. Rannacher. Adaptive finite element methods for optimal control of partial differential equations: Basic concept. *SIAM J. Control Optim.*, 39(1):113–132, 2000.
6. M. Bergounioux, M. Haddou, M. Hintermüller, and K. Kunisch. A comparison of a Moreau–Yosida-based active set strategy and interior point methods for constrained optimal control problems. *SIAM J. Optim.*, 11(2):495–521, 2000.
7. M. Bergounioux, K. Ito, and K. Kunisch. Primal-dual strategy for constrained optimal control problems. *SIAM J. Control Optim.*, 37(4):1176–1194, 1999.
8. S. C. Brenner and L. R. Scott. *The mathematical theory of finite element methods*. Springer, Berlin, 2nd edition, 2003.
9. E. Casas, J.-P. Raymond, and H. Zidani. Pontryagin's principle for local solutions of control problems with mixed control-state constraints. *SIAM J. Control Optim.*, 39(4):1182–1203, 2000.
10. K. Deckelnick and M. Hinze. Convergence of a finite element approximation to a state constrained elliptic control problem. *SIAM J. Numer. Anal.*, 45(5):1937–1953, 2007.
11. W. Dörfler. A convergent adaptive algorithm for Poisson's equation. *SIAM J. Numer. Anal.*, 33(3):1106–1124, 1996.
12. K. Eriksson, D. Estep, P. Hansbo, and C. Johnson. *Computational differential equations*. Cambridge University Press, Cambridge, 1995.
13. A. Gaevskaya, R. H. W. Hoppe, Y. Iliash, and M. Kieweg. A posteriori error analysis of control constrained distributed and boundary control problems. In O. Pironneau et al., editor, *Proc. Conf. Advances in Scientific Computing (Moscow)*, pages 85–108, Moscow, 2006. Russian Academy of Sciences.
14. A. Gaevskaya, R. H. W. Hoppe, Y. Iliash, and M. Kieweg. Convergence analysis of an adaptive finite element method for distributed control problems with control constraints. In K. Kunisch, G. Leugering, J. Sprekels, and F. Tröltzsch, editors, *Control of Coupled Partial Differential Equations (Oberwolfach, 2005)*, pages 47–68, Basel, 2007. Birkhäuser.
15. A. Gaevskaya, R. H. W. Hoppe, and S. Repin. A posteriori estimates for cost functionals of optimal control problems. In A. Bermúdez de Castro, D. Gómez, P. Quintela, and P. Salgado, editors, *Numerical Mathematics and Advanced Applications (ENUMATH 2005, Santiago de Compostela)*, pages 308–316, Berlin, 2006. Springer.
16. A. Günther and M. Hinze. A posteriori error control of a state constrained elliptic control problem. *J. Numer. Math.*, 16(4):307–322, 2008.
17. M. Hintermüller and R. H. W. Hoppe. Goal-oriented adaptivity in control constrained optimal control of partial differential equations. *SIAM J. Control Optim.*, 47(4):1721–1743, 2008.

18. M. Hintermüller and R. H. W. Hoppe. Adaptive finite element methods for control constrained distributed and boundary optimal control problems. In M. Heinkenschloss, L. N. Vicente, and L. M. Fernandes, editors, *Numerical PDE Constrained Optimization*, volume 73 of *Lecture Notes in Computational Science and Engineering*. Springer, Berlin, 2009. In press.
19. M. Hintermüller and R. H. W. Hoppe. Goal-oriented adaptivity in state constrained optimal control of partial differential equations. *SIAM J. Control Optim.*, 2009. Submitted.
20. M. Hintermüller, R. H. W. Hoppe, Y. Iliash, and M. Kieweg. An a posteriori error analysis of adaptive finite element methods for distributed elliptic control problems with control constraints. *ESAIM Control Optim. Calc. Var.*, 14(3):540–560, 2008. DOI 10.1051/cocv:2007057.
21. M. Hintermüller and K. Kunisch. Feasible and noninterior path-following in constrained minimization with low multiplier regularity. *SIAM J. Control Optim.*, 45(4):1198–1221, 2006.
22. M. Hintermüller and K. Kunisch. Path-following methods for a class of constrained minimization problems in function space. *SIAM J. Optim.*, 17(1):159–187, 2006.
23. R. H. W. Hoppe, Y. Iliash, C. Iyyunni, and N. H. Sweilam. A posteriori error estimates for adaptive finite element discretizations of boundary control problems. *J. Numer. Math.*, 14(1):57–82, 2006.
24. R. H. W. Hoppe and M. Kieweg. A posteriori error estimation of finite element approximations of pointwise state constrained distributed control problems. Submitted, 2007.
25. K. Kunisch and A. Rösch. Primal–dual active set strategy for a general class of constrained optimal control problems. *SIAM J. Optim.*, 13(2):321–334, 2002.
26. R. Li, W. Liu, H. Ma, and T. Tang. Adaptive finite element approximation for distributed elliptic optimal control problems. *SIAM J. Control Optim.*, 41(5):1321–1349, 2002.
27. W. Liu and N. Yan. A posteriori error estimates for distributed optimal control problems. *Adv. Comput. Math.*, 15(1–4):285–309, 2001.
28. C. Meyer, U. Prüfert, and F. Tröltzsch. On two numerical methods for state-constrained elliptic control problems. *Optim. Methods Softw.*, 22(6):871–899, 2007.
29. C. Meyer, A. Rösch, and F. Tröltzsch. Optimal control problems of PDEs with regularized pointwise state constraints. *Comput. Optim. Appl.*, 33(2–3):209–228, 2006.
30. P. Morin, R. H. Nochetto, and K. G. Siebert. Data oscillation and convergence of adaptive FEM. *SIAM J. Numer. Anal.*, 38(2):466–488, 2000.
31. P. Neittaanmäki and S. Repin. *Reliable methods for computer simulation. Error control and a posteriori estimates*. Elsevier, Amsterdam, 2004.
32. A. Rösch and F. Tröltzsch. Sufficient second-order optimality conditions for an elliptic optimal control problem with pointwise control-state constraints. *SIAM J. Optim.*, 17(3):776–794, 2006.
33. F. Tröltzsch. A minimum principle and a generalized bang-bang principle for a distributed optimal control problem with constraints on control and state. *Z. Angew. Math. Mech.*, 59(12):737–739, 1979.
34. F. Tröltzsch. Regular Lagrange multipliers for control problems with mixed pointwise control-state constraints. *SIAM J. Optim.*, 15(2):616–634, 2005.

35. R. Verfürth. *A review of a posteriori estimation and adaptive mesh-refinement techniques.* Wiley/Teubner, New York, 1996.
36. B. Vexler and W. Wollner. Adaptive finite elements for elliptic optimization problems with control constraints. *SIAM J. Control Optim.*, 47(1):509–534, 2008.

Feedback Solution and Receding Horizon Control Synthesis for a Class of Quantum Control Problems

Kazufumi Ito and Qin Zhang

Department of Mathematics, North Carolina State University, Raleigh, NC 27695-8205, USA, kito@unity.ncsu.edu

Summary. Control of quantum systems described by the linear Schrödinger equation are considered. Control inputs enter through coupling operators and results in a bilinear control system. Feedback control laws are developed for the orbit tracking and the performance of the feedback control laws is demonstrated by the stable and accurate numerical integrations of the closed-loop system. The receding horizon control synthesis is applied to improve the performance of the feedback law. The second order accurate numerical integrations via time-splitting and the monotone convergent iterative scheme are combined to solve the optimality system, i.e., the two-pint boundary value problem on a given time horizon. The feasibility of the proposed synthesis is demonstrated by numerical tests and the performance is greatly improved if we apply the receding horizon control.

1 Introduction

Consider a quantum system with internal Hamiltonian $\mathcal{H}_0$ prepared in the initial state $\Psi_0(x)$, where x denotes the relevant spatial coordinate. The state $\Psi(x,t)$ satisfies the time-dependent Schrödinger equation. In the presence of an external interaction taken as an electric field modeled by a coupling operator with amplitude $\varepsilon(t) \in \mathbb{R}$ and a time independent dipole moment operator μ results in the controlled Hamiltonian $\mathcal{H} = \mathcal{H}_0 + \varepsilon(t)\mu$ and the following dynamical system:

$$i\,\frac{\partial}{\partial t}\Psi(x,t) = (\mathcal{H}_0 + \varepsilon(t)\mu)\Psi(x,t), \quad \Psi(x,0) = \Psi_0(x). \tag{1}$$

where $\mathcal{H}_0$ is a positive, closed, self-adjoint operator in the Hilbert space H, $\mu \in \mathcal{L}(H)$ is self-adjoint, and $\varepsilon \in L^1(0,\infty)$ is the control input.

We consider the control problem of driving the state $\Psi(t)$ of (1) to an orbit $\mathcal{O}(t)$ of the uncontrolled dynamics

$$i\,\frac{d}{dt}\mathcal{O}(t) = \mathcal{H}_0\mathcal{O}(t), \tag{2}$$

W. Fitzgibbon et al. (eds.), *Applied and Numerical Partial Differential Equations*, Computational Methods in Applied Sciences 15, DOI 10.1007/978-90-481-3239-3_9,

specifically to the one that corresponds to an eigen-state or the manifold spanned by finite many eigen-states, in general. An element $\psi \in \mathrm{dom}(\mathcal{H}_0)$ is an eigen-state of $\mathcal{H}_0$ if $\mathcal{H}_0\psi = \lambda\psi$ for $\lambda > 0$. Then, the corresponding orbit is given by

$$\mathcal{O}(t) = e^{-i(\lambda t-\theta)}\psi, \tag{3}$$

where $\theta \in [0, 2\pi)$ is the phase factor. We have $|\mathcal{O}(t)|_H = 1$ if ψ is normalized as $|\psi|_H = 1$. We consider the discrete spectrum case: i.e. assume $\mathcal{H}_0$ only has discrete eigenvalues $\{\lambda_k\}$, the family of eigenfunctions $\{\psi_k\}_{k=1}^{\infty}$ forms an orthonormal basis of H and that $\{\lambda_k\}$ are arranged in increasing order.

We employ a variational approach based on the Lyapunov functional

$$V(t) = V(\Psi(t), \mathcal{O}(t)) = \frac{1}{2}|\Psi(t) - \mathcal{O}(t)|_X^2. \tag{4}$$

The variational approaches were previously discussed in [1,4,8], for example.

We shall see in Sections 2 and 3 that $|\Psi(t)|_H = 1$ for all $t \geq 0$. Together with $|\mathcal{O}(t)|_H = 1$ this implies that the functional V can equivalently be expressed as

$$V(\Psi(t), \mathcal{O}(t)) = 1 - \mathrm{Re}(\mathcal{O}(t), \Psi(t))_H. \tag{5}$$

It will be shown that

$$\frac{d}{dt}V(\Psi(t), \mathcal{O}(t)) = \varepsilon(t)\,\mathrm{Im}(\mathcal{O}(t), \mu\Psi(t))_H. \tag{6}$$

We propose the feedback law

$$\begin{aligned} \varepsilon(t) &= -\frac{1}{\alpha}(u(t) + \beta\,\mathrm{sign}(u(t))V(t)^{\gamma}) = F(\Psi(t), \mathcal{O}(t)), \\ u(t) &= \mathrm{Im}(\mathcal{O}(t), \mu\Psi(t))_X, \quad V(t) = V(\Psi(t), \mathcal{O}(t)), \end{aligned} \tag{7}$$

for $\alpha > 0$, $\beta \geq 0$, $\gamma \in (0, 1]$. The case $\beta = 0$ is analyzed in [4]. From (6)

$$\frac{d}{dt}V(\Psi(t), \mathcal{O}(t)) = -\frac{1}{\alpha}(|u(t)|^2 + \beta|u(t)|V(t)^{\gamma}). \tag{8}$$

Note that $u(t)$ is a linear in $\Psi(t)$. It will be shown in Section 5 that the performance of feedback laws significantly increases by incorporating the switching control term with $\beta > 0$.

The well-posedness of the closed loop system with the feedback law (7) is established in [5]. For the asymptotic tracking $V(\Psi(t), \mathcal{O}(t)) \to 0$ as $t \to \infty$, note that from (8) we have

$$\int_0^T |u(t+s)|^2\,ds \to 0, \quad \text{as } t \to \infty.$$

Our analysis of the asymptotic tracking is based on the LaSalle invariance principle, i.e. suppose

$$\Psi_\infty(\tau) = \lim_{t_n \to \infty} \Psi(t_n + \tau) = \sum_{k=0}^{\infty} A_k e^{i} \psi_k$$

be a ω-orbit of (1)–(7), we have the invariant set

$$\int_0^T |u(\tau))|^2 \, d\tau = 0 \quad \text{for all } T > 0$$

$$u(\tau) = \mathrm{Im}(\mathcal{O}_\infty(\tau), \mu\Psi_\infty(\tau)) = \mathrm{Im}\left(\sum_{k=1}^{\infty} A_k e^{i((\lambda_k - \lambda_{k_0})\tau - \theta_k + \tilde{\theta}_{k_0})}\right) \tag{9}$$

for the ω limit. Based on this and the Ingham theorem [2], one can prove a sufficient condition for the asymptotic tracking by a single control [4]:

Theorem 1. *Assume that all moments are non-vanishing:*

$$\mu_{k_0}^k = (\psi_{k_0}, \mu\psi)_H \neq 0.$$

If there exits a constant $\delta > 0$ such that $|\lambda_k + \lambda_\ell - 2\lambda_{k_0}| \geq \delta$ for all $k, \ell \geq 1$ with $\ell \neq k_0$, and $|\lambda_k - \lambda_\ell| \geq \delta$ for all $k \neq \ell$, then $\lim_{t \to \infty} V(\Psi(t), \mathcal{O}(t)) = 0$, for the feedback law (7).

For example, consider the harmonic oscillator case:

$$\mathcal{H}_0 \psi = -\frac{d^2}{dx^2}\psi + x^2\psi, \quad x \in \mathbb{R} = \Omega.$$

Then the eigen-pairs $\{(\lambda_k, \psi_k)\}_{k=1}^{\infty}$ are given by

$$\lambda_k = 2k - 1, \quad \psi_k(x) = \hat{c} H_{k-1}(x) e^{-\frac{x^2}{2}}$$

where H_k is the Hermite polynomial of degree k and $\hat{c}$ is a normalizing factor. In this case we have

$$\lambda_{k_0 - \ell} - \lambda_{k_0} = -(\lambda_{k_0 + \ell} - \lambda_{k_0}), \quad 1 \leq \ell \leq k_0 - 1,$$

and the gap condition $|\lambda_k + \lambda_\ell - 2\lambda_{k_0}| > \delta$ is not satisfied. The invariant set (9) implies

$$\mathrm{Im}\left(A_{k_0+\ell} e^{i(\lambda_\ell \tau - \theta_{k_0+\ell} + \tilde{\theta}_{k_0})} \mu_{k_0}^{k_0+\ell} + A_{k_0-\ell} e^{-i(\lambda_\ell \tau - \theta_{k_0-\ell} + \tilde{\theta}_{k_0})} \mu_{k_0}^{k_0-\ell}\right) = 0$$

for $1 \leq \ell < k_0$. That is, $A_{k_0-\ell}$ and $A_{k_0-\ell}$ are not necessary zero and thus $\Psi_\infty(\tau)$ is distributed over energy levels $1 \leq \ell \leq 2k_0 - 1$.

Consider the multiple control potentials of the form

$$\mu(t) = \sum_{j=1}^{m} \varepsilon_j(t) \mu_j. \tag{10}$$

The corresponding feedback law is given by

$$\varepsilon_j(t) = -\frac{1}{\alpha}(u_j(t) + \beta\,\mathrm{sign}(u_j(t))V(t)^\gamma), \quad u_j(t) = \mathrm{Im}(\mathcal{O}(t), \mu_j\Psi(t))_X.$$

Assume pairs (k_i, ℓ_i) are degenerated:

$$\lambda_{k_i} + \lambda_{\ell_i} - 2\lambda_{k_0} = 0$$

with $\ell_i \neq k_0$. It is shown in [4] that if the rank condition

$$\mathrm{rank}\begin{pmatrix} (\mu_1)_{k_0}^{k_i} & (\mu_1)_{k_0}^{\ell_i} \\ (\mu_2)_{k_0}^{k_i} & (\mu_2)_{k_0}^{\ell_i} \end{pmatrix} = 2 \tag{11}$$

holds for each i, then $A_{k_i} = A_{\ell_i} = 0$, and in particular $A_k = 0$ for all k, i.e., $\lim_{t\to\infty} V(\Psi(t), \mathcal{O}(t)) = 0$.

The nonlinear feedback method out-performs the linear one ($\beta = 0$) significantly (see Section 5). In our numerical tests numerical integrations of (1) and (7) via the Strang operator-splitting method is used and its convergence property is analyzed in Section 5. It is very efficient, second order accurate and unconditionally stable. The other contribution of this paper is to apply the receding control synthesis to further improve the tracking performance of the feedback law. Our implementation of the receding control uses the monotone convergent iterative scheme to solve the optimality system on a given time horizon. It is an iterative scheme for the two-pint boundary value problem and the monotone convergence property of fully discretized scheme is established. The effectiveness of the receding control synthesis is demonstrated via numerical tests in Section 5.

2 Control Formulation

Associated to the closed, positive, self-adjoint operator $\mathcal{H}_0$ densely defined in the Hilbert space H, we define the closed linear operator A_0 in $H \times H$ by

$$A_0 = \begin{pmatrix} 0 & \mathcal{H}_0 \\ -\mathcal{H}_0 & 0 \end{pmatrix}$$

with $\mathrm{dom}(A_0) = \mathrm{dom}(\mathcal{H}_0) \times \mathrm{dom}(\mathcal{H}_0)$. Here $\Psi = (\Psi_1, \Psi_2) \in H \times H$ is identified with $\Psi = \Psi_1 + i\,\Psi_2 \in X$. We note that

$$|(\Psi_1, \Psi_2)|_{H\times H} = |\Psi|_X \quad \text{and} \quad (\Phi, \Psi)_{H\times H} = \mathrm{Re}(\Phi, \Psi)_X$$

and that A_0 is skew-adjoint, i.e.

$$(A_0\Psi, \hat{\Psi})_{H\times H} = -(A_0\hat{\Psi}, \Psi)_{H\times H} \quad \text{for all } \Psi, \hat{\Psi} \in \mathrm{dom}(A_0).$$

Thus by the Stone theorem [9], A_0 generates C_0-group on X and $|S(t)\Psi_0|_X = |\Psi_0|_X$.

Associated to the self-adjoint operator $\mu \in \mathcal{L}(H)$ we define the skew-adjoint operator

$$B = \begin{pmatrix} 0 & \mu \\ -\mu & 0 \end{pmatrix}$$

Then for $\varepsilon \in L^2(0,T)$ there exists a unique mild solution $\Psi(t) \in C(0,T;X)$ to

$$\Psi(t) = S(t)\Psi_0 + \int_0^t S(t-s)\varepsilon(s)B\Psi(s)\,ds, \quad t \in [0,T], \tag{12}$$

and

$$\frac{d}{dt}\Psi = A_0\Psi(t) + \varepsilon(t)B\Psi(t) \quad \text{in } (\mathrm{dom}(A_0))^* \tag{13}$$

[3, Chapter 2], [9, Chapter 4]. Equivalently

$$\frac{d}{dt}\Psi(t) = -i\,(\mathcal{H}_0\Psi(t) + \varepsilon(t)\mu\Psi(t)).$$

Since $\mathcal{O}(t) \in C(0,T;\mathrm{dom}(A_0)) \cap C^1(0,T;X)$, we have

$$\frac{d}{dt}\mathcal{O}(t) = -i\,\mathcal{H}_0\mathcal{O}(t) \quad \text{in } H. \tag{14}$$

Thus,

$$\begin{aligned}
&\frac{d}{dt}\,\mathrm{Re}(\mathcal{O}(t),\Psi(t))_X \\
&= \mathrm{Re}\left((-i\,\mathcal{H}_0\mathcal{O}(t),\Psi(t))_X + (\mathcal{O}(t), -i(\mathcal{H}_0\Psi(t) + \varepsilon(t)\mu\Psi(t))_X)\right) \\
&= \mathrm{Re}\left(i\,\varepsilon(t)(\mathcal{O}(t),\mu\Psi(t))\right)_X = -\varepsilon(t)\,\mathrm{Im}(\mathcal{O}(t),\mu\Psi(t))_X,
\end{aligned}$$

which proves (6). Thus, we obtain the closed loop system of the form

$$\Psi(t) = S(t)\Psi_0 + \int_0^t S(t-s)F(\Psi(s),\mathcal{O}(s))B\Psi(s)\,ds. \tag{15}$$

3 Operator Splitting Method

Since the Hamiltonian is the sum of $\mathcal{H}_0$ and $\varepsilon(t)\mu$ it is very natural to consider time integration based on the operator splitting method. For the stepsize $h > 0$ consider the Strang splitting method:

$$\begin{aligned}
\frac{\hat{\Psi}_{k+1} - \hat{\Psi}_k}{h} = \varepsilon_k B \frac{\hat{\Psi}_{k+1} + \hat{\Psi}_k}{2}, \quad \hat{\Psi}_k = S\left(\frac{h}{2}\right)\Psi_k, \\
\Psi_{k+1} = S\left(\frac{h}{2}\right)\hat{\Psi}_{k+1},
\end{aligned} \tag{16}$$

where

$$\varepsilon_k = \frac{1}{h}\int_{kh}^{(k+1)h} \varepsilon(s)\,ds.$$

For time integration of the controlled Hamiltonian we employ the Crank–Nicholson scheme since it is a norm preserving scheme. In fact, since B is skew adjoint

$$\left(\frac{\Psi_{k+1}-\hat{\Psi}_k}{h}, \Psi_{k+1}+\hat{\Psi}_k\right)_X = 0,$$

and thus $|\Psi_{k+1}|_X^2 = |\hat{\Psi}_k|_X^2$. The Strang splitting is of second order as time-integration. We have the convergence of (16).

Theorem 2. *If we define $\Psi_h(t) = \Psi_k$ on $[kh,(k+1)h)$, then*

$$|\Psi_h(t) - \Psi(t)|_X \to 0 \quad \textit{uniformly in } t \in [0,T],$$

where $\Psi(t)$, $t \geq 0$, satisfies

$$\Psi(t) = S(t)\Psi_0 + \int_0^t S(t-s)\varepsilon(s)B\Psi(s)\,ds.$$

Proof. Define the one step transition operator

$$\Psi_{k+1} = T_h(t)\Psi_k$$

by

$$T_h(t) = S\left(\frac{h}{2}\right)\left(I - \frac{\varepsilon_k h}{2}B\right)^{-1}\left(I + \frac{\varepsilon_k h}{2}B\right)S\left(\frac{h}{2}\right)\Psi.$$

Then, $|T_h(t)\Psi|_X = |\Psi|_X$ and

$$A_h(t)\Psi = \frac{T_h(t)\Psi - \Psi}{h} = S\left(\frac{h}{2}\right)\frac{J_{h/2}(\varepsilon_k B) - I}{h/2}S\left(\frac{h}{2}\right)\Psi + \frac{S(h)\Psi - \Psi}{h}$$

where

$$J_{h/2}(\varepsilon_k B) = \left(I - \frac{\varepsilon_k h}{2}B\right)^{-1}.$$

Since for $\Psi \in X$

$$\lim_{h\to 0^+} \frac{J_{h/2}(\varepsilon_k B) - I}{h/2}\Psi = \varepsilon(t)B\Psi$$

and for $\Psi \in \mathrm{dom}(A)$

$$\lim_{h\to 0^+} \frac{S(h)\Psi - \Psi}{h} = A_0\Psi,$$

we have for $\Psi \in \mathrm{dom}(A)$ and $\varepsilon \in C(0,T)$

$$|A_h(t)\Psi - (A_0\Psi + \varepsilon(t)B\Psi)|_X \to 0 \quad \text{as } h \to 0^+.$$

It thus follows from the Chernoff theorem [3] that $|\Psi_h(t) - \Psi(t)|_X \to 0$ uniformly in $t \in [0,T]$.

Note that

$$\Psi_{k+1} = S(h)\Psi_k + hS\left(\frac{h}{2}\right)\varepsilon_k J_{h/2}(\varepsilon_k B)S\left(\frac{h}{2}\right)\Psi_k$$

and thus

$$\Psi^m = S(mh)\Psi_0 + \sum_{k=1}^{m} hS((m-k)h)S\left(\frac{h}{2}\right)\varepsilon_k B J_{h/2}(\varepsilon_k B)S\left(\frac{h}{2}\right)\Psi_{k-1}.$$

Thus, letting $h \to 0$ in this expression, $\Psi(t) \in C(0,T;X)$ satisfies (12).

For (16) there exists an ε_k on $[kh,(k+1)h)$ [5] such that for $\mathcal{O}_{k+1/2} = S(\frac{h}{2})\mathcal{O}_k$

$$\begin{aligned} \varepsilon_k &= F(\Psi_{k+1/2}, \mathcal{O}_{k+1/2}) = \frac{1}{\alpha}(u_{k+1/2} + \beta\,\mathrm{sign}(u_{k+1/2})V_k^{\gamma}), \\ u_{k+1/2} &= (\mathcal{O}_{k+1/2}, B\Psi_{k+1/2}), \quad \Psi_{k+1/2} = \frac{\hat{\Psi}_{k+1} + \hat{\Psi}_k}{2}. \end{aligned} \tag{17}$$

Then Ψ_k satisfies closed loop system

$$\begin{aligned} \frac{\hat{\Psi}_{k+1} - \hat{\Psi}_k}{h} &= \varepsilon_k B \frac{\hat{\Psi}_{k+1} + \hat{\Psi}_k}{2}, \quad \hat{\Psi}_k = S\left(\frac{h}{2}\right)\Psi_k, \\ \varepsilon_k &= F(\Psi_{k+1/2}, \mathcal{O}_{k+1/2}), \quad \Psi_{k+1} = S\left(\frac{h}{2}\right)\hat{\Psi}_{k+1}. \end{aligned} \tag{18}$$

Since

$$V\left(S\left(\frac{h}{2}\right)\hat{\Psi}_{k+1}, S\left(\frac{h}{2}\right)\mathcal{O}_{k+1/2}\right) = V(\hat{\Psi}_{k+1}, \mathcal{O}_{k+1/2}),$$

the discrete analog of (8)

$$V(\Psi_{k+1}, \mathcal{O}_{k+1}) = V(\Psi_k, \mathcal{O}_k) + \frac{1}{\alpha}\left(|u_k|^2 + \beta|u_k|V(\Psi_k, \mathcal{O}_k)^{\gamma}\right)$$

holds for the closed loop (18).

4 Receding Horizon Control Synthesis

In this section we consider the receding horizon control synthesis [6]. The receding horizon method is the time-decomposition technique for the longer horizon $[0, T_f]$. Consider the sequence of the finite horizon problem on $[T_t, T_i + T]$

$$\min \int_{T_i}^{T_i+T} \frac{1}{2}(|\Psi(t) - \mathcal{O}(t)|^2 + \alpha|\varepsilon(t)|^2)\,dt + \frac{1}{2}|\Psi(T_i+T) - \mathcal{O}(T_i+T)|^2 \tag{19}$$

subject to (13). We have the two types

(I) (Instantaneous Receding Horizon) Given $\Psi(T_i)$ we compute the optimal control ε_i on the horizon $[T_i, T_i + k\Delta T]$, $k \geq 1$ and then execute the control on $[T_i, T_i + \Delta]$, where $\Delta T > 0$ is small and denotes the execution duration.

(II) (Regular Receding Horizon) Given $\Psi(T_i)$ we compute the optimal control ε_i on the horizon $[T_i, T_i + T]$ and the execute the control on $[T_i, T_i + T]$, where T is relative large.

In any case the receding horizon synthesis is a feedback control in the sense that ε_i on $[T_i, T_i + T]$ is a function of $\Psi(T_i)$. It is shown in [3] that the necessary optimality condition for $\varepsilon = \varepsilon_i$ is given by

$$\begin{aligned} \frac{dt}{dt}\Psi(t) &= (A + \varepsilon(t)B)\Psi(t), & \varepsilon(t) &= \frac{1}{\alpha}(B\Psi(t), \chi(t)), \\ -\frac{dt}{dt}\chi(t) &= (A + \varepsilon(t)B)^*\chi(t) - \mathcal{O}(t), & \chi(T_i + T) &= \mathcal{O}(T_i + T). \end{aligned} \tag{20}$$

Consider the following iterative method to solve the two-point boundary value problem (20):

$$\begin{aligned} \frac{d}{dt}\Psi^{k+1} &= (A + \varepsilon^{k+1}B)\Psi^{k+1}, \\ \varepsilon^{k+1} &= (1-\delta)\tilde{\varepsilon}^k + \frac{\delta}{\alpha}(B\Psi^{k+1}, \chi^k), \\ -\frac{d}{dt}\chi^{k+1} &= (A + \tilde{\varepsilon}^{k+1}B)^*\chi^{k+1} + \mathcal{O}(t), \quad \chi^{k+1} = \mathcal{O}(T_i + T), \\ \tilde{\varepsilon}^{k+1} &= (1-\eta)\varepsilon^{k+1} + \frac{\eta}{\alpha}(B\Psi^{k+1}, \chi^{k+1}), \end{aligned} \tag{21}$$

where $\delta, \eta \in (0, 2)$ are the relaxation parameters. We let for $k = 1$

$$\tilde{\varepsilon}^0 = F(\Psi(t), \mathcal{O}(t)).$$

It can be proved [7] that

$$J(\varepsilon^{k-1}) - J(\varepsilon^k) = \frac{\alpha}{2}\int_0^T \left(\frac{2}{\delta} - 1\right)|\varepsilon^k - \tilde{\varepsilon}^{k-1}|^2 + \left(\frac{2}{\eta} - 1\right)|\varepsilon^{k-1} - \tilde{\varepsilon}^{k-1}|^2\, dt.$$

That is, one can improve the performance index as the number of iterates increases and (21) is called the monotone scheme.

4.1 Time-Discretization

In this section we discuss the time-discretization of (19) using the splitting method (16). Consider the problem

$$J_h(\varepsilon) = \frac{1}{2}|\Psi_m - \mathcal{O}_m|^2 \Delta t \sum_{j=0}^{m-1}\left(\frac{\alpha}{2}|\varepsilon_j|^2 + \frac{1}{2}(\hat{\Psi}_{j+1} - \mathcal{O}_{j+1/2}|^2 + |\hat{\Psi}_j - \mathcal{O}_{j+1/2}|^2)\right), \tag{22}$$

subject to (16);

$$\hat{\Psi}_j = S\left(\frac{h}{2}\right)\Psi_j, \quad \frac{\hat{\Psi}_{j+1} - \hat{\Psi}_j}{\Delta t} = \varepsilon_j B \frac{\hat{\Psi}_{j+1} + \hat{\Psi}_j}{2}, \quad \Psi_{j+1} = S\left(\frac{h}{2}\right)\hat{\Psi}_{j+1}.$$

Define the Lagrangian

$$L(\varepsilon, \Psi, \chi) = J_h(\varepsilon) + \sum_{j=0}^{m-1} \frac{\hat{\chi}_j + \hat{\chi}_{j+1}}{2}\left(\frac{\hat{\Psi}_{j+1} - \hat{\Psi}_j}{\Delta t} - \varepsilon_j B \frac{\hat{\Psi}_{j+1} + \hat{\Psi}_j}{2}\right)\Delta t.$$

Then, we obtain the necessary optimality condition

$$\hat{\chi}_{j+1} = S\left(\frac{h}{2}\right)^* \chi_{j+1}, \quad -\frac{\hat{\chi}_j - \hat{\chi}_{j+1}}{\Delta t} = \varepsilon_j B^* \frac{\hat{\chi}_j + \hat{\chi}_{j+1}}{2} + \mathcal{O}_{j+1/2},$$

$$\chi_j = S\left(\frac{h}{2}\right)^* \hat{\chi}_j, \quad \varepsilon_j = \frac{1}{\alpha}\left(\frac{\hat{\chi}_j + \hat{\chi}_{j+1}}{2}, B\frac{\hat{\Psi}_{j+1} + \hat{\Psi}_j}{2}\right).$$

The corresponding monotone scheme is given by

$$\hat{\Psi}_j^{k+1} = S\left(\frac{h}{2}\right)\Psi_j^{k+1}, \quad \frac{\hat{\Psi}_{j+1}^{k+1} - \hat{\Psi}_j^{k+1}}{\Delta t} = \varepsilon_j^{k+1} B \frac{\hat{\Psi}_{j+1}^{k+1} + \hat{\Psi}_j^{k+1}}{2},$$

$$\Psi_{j+1}^{k+1} = S\left(\frac{h}{2}\right)\hat{\Psi}_{j+1}^{k+1},$$

$$\varepsilon_j^{k+1} = (1-\delta)\tilde{\varepsilon}_j^k + \frac{\delta}{\alpha}\left(B\frac{\hat{\Psi}_j^{k+1} + \hat{\Psi}_{j+1}^{k+1}}{2}, \frac{\hat{\chi}_j^k + \hat{\chi}_{j+1}^k}{2}\right),$$

$$\hat{\chi}_{j+1}^{k+1} = S\left(\frac{h}{2}\right)^* \chi_{j+1}^{k+1}, \quad -\frac{\hat{\chi}_j^{k+1} - \hat{\chi}_{j+1}^{k+1}}{\Delta t} = \tilde{\varepsilon}_j^{k+1} B^* \frac{\hat{\chi}_j^{k+1}, + \hat{\chi}_{j+1}^{k+1}}{2} + \mathcal{O}_{j+1/2},$$

$$\chi_j^{k+1} = S\left(\frac{h}{2}\right)^* \hat{\chi}_j^{k+1},$$

$$\tilde{\varepsilon}_j^{k+1} = (1-\eta)\varepsilon_j^{k+1} + \frac{\eta}{\alpha}\left(B\frac{\hat{\Psi}_j^{k+1} + \hat{\Psi}_{j+1}^{k+1}}{2}, B\frac{\hat{\chi}_{j+1}^{k+1} + \hat{\chi}_j^{k+1}}{2}\right),$$

where $\Psi_0^{k+1} = \Psi_0$ and $\chi_m^{k+1} = \mathcal{O}_m$

Theorem 3

$$J_h(\varepsilon^{k-1}) - J_h(\varepsilon^k) = \frac{\alpha \Delta t}{2}\sum_{j=1}^{m}\left(\frac{2}{\delta} - 1\right)|\varepsilon_j^k - \tilde{\varepsilon}_j^{k-1}|^2 + \left(\frac{2}{\eta} - 1\right)|\varepsilon_j^{k-1} - \tilde{\varepsilon}_j^{k-1}|^2.$$

Proof. Since $|\Psi_j^k| = 1$, $|\mathcal{O}_j| = 1$, we have

$$J(\varepsilon^{k+1}) - J(\varepsilon^k) = -(\Psi_m^{k+1} - \Psi_m^k, \mathcal{O}_m)$$
$$- \Delta t \sum_{j=0}^{m-1} \left(\frac{\hat{\Psi}_{j+1}^{k+1} + \hat{\Psi}_j^{k+1}}{2} - \frac{\hat{\Psi}_{j+1}^k + \hat{\Psi}_j^k}{2}, \mathcal{O}_{j+\frac{1}{2}} \right) + \frac{\alpha \Delta t}{2} \sum_{j=0}^{m-1} [|\varepsilon_j^{k+1}|^2 - |\varepsilon_j^k|^2].$$

Note that $\Psi_0^k = \Psi_0^{k+1} = \Psi_0$ and we use the following equality:

$$\sum_{j=0}^{m-1} \left(\frac{\widehat{\Psi}_{j+1} - \widehat{\Psi}_j}{\Delta t}, \frac{\widehat{\chi}_{j+1} + \widehat{\chi}_j}{2} \right) + \left(\frac{\widehat{\Psi}_{j+1} + \widehat{\Psi}_j}{2}, \frac{\widehat{\chi}_{j+1} - \widehat{\chi}_j}{\Delta t} \right)$$
$$= \sum_{j=0}^{m-1} \left(\frac{\widehat{\Psi}_{j+1}}{\Delta t}, \widehat{\chi}_{j+1} \right) - \left(\frac{\widehat{\Psi}_j}{\Delta t}, \widehat{\chi}_j \right)$$
$$= \frac{1}{\Delta t}((\widehat{\Psi}_m, \widehat{\chi}_m) - (\widehat{\Psi}_0, \widehat{\chi}_0)) = \frac{1}{\Delta t}((\Psi_m, \chi_m) - (\Psi_0, \chi_0)).$$

Thus, we have

$$- (\Psi_m^{k+1} - \Psi_m^k, \mathcal{O}_m) - \Delta t \sum_{j=0}^{m-1} \left(\frac{\hat{\Psi}_{j+1}^{k+1} + \hat{\Psi}_j^{k+1}}{2} - \frac{\hat{\Psi}_{j+1}^k + \hat{\Psi}_j^k}{2}, \mathcal{O}_{j+\frac{1}{2}} \right)$$
$$= \Delta t \sum_{j=0}^{m-1} \left\{ B \left(\left[\varepsilon_j^{k+1} \frac{\widehat{\Psi}_j^{k+1} + \widehat{\Psi}_{j+1}^{k+1}}{2} - \varepsilon_j^k \frac{\widehat{\Psi}_j^k + \widehat{\Psi}_{j+1}^k}{2} \right], \frac{\widehat{\chi}_j^k + \widehat{\chi}_{j+1}^k}{2} \right) \right.$$
$$\left. + \left(\frac{\widehat{\Psi}_j^{k+1} + \widehat{\Psi}_{j+1}^{k+1}}{2} - \frac{\widehat{\Psi}_j^k + \widehat{\Psi}_{j+1}^k}{2}, B \widetilde{\varepsilon}_j^k \frac{\widehat{\chi}_j^k + \widehat{\chi}_{j+1}^k}{2} \right) \right\}$$
$$= -\alpha \Delta t \sum_{j=0}^{m-1} \frac{1}{\delta} (\varepsilon_j^{k+1} - \widetilde{\varepsilon}_j^k, \varepsilon_j^{k+1} - (1-\delta) \widetilde{\varepsilon}_j^k) + \frac{1}{\eta} (\varepsilon_j^k - \widetilde{\varepsilon}_j^k, \widetilde{\varepsilon}_j^k - (1-\eta) \varepsilon_j^k)$$

and hence we have

$$J(\varepsilon^{k+1}) - J(\varepsilon^k) = -\alpha \Delta t \sum_{j=0}^{m-1} \frac{1}{\delta} (\varepsilon_j^{k+1} - \widetilde{\varepsilon}_j^k, \varepsilon_j^{k+1} - (1-\delta) \widetilde{\varepsilon}_j^k)$$
$$+ \frac{1}{\eta} (\varepsilon_j^k - \widetilde{\varepsilon}_j^k, \widetilde{\varepsilon}_j^k - (1-\eta) \varepsilon_j^k) - \frac{1}{2} |\varepsilon_j^{k+1}|^2 + \frac{1}{2} |\varepsilon_j^k|^2$$
$$= -\frac{\alpha \Delta t}{2} \sum_{j=0}^{m-1} \left(\frac{2}{\delta} - 1 \right) |\widetilde{\varepsilon}_j^k - \varepsilon_j^{k+1}|^2 + \left(\frac{2}{\eta} - 1 \right) |\varepsilon_j^k - \widetilde{\varepsilon}_j^k|^2 \leq 0.$$

5 Numerical Tests

In this section we demonstrate the feasibility of our proposed feedback laws using a test example. We set $H = L^2(0,1)$ and

$$\mathcal{H}_0\psi = \sum_{k=1}^{\infty} \lambda_k (\psi, \psi_k)_H \psi_k,$$

where

$$\psi_k(x) = \sqrt{2}\sin(k\pi x) \quad \text{and} \quad \lambda_k = k\pi.$$

The control Hamiltonians are given by

$$(\mu_i \Psi)(x) = b_i(x)\Psi(x), \quad x \in (0,1),$$

with $i = 1, 2$. For computations we truncated the expansion of $\mathcal{H}_0$ at $N = 99$, so that

$$S_N(h)\Psi_0 = \sum_{k=1}^{N} e^{-i\lambda_k h}(\Psi_0, \psi_k)\psi_k.$$

To integrate the control Hamiltonian term, the collocation method was used in the form

$$(B_i^N \psi)(x_n^N) = b_i(x_n^N)\psi(x_n^N), \quad i = 1, 2,$$

where $x_n^N = \frac{n}{N}$, $1 \le n \le N-1$. Thus, we implemented the feedback law based on the Strang splitting method in the form

$$\Psi^{k+1} = S_N\left(\frac{h}{2}\right) \mathcal{F}_N \left(I - \varepsilon_1^k \frac{h}{2} B_1^N - \varepsilon_2^k \frac{h}{2} B_2^N\right)^{-1} \left(I + \varepsilon_1^k \frac{h}{2} B_1^N + \varepsilon_2^k \frac{h}{2} B_2^N\right) S_N\left(\frac{h}{2}\right)$$

$$\varepsilon_i^k = F_i(\Psi^{k+1/2}, \mathcal{O}^{k+1/2}), \quad i = 1, 2,$$

where $\mathcal{F}_N$ and $\mathcal{F}_N^{-1}$ are the discrete Fourier sine transform and its inverse transform, respectively and B_i^N is the diagonal matrix with diagonal

$$b_i(x_1^N), \cdots, b_i(x_{N-1}^N) \quad \text{for each } i = 1, 2.$$

This is an implicit method and its well-posedness is discussed in Section 3 for given $\beta > 0$ and $\gamma \in [0, 1]$. The numerical tests that we report on are computed with $h = 0.01$, $\alpha = 1/500$ and

$$b_1(x) = (x - .5) + 1.75(x - .5)^2, \quad b_2(x) = 2.5(x - .5)^3 - 2.5(x - .5)^4.$$

These control potentials satisfy the rank condition in Section 1 and are selected by minimizing the tracking time by trial and error tests. Figure 1 shows the orbit tracking performance $V = \frac{1}{2}|\Psi(t) - \mathcal{O}(t)|_X^2$ comparison between different β and different power γ of V. As β increases, the performance V is significantly improved and the 10 % performance level is achieved in much shorter horizon. By decreasing the power of V, the performance V improves also and more rapidly in the beginning of the time horizon.

Figure 2 shows the tracked state (real and imaginary parts) after 50 time units compared to the desired orbit. The imaginary part of the desired state is zero at T and there remains some tracking error. On the right the tracking error in terms of $V_1(\Psi^k, \mathcal{O}^k)$ is shown.

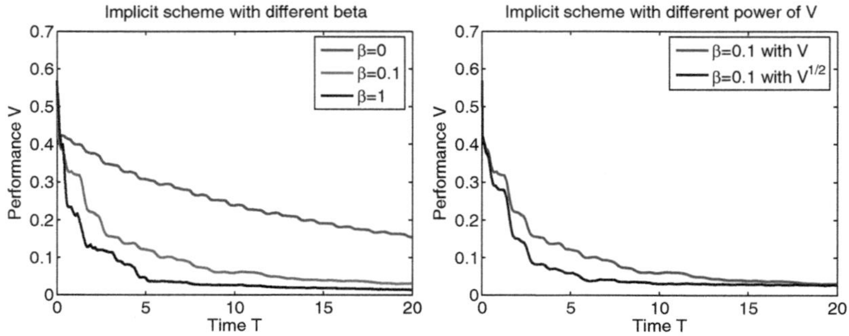

Fig. 1. Performance comparison for implicit scheme.

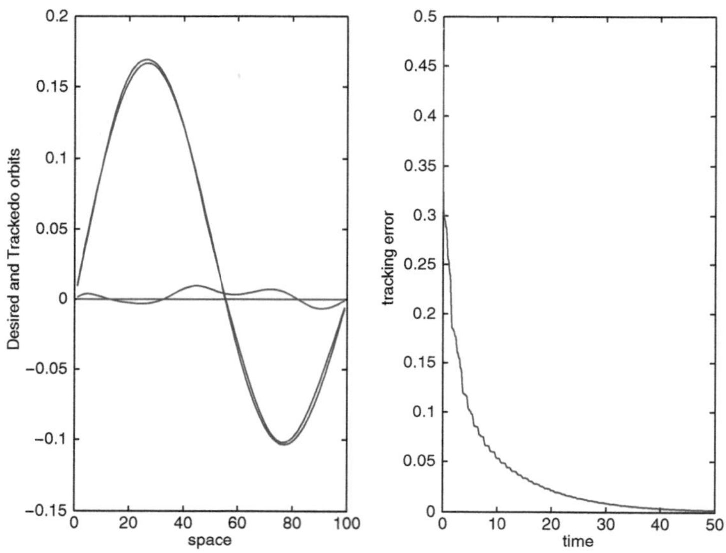

Fig. 2.

In Section 4.1, we describe applying the monotone scheme with relaxation constant $\delta, \eta \in (0, 2)$ for the receding horizon control synthesis. As shown in Figure 3, we observe that the under-relaxation performs better than the over-relaxation.

In Figure 4 we show numerical results for the instantaneous receding horizon method with $\Delta T = .01$. We observe that the receding horizon synthesis improves the performance of the tracking feedback law significantly.

The end performance at $T_f = 200$ for the instantaneous receding horizon control is $V(T_f) = 6.95 \cdot 10^{-4}$ compared with $V(T_f) = 1.49 \cdot 10^{-3}$ for original feedback law. Also it is observed that the performance improves with increasing receding step k.

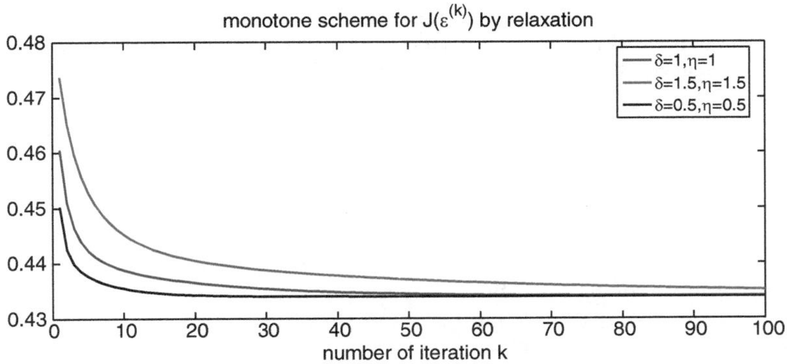

Fig. 3. $J(\varepsilon^{(k)})$ of monotone scheme with different relaxations (δ, η).

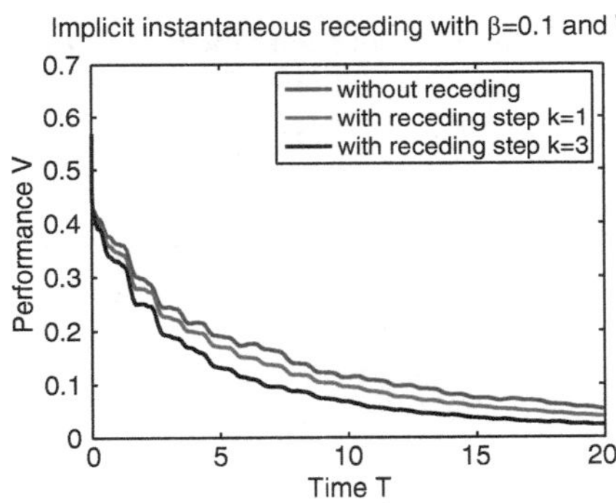

Fig. 4. Performance of instantaneous receding horizon control.

Table 1. Performance of regular receding horizon control, $T_f = 200$, $\beta = 0.1$

T	m	$\lvert\Psi(T_f) - O(T_f)\rvert^2$
1	1	$1.60 \cdot 10^{-4}$
1	2	$8.66 \cdot 10^{-5}$
0.5	1	$3.14 \cdot 10^{-4}$

In Table 1 we show numerical results for the regular receding horizon for different numbers of iteration m for the monotone scheme in Section 4.1. It is observed that the performance increases with m as we expected. A larger horizon $T = 1$ performs much better than a shorter horizon $T = .5$ as should be. The regular receding horizon control performs better than the instantaneous receding horizon control but has more computational cost.

Acknowledgement. The authors express their appreciation to Prof. Karl Kunisch for discussions on various aspects of this paper. Research is partially supported by the Army Research Office under DAAD19-02-1-0394 and the Air Force Office of Scientific Research under FA 9550-06-01-0241.

References

1. K. Beauchard, J. M. Coron, M. Mirrahimi, and P. Rouchon. Implicit Lyapunov control of finite dimensional Schrödinger equations. Preprint.
2. A. E. Ingham. Some trigonometrical inequalities with applications to the theory of series. *Math. Z.*, 41(1):367–379, 1936.
3. K. Ito and F. Kappel. *Evolution equations and approximations.* World Scientific, River Edge, NJ, 2002.
4. K. Ito and K. Kunisch. Asymptotic properties of feedback solutions for a class of quantum control problems. *SIAM J. Control Optim.* Submitted.
5. K. Ito and K. Kunisch. Feedback solutions for a class of quantum control problems. In *Oberwolfach Proceedings.* Submitted.
6. K. Ito and K. Kunisch. Asymptotic properties of receding horizon optimal control problems. *SIAM J. Control Optim.*, 40(5):1585–1610, 2002.
7. K. Ito and K. Kunisch. Optimal bilinear control of an abstract Schrödinger equation. *SIAM J. Control Optim.*, 46(1):274–287, 2007.
8. M. Mirrahimi, P. Rouchon, and G. Turinici. Lyapunov control of bilinear Schrödinger equations. *Automatica J. IFAC*, 41(11):1987–1994, 2005.
9. A. Pazy. *Semigroups of linear operators and applications to partial differential equations.* Springer, Berlin, 1983.

Fluid Dynamics of Mixtures of Incompressible Miscible Liquids

Daniel D. Joseph

University of Minnesota, Minneapolis University of California, Irvine, USA,
joseph@aem.umn.edu

Summary. The velocity field of binary mixture of incompressible miscible liquids is non-solenoidal when the densities of the two liquids are different. If the mixture density is linear in the volume fraction, as is the case of liquids which satisfy the law of additive volumes, then the velocity can be decomposed into a solenoidal and expansion part. Here we propose a theory for liquids which do not satisfy the law of additive volumes. In this theory the mixture density is again given by a linear form but the densities of the liquids are scaled by the factor expressing the change of the volume of the mixture upon mixing. The dynamical theory of simple mixtures of incompressible liquids can be formed as the correct form of the Navier–Stokes equations in which the compressibility of the mixture is recognized. A rigorous form of the diffusion equation, different than the usual one, is also derived from first principles. The diffusion equation is based a non-linear form of Fick's law, expressed in terms of gradients of the chemical potential. It is argued that the diffusion of species (of heat and in general) is impossible; signals must move with a finite speed though they may rapidly decay to diffusion. The underlying equation for the evolution of species and heat in the linear case is a damped wave equation rather than the conventional diffusion equation. The Navier–Stokes theory can be identified as a mass transport theory. The solenoidal part of the velocity satisfies an equation which can be shown to govern the transport of volume; it differs from the mass transport velocity by an irrotational expansion velocity associated with the dilitation of the mixture. The equations governing the transport of mass and volume differ from one another by well-defined mathematical transformations; the choice of one or the other is a matter of convenience. However, a genuine difference is associated with boundary conditions. The conventional assumption that the mass transport velocity vanishes is supported by calculations from molecular dynamics but these calculations employ entirely different assumptions and, hence, lack authority. The idea that gradients of composition ought to induce stresses and not just diffusion has been considered and is modeled by a second-order theory introduced by Korteweg 1901. There is not strong evidence that these stresses are important except in regions of strong gradients where a relaxation theory rather than a second-order theory ought to apply. A relaxation theory for stresses due to gradients of composition which relaxes into the second-order theory when the gradients are small is proposed and applied to explain observations of a transient interfacial tension which may be traced to a difference between the relaxation times for diffusion and stresses.

W. Fitzgibbon et al. (eds.), *Applied and Numerical Partial Differential Equations*, Computational Methods in Applied Sciences 15, DOI 10.1007/978-90-481-3239-3_10,

1 Introduction

Joseph [18], Galdi et al. [13], Hu and Joseph [16], Joseph et al. [19, 22] and Camacho and Brenner [6] developed a theory of non-solenoidal velocity effects and Korteweg stresses in simple mixtures of incompressible liquids. The theory may be framed in terms of a mass transport velocity **u**, which is not solenoidal; and a volume transport velocity **w**, which is solenoidal. The equations which govern the mass transport are appropriately described as Navier–Stokes equations for the mixture; they may also be described in terms of molecular modeling as mass averaged whilst the velocity **w** is volume averaged.

The effects of diffusion in miscible liquids are believed to be well understood. It is not so well known but obvious that diffusion is impossible; the propagation of changes of composition, heat and other diffusing scalars must initially occur as a wave, which typically rapidly decays to diffusion. These kinds of effects can be modeled (Section 7) by assuming that the flux of species depends on the time rate of change of flux as well as the gradients of the concentration of species.

The possibility that motions can be driven by additional stresses associated with gradients of composition can also be considered. Dynamical effects can arise in thin mixing layers where the gradients of composition are large. This possibility was already recognized in discussions given by Korteweg [25] following earlier work by Van der Waals [45] in which he proposes a constitutive equation includes the stresses induced by gradients of density and by gradients of composition which could give rise to effects which mimic surface tension in regions where the gradients are large. Various theories based on thermodynamic arguments in which the consequences of the assumption that density gradients give rise to stresses, even when these stresses are induced in single component liquids by temperature gradients have been put forward by Brenner (see [4] and the references there). Serrin [43] has considered the form of interfacial surfaces produced by density variations in Korteweg's theory of phase equilibria. He shows "...that, *unless rather special conditions are satisfied*, the only geometric phase boundaries which are consistent with Korteweg's theory are spherical, cylindrical, or planar".

A review of literature about effects which mimic interfacial tension between diffusing liquids is given by Joseph [18] and Joseph and Renardy [22]. Experiments show that the shape of sharp interfaces in the presence of slow diffusion resemble familiar shapes which can be seen in immiscible liquids with real interfacial tension.

Attempts to model the aforementioned effects with Korteweg stresses use the fact that these stresses are large when and where the gradients are large and they are infinitely large at surfaces of discontinuity. These models are very difficult to evaluate since the model parameters are not known and even when the parameters are chosen to fit, the fits are far from perfect.

One flaw of the theory is that higher tensions at the early times when diffusion starts are destroyed rapidly by diffusion, even when diffusion is slow. The Korteweg stresses arise from the simplest constitutive assumption in which the compositional stress is determined by the present value of compositional gradients. A properly invariant theory of this kind leads to the quadratic expression (23); in the sense and in analogy with facts well known in rheological modeling, this can be called a second order model. Relaxation effects which may be expected for discontinuous initial data are not present in Korteweg's theory.

The modeling of interfacial stresses for miscible liquids should be framed as a small part of the general problem of modeling stresses which arise from gradients of composition. The Korteweg model is one among possibly many. It is not yet clear what may be the observable consequences of such stresses.

2 Simple Mixtures and the Law of Additive Volume

Suppose there are two species, for example, glycerin and water, designated with subscripts g and w. If the volume V of a mixture of the two liquids does not change, then $V = V_w + V_g$ and the mixture density can be expressed in terms of the volume fraction of one of them, say $\phi = V_w/V$,

$$\rho(\phi) = \rho_w \phi + \rho_g(1 - \phi), \tag{1}$$

where ρ_w and ρ_g are the handbook values of water and glycerin and

$$\begin{aligned} m_w &= \rho_w V_w, \\ m_g &= \rho_g V_g \end{aligned} \tag{2}$$

are the mass of water and the mass of glycerin in the mixture. The equation $V = V_w + V_g$ which states that the total volume is the sum of the volumes of the two constituents is called the law of additive volumes.

In general, the volume of the mixture is not the same as the volume of the mixture. The new volume

$$U = Vf(\phi) \tag{3}$$

is more or less than the original one and the dilation factor f depends on ϕ with

$$f(0) = f(1) = 1. \tag{4}$$

The dilation factor is nearly one for glycerin and water with a maximum near

$$f(1/2) = 1.01.$$

The maximum deviation of ethanol and water solutions is about 3%.

Noting now that in the mixtures the masses m_w and m_g do not change, we may assume that the same dilation $f(\phi)$ of V, applies also to V_w and V_g, so that

$$\begin{aligned} U_w &= V_w f(\phi), \\ U_g &= V_g f(\phi). \end{aligned} \tag{5}$$

Hence

$$\begin{aligned} m_w &= \rho_w^* U_w, \\ m_g &= \rho_g^* U_g \end{aligned} \tag{6}$$

and

$$U = U_w + U_g \tag{7}$$

is a restatement of the "law of additive volumes". The new volume fraction

$$\phi^* = U_w/U = V_w/V = \phi \tag{8}$$

is the same as the old one.

The new mixture density is the same as the old mixture density

$$\rho^*(\phi) = \rho_w^* \phi + \rho_g^*(1-\phi) = \rho(\phi)/f(\phi) \tag{9}$$

with the caveat that $\rho_w^* = \rho_w/f(\phi)$ and $\rho_g^* = \rho_g/f(\phi)$ given by (2), (5) and (6) are no longer table values.

The continuity equation is

$$\frac{\mathrm{d}\rho^*}{\mathrm{d}t} = \frac{\mathrm{d}\rho^*}{\mathrm{d}\phi}\frac{\mathrm{d}\phi}{\mathrm{d}t} = -\rho^* \operatorname{div} \mathbf{u}, \tag{10}$$

where

$$\frac{\mathrm{d}\rho^*}{\mathrm{d}t} = \bar{\rho}(\phi)\frac{\mathrm{d}\phi}{\mathrm{d}t} \tag{11}$$

and

$$\bar{\rho}(\phi) = d[\rho(\phi)/f(\phi)]/d\phi \tag{12}$$

is not zero. If

$$\frac{\mathrm{d}\phi}{\mathrm{d}t} = \frac{\partial \phi}{\partial t} + \mathbf{u}\cdot\nabla\phi \neq 0, \tag{13}$$

the flow cannot be solenoidal.

Volume changes due to mixing require that distances between molecules change; such molecular rearrangements produce work and energy in the form of exothermic reactions. Thermodynamics of miscible mixtures which do not satisfy the law of additive volumes should be considered.

3 Mass Transport Equations for Simple Mixtures of Incompressible Miscible Liquids

The governing equations ([22, Vol. 2, Chapter X], [19, 28]) expressing the diffusion of species, balance of mass and momentum for simple incompressible binary mixtures can be formulated as

$$\frac{\partial \phi}{\partial t} + \nabla \cdot (\phi \mathbf{u}) = \nabla \cdot \left(\frac{D}{1 - \zeta \phi} \nabla \phi \right), \tag{14}$$

$$\nabla \cdot \left(\mathbf{u} - \frac{\zeta D}{1 - \zeta \phi} \nabla \phi \right) = 0, \tag{15}$$

and

$$\rho \left(\frac{\partial \mathbf{u}}{\partial t} + (\mathbf{u} \cdot \nabla) \mathbf{u} \right) = \rho \mathbf{g} - \nabla p + \mu \left\{ \frac{1}{3} \nabla (\nabla \cdot \mathbf{u}) + \nabla^2 \mathbf{u} \right\} - \frac{2}{3} (\nabla \mu)(\nabla \cdot \mathbf{u}) + 2 \nabla \mu \cdot \mathbf{D}[\mathbf{u}], \tag{16}$$

where ϕ is the volume fraction, $\mathbf{u}$ is the mass averaged velocity, $\zeta = 1 - \rho_\gamma/\rho_v > 0$ is the normalized density difference, D is the diffusion coefficient, ρ is the density and μ is the viscosity. Note that D, ρ and μ are functions of ϕ. Equation (16) is the Navier–Stokes equation for a compressible mixture of two miscible incompressible liquids in which the acceleration and viscous stress are expressed in terms of the mass averaged velocity and the bulk viscosity is chosen according to Stokes hypothesis which forces the stress deviator to have a zero trace.

The expansion velocity $\mathbf{u}_e$

$$\mathbf{u}_e \overset{\text{def}}{=} \nabla h = \frac{\zeta D}{1 - \zeta \phi} \nabla \phi \tag{17}$$

and the solenoidal velocity $\mathbf{w}$

$$\mathbf{w} \overset{\text{def}}{=} \mathbf{u} - \mathbf{u}_e. \tag{18}$$

The expansion velocity $\mathbf{u}_e$ has a zero curl and a non-zero divergence and $\nabla \wedge \mathbf{u} = \nabla \wedge \mathbf{w}$. The function

$$h = h(\phi) = \int_0^\phi \frac{\zeta D}{1 - \zeta \phi} d\phi \tag{19}$$

is a chemical potential derived in [22, eq. (4e.7)] using the theory presented in [26, p. 357]. The equation (15) is derived (see [13]) from manipulations using the continuity equation (10) written for simple mixtures satisfying the law of additive volumes and the diffusion equation (14). The decomposition of the velocity into a solenoidal part and a gradient is a realization the Helmholtz decomposition. The same manipulations do not lead to a similar decomposition when the theory (5) of volume changes due to mixing is applied. Here ζ is a primary parameter. The expansion velocity $\mathbf{u}_e$ is proportional to ζ and D, the diffusion coefficient, and is zero for two species with the same density.

If D is constant, then

$$h = -D \log[1 - \zeta]. \tag{20}$$

The viscosity $\mu = \mu(\phi)$ is a rapidly varying function, in general. We could think of ϕ as the water fraction of glycerin–water mixture, then empirically [42] $\mu(\phi)$ may be approximated by $\mu_G \exp[\alpha_1 \phi + \alpha_2 \phi^2 + \alpha_3 \phi^3]$ and, for example, at 60°C, the coefficients are $\alpha_1 = -10.8$, $\alpha_2 = 9.47$, and $\alpha_3 = -3.83$. And, according to the simple mixture assumption, the density is given by $\rho = \rho_G(1 - \zeta\phi)$. In the case of glycerin–water mixtures, the model gives less than 1 % error with the maximum error near $\phi = 0.5$. Values of $D(\phi)$ for glycerin–water mixtures may be obtained from the paper on miscible displacement in capillary tubes by Petitjeans and Maxworthy [37, Table 1]. (They measured $D(C_g)$, where C_g is the percentage of glycerin by weight, over the whole range $0 \leq C_g \leq 1$.) Their paper and the companion papers on numerical simulation of miscible displacement by Chen and Meiburg [9, 10] make some comparisons between the usual solenoidal theories in which the weight difference is neglected and the non-solenoidal theory under study here. They appear to have neglected the effects of the dilatational stress which is important when $\operatorname{div} \mathbf{u}$ does not vanish and the viscosity and viscosity gradients are large as in glycerin–water mixtures.

4 Volume Transport Equations for Simple Mixtures of Incompressible Miscible Liquids

In terms of $\mathbf{w}$, h and ϕ, the governing equations (14), (15), (16) can be written as

$$\frac{\partial h}{\partial t} + (\mathbf{w} \cdot \nabla) h = D \nabla^2 h - \nabla h \cdot \nabla h, \tag{21}$$

$$\frac{\partial \phi}{\partial t} + (\mathbf{w} \cdot \nabla) \phi = \nabla \cdot (D \nabla \phi), \tag{22}$$

$$\nabla \cdot \mathbf{w} = 0 \tag{23}$$

and

$$\rho \left(\frac{\partial \mathbf{w}}{\partial t} + (\mathbf{w} \cdot \nabla)\mathbf{w} + (\nabla \mathbf{w}) \cdot \nabla h - \nabla h \cdot (\nabla \mathbf{w}) \right) + \rho \nabla \left\{ D \nabla^2 h - \frac{1}{2} (\nabla h \cdot \nabla h) \right\}$$
$$= \rho g - \nabla p + \mu \nabla^2 w + \frac{1}{3} \mu \nabla (\nabla^2 h) - \frac{2}{3} (\nabla \mu) \nabla^2 h + 2 \nabla \mu \bullet D[w] + 2 \nabla \mu \bullet D[\nabla h]. \tag{24}$$

5 Boundary Conditions at a Solid Wall

The diffusive flux of any species across an impermeable, bounding surface vanishes. If $\mathbf{n}$ is the outward at such a surface (from the fluid to solid), we have

$$\mathbf{n} \cdot \nabla \phi = 0 \tag{25}$$

at an impermeable boundary.

The nature of the boundary condition for the velocity at a solid wall can be considered [28]. For miscible liquids, like glycerin and water, the mixture looks and feels like any other liquid and it is natural to think that the no-slip condition $\mathbf{u} = 0$ which applies to solutions of the Navier–Stokes equation, like (3) ought to apply. This is the point of view adopted by Landau and Lifshitz [27], by Camacho and Brenner [6] and in our earlier work (see [22]) and here. However, in mixtures we do not know, at present, what is the appropriate average of the species velocities to insert in the viscous stress terms in the momentum balance, or in the no-slip condition at a solid boundary.

Gases are different than liquids, because the molecules between species of different types are not held together by short range forces at a distance; collisions are the dominating dynamical process. It is perhaps more natural to consider averages over the two species of a binary mixture of gases, which unlike the constituents of miscible liquids, are not tied together in lock step by molecular fields of force. When viewed in this way, we may identify $\mathbf{u}$ with the mass averaged velocity and the solenoidal part $\mathbf{w}$ with the volume averaged velocity. We may then consider whether $\mathbf{w}$, $\mathbf{u}$ or some combination of these ought to vanish at a solid wall.

Careful experiments on isobaric interdiffusion of binary gases in porous plugs by Graham [15] and others lead to the conclusion that the total mass flux does not vanish even though the pressure is the same at either end of a capillary tube. Jackson [17] has shown that Graham's law which implies the existence of a mass flux in isobaric conditions holds from free molecule to continuum flow. Jackson [17, pp. 25–33] generalized a kinetic theory argument of Maxwell for a pure gas to a gas consisting of a mixture of two substances to show that a weighted mass averaged velocity, which is neither the mass or volume averaged velocity ought to vanish at a solid wall.

Mo and Rosenberger [34] did molecular-dynamics simulations of flow of gases with binary diffusion in a two-dimensional channel with atomically rough walls. They found that the no-slip condition for the mass averaged velocity arises when the mean free path in the gas mixture is of the same order of magnitude or smaller than the atomic-wall-roughness amplitude. However, if there are concentration gradients along the wall, the component velocities at the wall do not vanish. Thus, the no-slip condition is established via the mutual cancellation of the non-vanishing, opposing slip velocities of the components. Mo and Rosenberger note that their work does not settle the apparent contradiction between the results of isobaric interdiffusion experiments and the expected vanishing of the mass averaged velocity at all locations; they speculate about possible reasons for the discrepancy.

On the other hand, Koplick and Banavar [24] did molecular dynamic simulations of the flow of liquids with binary diffusion in a two-dimensional channel with atomically rough walls. Their simulations indicate that the velocity of each individual liquid species satisfies the no-slip condition and, therefore, so do mass and volume averages. Certain mathematical problems are associated with this approach; if the velocity of each species of a binary mixture vanishes, then the second order diffusion equation is well-posed with one boundary condition, (25), but not with two.

The sidewall effects, would disappear if the volume averaged velocity were to vanish at solid wall. This possibility seems to have been rejected by all workers in this subject, but the nature of the boundary conditions at a solid wall still needs clarification.

6 Korteweg Stresses

Korteweg [25], following his teacher Van der Waals [45], gave a theory of *equilibrium* surface tension in which the surface of discontinuity between a liquid and its vapor is replaced by a transition layer. He proposed that the stress in a compressible fluid is the usual one plus another stress which depends on gradients $\partial\rho/\partial\mathbf{x}$ of the density. Van der Waals reduced the form of the added stress to a quadratic form by requiring that the relation between the density gradient and the stress be form invariant under rigid body transformations. Korteweg suggested in [25, footnote] that his theory could be adapted for miscible mixtures using ϕ instead of ρ. This theory is different than his equilibrium theory because it is tied to diffusion and ultimately to motion.

The stress is given by

$$\mathbf{T} = \mathbf{T}^{(1)} + \mathbf{T}^{(2)}, \tag{26}$$

where

$$\mathbf{T}^{(1)} = -p\mathbf{1} + 2\mu\mathbf{D}[\mathbf{u}] - \frac{2}{3}\mathbf{1}\,\mathrm{div}\,\mathbf{u} \tag{27}$$

and

$$\mathbf{T}^{(2)} = \hat{\delta}\nabla\phi\otimes\nabla\phi + \hat{\gamma}\nabla\otimes\nabla\phi, \tag{28}$$

$$T^{(2)}_{ij} = \hat{\delta}\frac{\partial\phi}{\partial x_i}\frac{\partial\phi}{\partial x_j} + \hat{\gamma}\frac{\partial^2\phi}{\partial x_i\partial x_j} \tag{29}$$

is the Korteweg stress. Invariance requires that $T^{(2)}_{ij}$ is invariant to a change from $\partial\mathbf{x}$ to $-\partial\mathbf{x}$. This symmetry means those linear terms in $\nabla\phi$ cannot appear. The coefficients $\hat{\delta}$ and $\hat{\gamma}$ are unknown and experiments in which they may be measured are also not known.

Here is a simple way to think about the Korteweg stresses. Since terms linear in $\nabla\phi$ are excluded, we suppose that

$$T_{ij}^{(2)} = \frac{\partial^2 F(\phi)}{\partial x_i \partial x_j} = F''(\phi)\frac{\partial \phi}{\partial x_i}\frac{\partial \phi}{\partial x_j} + F'\frac{\partial^2 \phi}{\partial x_i \partial x_j}. \tag{30}$$

The Navier–Stokes equation for simple mixtures of incompressible miscible liquids with the Korteweg stresses are given in Section 2 and

$$\rho\frac{\mathrm{d}\mathbf{u}}{\mathrm{d}t} = -\nabla p + \mathrm{div}(\mathbf{T}^{(1)} + \mathbf{T}^{(2)}). \tag{31}$$

7 Relaxation Effects

Diffusion of initially discontinuous data is impossible. This data must propagate as a damped wave governed by a damped wave equation rather than a diffusion equation. If the damping is rapid as is usually true for heat conduction; this is just as true for the propagation of species concentration; a change of concentration at some point cannot be felt instantly at distant points.

A simple model for the propagation of heat by damped waves was first given by Cattaneo [8] who assumed that the temperature gradient depends on the rate of change of the heat flux as well as its present value. A similar assumption was made for gases, but not pursued, by Maxwell [32]. The literature on heat waves was reviewed and evaluated by Joseph and Preziosi [20,21]. These reviews have stimulated a topic of heat transfer research called hyperbolic heat conduction. The goals of research on hyperbolic heat conduction are to identify the applications in which relaxation effects are important, to study different models relating the heat flux to the temperature gradient, to investigate effects of a spectrum of relaxation times rather than a single one, to determine the conditions which lead to the smoothing of waves and to develop analytic and numerical methods to solve outstanding problems. Some of these research directions are discussed in papers by Joseph and Preziosi and others are readily found in internet search. Maddox [30] commented on issues raised in the papers by Joseph and Preziosi in the "News and Views" section of Nature; he wrote "Heat conduction is a can of worms". In commenting on Fick's law, Malone and Wheatley [31] have written a similar note "A bigger can of worms". Heat conduction satisfying Fourier's law has the same problems and possibly the same remedies as the diffusion of species following Fick's law.

The basic idea is to replace Fourier's law $\mathbf{q} = -k\nabla T$ with a relaxation law

$$\lambda\frac{\mathrm{d}\mathbf{q}}{\mathrm{d}t} + \mathbf{q} = -k\nabla T, \tag{32}$$

where $\mathbf{q}$ is the heat flux, k is the conductivity, T the temperature, and λ a relaxation time. The energy equation is

$$\rho C_p\frac{\mathrm{d}T}{\mathrm{d}t} = -\,\mathrm{div}\,\mathbf{q} \tag{33}$$

if $\mathbf{q} = -k\nabla T$, then

$$\frac{\mathrm{d}T}{\mathrm{d}t} = \frac{k}{\rho C_p}\nabla^2 T \tag{34}$$

giving rise to diffusion.

On the other hand, using the relaxation equation (32), we get

$$\lambda\frac{\mathrm{d}^2 T}{\mathrm{d}t^2} + \frac{\mathrm{d}T}{\mathrm{d}t} = \frac{k}{\rho C_p}\nabla^2 T. \tag{35}$$

The equation (35) is a damped wave equation. When λ is small (35) reduces to diffusion. When λ is large (35) reduces to a wave equation. Waves of the form $T(x - ct)$ propagate with a speed

$$c = \sqrt{k/\rho C_p \lambda}. \tag{36}$$

Suppose that the temperature $T(0,t)$ of a semi-infinite solid, initially at a uniform temperature, is suddenly raised to $T_0 > 0$. In the classical case, the news of this change is felt immediately at infinitely distant points; the temperature distribution is self similar and is described by an error function of argument proportional $x/\sqrt{t}$ (see [7, p. 94]). In the hyperbolic case (35), the speed of the wave (36) is finite and the temperature in front of the wave is identically zero with a diffusion like profile behind. A dimensionless formulation of the initial value problem just described is

$$\frac{\partial\theta}{\partial\hat{t}} + \frac{\partial^2\theta}{\partial\hat{t}^2} = \frac{\partial^2\theta}{\partial\hat{x}^2}, \quad \theta = T/T_0,\ \theta(0,\hat{t}) = H(t),\ \theta(\hat{x},0) = \frac{\partial\theta(\hat{x},0)}{\partial\hat{t}} = 0,$$

where θ is bounded as $\hat{x} \to \infty$, $\hat{x} = x\sqrt{\rho C_p/\lambda k}$, $\hat{t} = t/\lambda$.

The solution (see Figure 1) of this problem is well known [36].

The diffusion of species is typically modeled in the same way as the diffusion of heat. In the classical case of diffusion we have

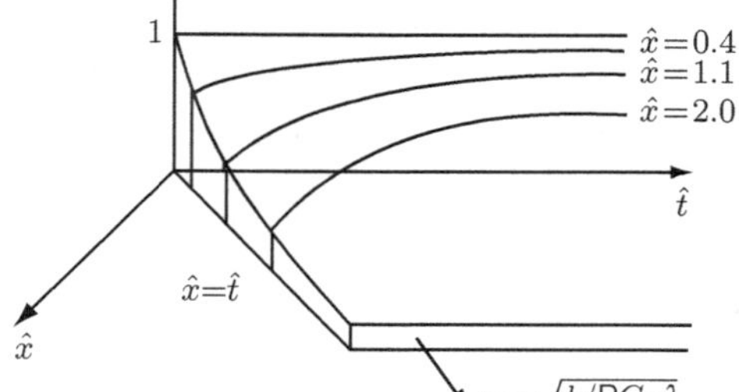

Fig. 1. Heat conduction into a semi-infinite region $\hat{x} > 0$ initially at a temperature $\theta = 0$. At $\hat{t} = 0_+$, the temperature at $\hat{x} = 0$ is raised to 1 and, therefore, propagates as a decaying wave, with $\theta = 0$ in front and diffusion behind.

$$\frac{\partial \phi}{\partial t} = -\operatorname{div} \mathbf{q}_\phi, \qquad \text{diffusion equation} \tag{37}$$

$$q_\phi = -D\nabla\phi, \qquad \text{Fick's law} \tag{38}$$

$$\frac{\partial \phi}{\partial t} = D\nabla^2\phi. \qquad \text{parabolic equation} \tag{39}$$

Suppose now that we have relaxation instead of Fick's law. Then

$$\lambda_3 \frac{\partial q_\phi}{\partial t} + q_\phi = -D\nabla\phi, \tag{40}$$

and

$$\lambda_3 \frac{\partial^2 \phi}{\partial t^2} + \frac{\partial \phi}{\partial t} = D\nabla^2\phi. \tag{41}$$

A typical problem is the smoothing an initial discontinuity of composition as studied by Liao and Joseph [28] but with effects of sidewalls neglected. The composition for $x > 0$ is initially, $\phi = 1$ (all water) and $\phi = 0$ for $x < 0$ (all glycerin). Then, the species mix so that the final composition tends to $\frac{1}{2}$. Of course, the uniform concentration cannot be established on the infinite domain but more and more of the fluid comes under the influence of diffusion in the region around $x = 0$ behind the waves composition propagating to the left and right. In a bounded domain we eventually get complete mixing, by diffusion alone when $\lambda_3 = 0$ and, in principle, by repeated reflection of waves of composition off the bounding walls. In practice, the decay of the wave amplitude may be so rapid that the waves could not be observed (see Figure 2).

The modeling of the relaxation between the species flux $\mathbf{q}_\phi$ and the species concentration ϕ in the nonlinear case is complicated and, in fact, there are many different models which satisfy the requirement of form invariance. For example,

$$\lambda_3 \overset{\nabla}{\mathbf{q}}_\phi + \mathbf{q}_\phi = -D_\gamma(\phi)\nabla\phi, \tag{42}$$

where $\overset{\nabla}{\mathbf{q}}$ is an invariant derivative of a vector discussed in [18, pp. 13–44].

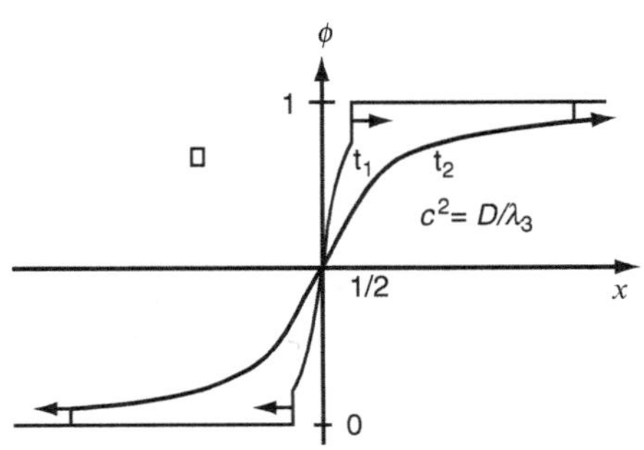

Fig. 2. Mixing of initial discontinuity by a damped wave. The discontinuity propagates as wave; faraway there is no mixing. Diffusion acts behind the wave. The amplitude of the wave decays. After some time the wave collapses to diffusion.

8 Relaxation Effects in the Modeling of Gradient Stresses

The Korteweg stress (28) is large where and when the gradient of composition is large. The stresses are infinite across surfaces of discontinuity such as when fresh water is put into contact with pure glycerin. The discontinuity is immediately smoothed by diffusion; this smoothing occurs so rapidly that the molecular rearrangement of glycerin and water could not occur. As in other problems eventually governed by diffusion, the initial response is momentarily elastic, the glycerin and water relax into diffusion.

A template for modeling relaxation effects is provided by the vast literature modeling the relation of stress and deformation in fluids. Many models are possible, these models are required to be form invariant under rigid body motions. This kind of invariance was established by Van der Waals for the Korteweg stresses depending on the gradients of density. His analysis is rigorous and modern. The modeling of relaxation effects satisfying the requirements of invariance is more complicated than the simple theory given in section. The modeling of relaxation effects for heat and temperature was treated in a comprehensive manner by Joseph and Preziosi [20, 21].

To describe the relaxation of stresses due to gradients of composition to the Korteweg stresses a templet leading to an upper converted Maxwell model can be considered. The mass transport equations for this model are (14) and (15) and

$$\rho\frac{\mathrm{d}\mathbf{u}}{\mathrm{d}t} = \rho\mathbf{g} - \nabla p + \mathrm{div}[\boldsymbol{\tau}^{(1)} + \boldsymbol{\tau}^{(2)}], \tag{43}$$

where

$$\lambda_1 \overset{\triangledown}{\boldsymbol{\tau}}{}^{(1)} + \boldsymbol{\tau}^{(1)} = 2\mu\mathbf{D}[\mathbf{u}] - \frac{2}{3}\mathbf{1}\,\mathrm{div}\,\mathbf{u} \tag{44}$$

for the stresses due to motion and

$$\lambda_2 \overset{\triangledown}{\boldsymbol{\tau}}{}^{(2)} + \boldsymbol{\tau}^{(2)} = \hat{\delta}\nabla\phi \otimes \nabla\phi + \nabla \otimes \nabla\phi \tag{45}$$

for the Korteweg stress where

$$\overset{\triangledown}{\boldsymbol{\tau}} = \frac{\partial \boldsymbol{\tau}}{\partial t} + (\mathbf{u}\cdot\nabla)\boldsymbol{\tau} - 2\mathbf{L}\boldsymbol{\tau} - 2\boldsymbol{\tau}\mathbf{L}^T \tag{46}$$

and $\mathbf{L} = \nabla\mathbf{u}^T$. If the relaxation time λ_1 for stresses is different than the relaxation time λ_2 for composition, very interesting short time effects can be expected.

9 Transient Interfacial Tension

The possibility that forces associated with steep gradient of composition can lead to transient effects mimicking interfacial tension has been considered. It is necessary to recognize the difference between liquids which mix in all proportions and those which only partially mix. Liquids which partially mix are immiscible, but the partially mixed fluids are not pure. Interfacial tension in the classical sense applies to them but the tension between the pure liquids will be greater than the final tension between the partially mixed liquids. Dynamic tension describes the reduction in tension during mixing.

Freundlich [12] says that

> We have only to remember here we are in the end always dealing with solutions. For the one liquid will always be soluble in the other to some degree, however small. Hence the *dynamic* tension of liquids, when first brought into contact, is to be distinguished from the *static* tension, when the two liquids are mutually saturated. Not only do liquids which are not miscible in all proportions have a mutual surface tension; even two completely miscible liquids, before they have united to form one phase, exhibit a dynamic interfacial tension. For we get by careful overlaying of any two liquids a definite meniscus, a jet of one liquid may be generated in another, and so on. The tension decreases rapidly during the process of solution, and becomes zero as soon as the two liquids have mixed completely.

A few attempts were made to measure the tension at early times t. Quinke [40] got 0.8–3.0 dyn/cm for ethyl alcohol/salt water. Smith et al. [44] got 1 dyn/cm for 1 cs/2,000 of silicone oil.

Davis [11] used the Irving–Kirkwood pressure tensor to evaluate the jump of pressure across a plane mixing layer. The Irving–Kirkwood pressure tensor gives the pressure in terms of the square of the concentration, something like the Korteweg stresses. The tension is expressed as a jump of pressure across the layer; it is a function of time due to the diffusive spreading of the front. He concludes that the tension of a diffusing mixing zone between miscible liquids while small, is nevertheless not zero.

Joseph [18] and Joseph et al. [19] used the full set of basic equations for mixing liquids to study transient tension. They found that the Korteweg stresses do not enter into the jump of the normal stress ΔP at a plane layer. At a spherical layer centered on r_0

$$\Delta P = \frac{2}{r_0}\sqrt{\frac{D}{t}}\left(164\frac{-\hat{\delta}}{D} - 429\right).$$

To get positive tension

$$-164\frac{\hat{\delta}}{D} - 429 > 0.$$

Lowengrub and Truskinovsky [29] presented a model for a binary miscible mixture in a narrow transition layer in which Korteweg-like dynamics involving motion and diffusion determine the internal structure of the layer. The effects of surface tension are determined by the pressure drop across the layer.

Pojman et al. [38] and Zoltowski et al. [46] presented evidence for the existence of effective interfacial tension between several binary mixtures of miscible fluids. These liquids mix in certain properties at one temperature and in all proportions at another. They used a spinning drop tensionmeter to measure the tension after erasing the solubility threshold. A rather complete review of the literature on this topic is given in these papers.

The effects of relaxation on transient tension have not been systematically studied. The effects of different rates of relaxation for stresses due to motion and stresses due to composition are of interest. These effects would be magnified in binary mixtures of fluids with high viscosity and low diffusion, as in the experiments of Mungall [35] and Runge and Frischat [41].

Mungall [35] developed a relaxation theory for gradient stresses between silicate melts. He models these fluids as viscoelastic compressible solids in which the compressibility may be induced by density gradients of heat or composition. His viscoelastic compressible Maxwell solid presumably reduces to a viscous fluid after the stress relaxes but a description of the reduction to diffusion is not presented. His model does not relax to Korteweg's and may not be an appropriate description of mixtures of incompressible miscible liquids. The diffusion of species in Mungall's theory is classical without relaxation effects. He gets interfacial effects by comparing the classical diffusion time with the Maxwell relaxation time. He says that

> I present observations of interfacial tension between miscible pairs of silicate liquid and propose a theoretical model that quantitatively accounts for them. Viscoelastic rheology of the liquids permits the establishment of gradient stress completely analogous to thermal stress in the Maxwell solids; this is expected whenever the time scale of diffusion is shorter than the Maxwell relaxation time. The existence of gradient stress may profoundly affect interface processed during the mixing of miscible fluids.

He presents data from an experiment which appears to support his ideas about transient interfacial tension (see Figure 3):

> I observed interfacial tension in molten silicates at 1300°C and 1 bar pressure ... they correspond to the natural lava types trachyte (P16a) and basalt (146). Each experiment consisted of a block of basaltic glass situated underneath a block of trachytic glass, in contact at a polished planar horizontal interface, in an open Pt crucible. The finished assembly was hung with Pt wire within the hotspot of the 1 atm tube furnace and brought to temperature within 2 min. One experiment was quenched immediately upon reaching the run temperature of 1300°C whereas the others had durations of 17 and 84 min after reaching 1300°C...

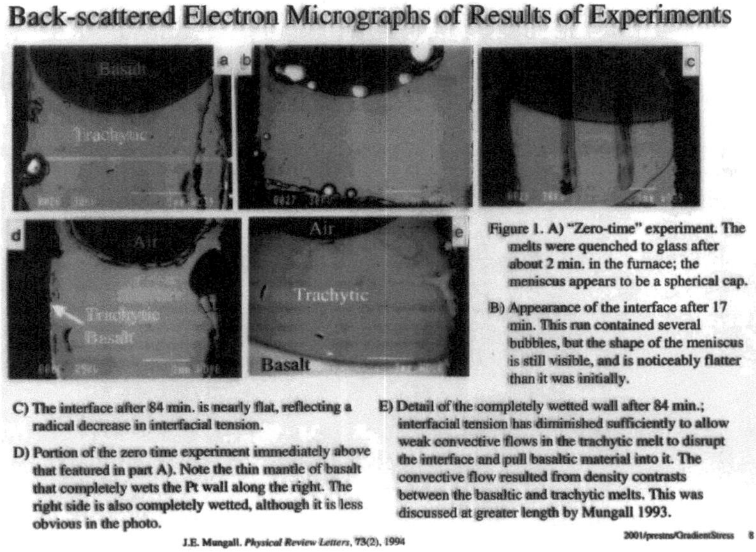

Fig. 3. Back-scattered electron micrographs of results of experiments.

> Backscattered scanning electron micrographs of vertical sections through the run products are shown in Figure 1 [in the next slide]. ...At the time of first melting of the glasses the interface was perfectly planar. The interface as it appeared after quench is clearly visible and in all four experiments forms a meniscus (see figure caption). In the reversed experiment the sense of curvature of the meniscus is also reversed [our emphasis]. Since basaltic and trachytic liquids are completely miscible, the classic concept of a surface tension cannot be applied to this case.

Runge and Frischat [41] presented data on glass melts containing Al_2O_3 which can be interpreted as giving rise to strong effects of transient tension (see Figures 4 and 5). Their binary system is high viscosity, slow diffusion and surely viscoelastic:

> Al_2O_3-containing droplets ("model cords") were placed on an Al_2O_3-free matrix glass of the system Na_2O–CaO–SiO_2 and were allowed to sink in and react with the glass melt. From a comparison using Al_2O_3-free substances it could be shown that the Al_2O_3-containing droplets achieved only by assuming an acting interfacial energy in the contact zone between the two glass melts. From the droplet contours an effective interfacial tension between 0.4 and 1.1 mNm^{-1} could be calculated. The interdiffusion profiles in the contact zone showed that although Al_2O_3 and SiO_2 equilibriated slowly by diffusion, both Na_2O

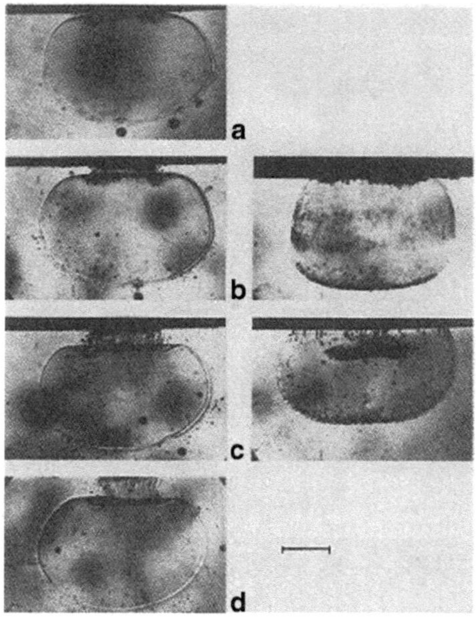

Fig. 4. Droplets after nearly totally dipped cord performs at $\nu_{\max} = 1{,}045°C$, cylindric droplet performs of 3 mm in height and 3 mm in diameter, matrix Al_2O_3-free glass, droplet Al_2O_3-containing glass (left column) and Al_2O_3-free glass (right column), respectively. Before dipping the matrix melt was held at $\nu_{\max}$. (**a**) $t_h = 15$, (**b**) 30, (**c**) 45, and (**d**) 60 min, respectively. Bar = 1 mm [41].

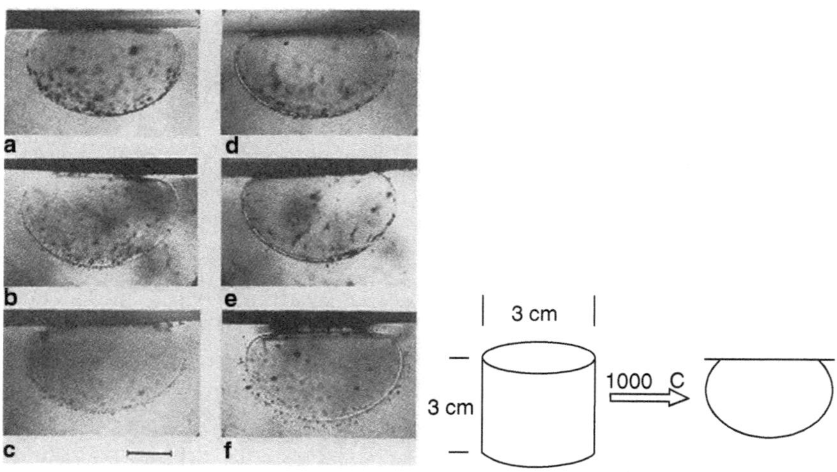

Fig. 5. Droplets after sink-in experiments at $u_{\max} = 996°C$, similar conditions as in [their] Figure 2; (**a**) $t_h = 0$, (**b**) 7.5, (**c**) 15, (**d**) 30, (**e**) 45, and (**f**) 60 min, respectively. Bar = 1 mm [41].

and CaO withstand equilibriation by uphill diffusion. *In conclusion, both the effective interfacial energy and the uphill diffusion prevent a rapid dissolution of Al_2O_3-containing cords* [our emphasis].

10 Discussion

A brief summary of the issues raised in generalizing the Navier–Stokes equations for mixtures of incompressible miscible liquids to accommodate for the effects of a nonzero irrotational expansion and diffusion is given in the abstract to this paper. We have argued that diffusion is impossible because the news of a finite change of composition cannot be felt everywhere instantly; the news of such a change must travel to distant places with a finite speed. Typically the amplitude of the wave front decays to zero, and the species profile behind the wave decays to diffusion. In the linear case, the diffusion of species is governed by a damped wave equation. Elements of a theory of mixtures which do not satisfy the law of additive volumes were proposed; the energy equation for non-simple mixtures ought to considered because changing the distance between molecules will generate large amounts of heat. Gradients of composition do not only enter into diffusion but they can be expected to generate stresses which may be modeled by the second-order theory of Korteweg [25]. The central question is the nature of stresses induced by gradients of composition. Korteweg's model is one constitutive assumption relating stresses to gradients of composition. Other constitutive assumptions could be considered. The effects of the Korteweg stresses seem to be negligible except when the gradients are large and then the effects of stress relaxation should be considered. We introduced a viscoelastic model which reduces to Korteweg's after stress relaxation and a Cattaneo type of diffusion model which reduces to Fick's law and has a different time of relaxation. It remains to see if this model can be tuned to explain the pseudo-transient interfacial effects observed in experiments.

References

1. F. T. Arecchi, P. K. Buah-Bassuah, F. Francini, C. Pérez-Garcia, and F. Quercioli. An experimental investigation of the break-up of a liquid drop falling in a miscible fluid. *Europhys. Lett.*, 9(4):333–338, 1989.
2. D. P. Barkey. Morphology selection and the concentration boundary layer in electrochemical deposition. *J. Electrochem. Soc.*, 138(10):2912–2917, 1991.
3. N. Baumann, D. D. Joseph, P. Mohr, and Y. Renardy. Vortex rings of one fluid in another in free fall. *Phys. Fluids A*, 4(3):567–580, 1991.
4. H. Brenner. Navier–Stokes revisited. *Phys. A*, 349(1–2):60–132, 2005.
5. J. Cahn and J. Hilliard. Free energy of s nonuniform system I: Interfacial free energy. *J. Chem. Phys.*, 28:258–267, 1958.

6. J. Camacho and H. Brenner. On convection induced by molecular diffusion. *Ind. Eng. Chem. Res.*, 34:3326–3335, 1995.
7. H. S. Carslaw and J. C. Jaeger. *Operational methods in applied mathematics.* Dover, New York, 1963.
8. C. Cattaneo. Sulla conduzione del calore. *Atti Sem. Mat. Fis. Univ. Modena*, 3:83–101, 1949.
9. C. Chen and E. Meiburg. Miscible displacements in capillary tubes. Part 2: Numerical simulations. *J. Fluid Mech.*, 326:57–90, 1996.
10. C. Y. Chen and E. Meiburg. Miscible displacements in capillary tubes: Influence of Korteweg stresses and divergence effects. *Phys. Fluids*, 14(7):2052–2058, 2002.
11. H. T. Davis. A theory of tension at a miscible displacement front. In *Numerical Simulation in Oil Recovery (Minneapolis, MN, 1986)*, pages 105–110, Berlin, 1988. Springer.
12. H. Freundlich. *Colloid and capillary chemistry.* Mathuen & Co. Ltd, London, 1926.
13. P. Galdi, D. D. Joseph, L. Preziosi, and S. Rionero. Mathematical problems for miscible incompressible fluids with Korteweg stresses. *European J. Mech. B Fluids*, 10(3):253–267, 1991.
14. P. Garik, J. Hetrick, B. Orr, D. Barkey, and E. Ben-Jacob. Interfacial cellular mixing and a conjecture on global deposit morphology. *Phys. Rev. Lett.*, 66(12):1606–1609, 1991.
15. T. Graham. On the law of diffusion of gases. *Philos. Mag. 2*, 175, 194:269–276, 351–358, 1833. Reprinted in *Chemical and Physical Researches*, Edinburgh University Press, 1876, pp. 44–70.
16. H. H. Hu and D. D. Joseph. Miscible displacement in a Hele-Shaw cell. *Z. Angew. Math. Phys.*, 43(4):626–644, 1992.
17. R. Jackson. *Transport in porous catalysts.* Elsevier, New York, 1977.
18. D. D. Joseph. Fluid dynamics of two miscible liquids with diffusion and gradient stresses. *European J. Mech. B Fluids*, 9(6):565–596, 1990.
19. D. D. Joseph, A. Huang, and H. Hu. Non-solenoidal velocity effects and Korteweg stresses in simple mixtures of incompressible liquids. *Phys. D*, 97 (1–3):104–125, 1996.
20. D. D. Joseph and L. Preziosi. Heat waves. *Rev. Mod. Phys.*, 61(1):41–73, 1989.
21. D. D. Joseph and L. Preziosi. Addendum to the paper "Heat waves" [Rev. Mod. Phys. 61, 41 (1989)]. *Rev. Mod. Phys.*, 62(2):375–391, 1990.
22. D. D. Joseph and Y. Renardy. *Fundamentals of Two-Fluid Dynamics, Part II.* Springer, New York, 1992.
23. M. Kojima, E. J. Hinch, and A. Acrivos. The formation and expansion of a toroidal drop moving in a viscous fluid. *Phys. Fluids*, 27(1):19–32, 1984.
24. J. Koplik and J. R. Banavar. No-slip condition for a mixture of two liquids. *Phys. Rev. Lett.*, 80(23):5125–5128, 1998.
25. D. Korteweg. Sur la forme que prennant les équations du mouvement des fluides si l'on tient compte des forces capillaires causées par les variations de densité. *Arch. Neerl. Sci. Ex. Nat., Ser. II*, 6:1–24, 1901.
26. L. D. Landau and E. M. Lifshitz. *Fluid Mechanics*, volume 6 of *Course of Theoretical Physics.* Pergamon Press, Addison-Wesley Publishing Co., London, Reading, Mass., 1959.
27. L. D. Landau and E. M. Lifshitz. *Course of theoretical physics. Vol. 6. Fluid mechanics.* Pergamon Press, Oxford, 2nd edition, 1987.

28. T. Y. Liao and D. D. Joseph. Sidewall effects in the smoothing of an initial discontinuity of concentration. *J. Fluid Mech.*, 342:37–51, 1997.
29. J. Lowengrub and L. Truskinovsky. Quasi-incompressible Cahn–Hilliard fluids and topological transitions. *R. Soc. Lond. Proc. Ser. A Math. Phys. Eng. Sci.*, 454(1978):2617–2654, 1998.
30. J. Maddox. Heat conduction is a can of worms. *Nature*, 338:373, 1989.
31. P. C. Malone and D. N. Wheatley. A bigger can of worms. *Nature*, 349:373, 1991.
32. J. C. Maxwell. On the dynamical theory of gases. *Philos. Trans Roy. Soc. London*, 157:49, 1867.
33. S. E. May and J. V. Maher. Capillary-wave relaxation for a meniscus between miscible liquids. *Phys. Rev. Lett.*, 67(15):2013–2016, 1991.
34. G. Mo and F. Rosenberger. Molecular-dynamics simulations of flow with binary diffusion in a two-dimensional channel with atomically rough walls. *Phys. Rev. A*, 44(8):4978–4985, 1991.
35. J. E. Mungall. Interfacial tension in miscible two-fluid systems with linear viscoelastic rheology. *Phys. Rev. Lett.*, 73(2):288–291, 1994.
36. A. Narain and D. D. Joseph. Linearized dynamics for step jumps of velocity and displacement of shearing flows of a simple fluid. *Rheol. Acta*, 21(3):228–250, 1982.
37. P. Petitjeans and T. Maxworthy. Miscible displacements in capillary tubes. Part 1. Experiments. *J. Fluid Mech.*, 326:37–56, 1996.
38. J. A. Pojman, C. Whitmore, M. L. Turco Liveri, R. Lombardo, J. Marszalek, R. Parker, and B. Zoltowski. Evidence for the existence of an effective interfacial tension between miscible fluids: Isobutyric acid-water and 1-butanol-water in a spinning-drop tensiometer. *Langmuir*, 22:2569–2577, 2006.
39. V. V. Pukhnachov. Mathematical model of natural convection under low gravity. Preprint 796, Institute for Mathematics and its Applications, Univ. of Minnesota, 1991.
40. G. Quinke. Die oberfachenspannung an der Grenze von Alkohol mit wasserigen Salzlosungen. *Ann. Phys.*, 9(1), 1902.
41. S. Runge and G. H. Frischat. Stability of al₂o₃-containing droplets in glass melts. *J. Non-Cryst. Solids*, 102(1–3):157–164, 1988.
42. J. B. Segur. Physical properties of glycerol and its solutions. In C. S. Miner and N. N. Dalton, editors, *Glycerol*, pages 238–334. Reinhold, 1953.
43. J. Serrin. The form of interfacial surfaces in Korteweg's theory of phase equilibria. *Quart. Appl. Math.*, 41(3):357–364, 1983/84.
44. P. G. Smith, M. Van Den Ven, and S. G. Mason. The transient interfacial tension between two miscible fluids. *J. Colloid Interface Sci.*, 80(1):302–303, 1981.
45. M. Van der Waals. Theorie thermodynamique de la capillarité dans l'hypothése d'une variation de densité. *Arch. Neerl. Sci. Ex. Nat.*, 28:121–201, 1895.
46. B. Zoltowski, Y. Chekanov, J. Masere, J. A. Pojman, and V. Volpert. Evidence for the existence of an effective interfacial tension between miscible fluids. 2. Dodecyl acrylate-poly(dodecyl acrylate) in a spinning drop tensiometer. *Langmuir*, 23:5522–5531, 2007.

Demand Forecasting Method Based on Stochastic Processes and Its Validation Using Real-World Data

Yinggao Zheng[1], Hiroshi Suito[1], and Hideo Kawarada[2]

[1] Graduate School of Environmental Sciences, Okayama University, 3-1-1, Tsushima-naka, Okayama, 700-8530, Japan, zheng@s.ems.okayama-u.ac.jp, suito@ems.okayama-u.ac.jp

[2] Iwaki Credit Bank, 74-2, Shimokajiro-Yamakamisawa, Onahama, Iwaki, 970-0316, Japan, kawarada0@nifty.com

Summary. Demand forecasting problems frequently arise in logistics and supply chain management. The Newsboy Problem is one such problem. In this paper, we present an improved solution method using application of the Black–Scholes model incorporating stochastic processes used in financial engineering for option pricing. Through numerical experiments using real-world data, the proposed model is demonstrated to be effective.

1 Introduction

In demand forecasting problems, the quantity of expected demand and the most suitable wholesale quantity are sought. They are important problems in Effective Demand Management; the problems often appear in the fields of logistics and supply chain management. Demand forecasting systems are expected to support the purchasing/procurement and sales departments of companies. They are intended to reduce differences between sales and demand. In recent years, various means have been adopted to determine quantities of demand and optimal wholesale size. Makridakis and Wheelwright [3] classified forecast methods into quantitative forecasting, judgment forecasting, and technological forecasting. Sekine [6] presented a forecasting system used by KAO Corp., which includes forecasting according to season, daily goods, new products, commodity switching, and similar goods. Munakata and Saito [5] proposed a forecast technique for new products based on short-term time series data, quantities of accumulated aggregate demand, and product life cycles. In contrast, few studies have examined Newsboy Problems (NBPs). Some commodities become worthless after a certain time passes: newspapers, perishable foods, etc. For stock control of such commodities, the daily decision for wholesale quantity becomes an important problem. Such problems are

W. Fitzgibbon et al. (eds.), *Applied and Numerical Partial Differential Equations*, Computational Methods in Applied Sciences 15, DOI 10.1007/978-90-481-3239-3_11,

designated simply as Problems herein. Masui [4] proposed a method of finding optimal wholesale size that minimizes expected values of the loss and maximizes expected profit for various wholesale quantities. Kawarada and Hachiya [2] reformulated it using the Martingale theory and the Black–Scholes model. In this formulation, the Black–Scholes model, which is widely used in financial engineering, is adopted. The opportunity and disposal losses correspond respectively to call and put options.

In this study, we present practical methods of finding an optimal wholesale quantity and evaluate it through numerical experiments using time series data in a retail shop. In Section 2, the Newsboy Problem is described. Section 3 presents an outline of the formulation of Kawarada and Hachiya. Section 4 describes validation data. Section 5 shows the verification method and a result obtained using real-world data. Section 6 explains our salient conclusions.

2 Outline of the Newsboy Problem

We respectively denote the wholesale price and profit margin of the commodity as α, and β, so the retail price of the commodity is

$$\gamma = (\alpha + \beta) \times (1 + r_c), \tag{1}$$

where r_c is a consumption tax rate.

Variables v and u respectively represent demand and wholesale quantities. In the case where a commodity remains unsold, i.e. $v \leq u$, a disposal loss of

$$l_d(v, u) = \alpha(u - v)_+ \quad (v, u \in Z_+) \tag{2}$$

is incurred, where

$$(x)_+ = \begin{cases} x, & x > 0, \\ 0, & x \leq 0, \end{cases} \tag{3}$$

and Z_+ represents the set of positive integers.

For a sold out condition, i.e. $v > u$, the firm realizes an opportunity loss of

$$l_o(v, u) = \beta(v - u)_+, \quad v, u \in Z_+. \tag{4}$$

Consequently, the total loss $l(v, u)$ is expressed as

$$l(v, u) = l_d(v, u) + l_o(v, u), \quad v, u \in Z_+. \tag{5}$$

Figure 1 portrays this relation.

We regard v as a random variable that represents a demand quantity distribution subject to a probability density function $f(v)$, $v > 0$. Therefore, we can define the expected value $E(l(\cdot, u))$ of loss function $l(v, u)$. An optimal wholesale value is defined as the wholesale value u^* that gives the minimum

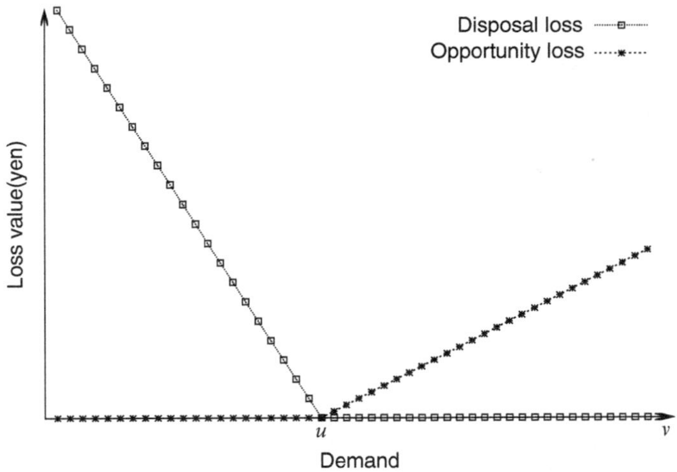

Fig. 1. Loss function.

of the expected loss. The demand quantity distribution $f(v)$ is assumed to be a continuous type. The expected value $L(u)$ of the loss is written as

$$L(u) = E(l(\cdot, u)) = \alpha \int_0^u (u - v)_+ f(v)dv + \beta \int_u^\infty (v - u)_+ f(v)dv. \quad (6)$$

The derivative of the equation (6) with respect to u is

$$\frac{dL}{du} = \alpha \int_0^u f(v)dv - \beta \int_u^\infty f(v)dv. \quad (7)$$

Then, by setting (7) to zero, the optimal wholesale quantity u^* is characterized as

$$\int_0^{u^*} f(v)dv = \frac{\beta}{\gamma/(1 + r_c)}. \quad (8)$$

3 Black–Scholes Models

Black and Scholes described an uncertain time-variation of stock value by Brownian motion [1]. They obtained an equation representing the valuation of option prices using the probability process theory. Ito's Lemma played an important role in this procedure.

The Black–Scholes equation for the value of the call option $c(t, S)$ is written as

$$\frac{\partial c}{\partial t} + rS\frac{\partial c}{\partial S} + \frac{1}{2}\sigma^2 S^2 \frac{\partial^2 c}{\partial S^2} - rc = 0, \quad 0 < S < \infty, \ 0 \leq t < T, \quad (9)$$

where t, S, r, T, and σ respectively signify time, the current price of a stock, the current continuously compounded risk-free interest rate, the end of a time period, and the standard deviation of the continuously compounded annual rate of return $P_r(t)$, which is defined as

$$P_r(t) = \frac{S_{(t+1)} - S_{(t)}}{S_{(t)}} = \frac{\Delta S_{(t)}}{S_{(t)}}. \tag{10}$$

Boundary and initial conditions are

$$\begin{aligned} c(t,0) &= 0, && 0 \le t < T, && (11)\\ c(t,S) &\to S, && S \to \infty,\ 0 \le t < T, && (12)\\ c(T,S) &= (S-X)_+, && 0 < S < \infty, && (13) \end{aligned}$$

where X indicates the strike price of the option. The equation (13) denotes the payoff of a European call option at $t = T$.

In the model, the call option price $c(t, S)$ depends on the volatility, which describes a random characteristic of the asset price. The volatility influences the distribution of the asset price at the end of a time period. Consequently, it influences the expected earnings from the option.

In fact, the equation (9) can be transformed to a heat equation through variable redefinitions. Then, using the solution of the heat equation, the solution of the equation (9) at time $t = 0$ is obtained as

$$c\,(0,S) = S \cdot N(d_1) - X \cdot e^{-rT} N(d_2), \tag{14}$$

where

$$N(x) = \frac{1}{\sqrt{2\pi}} \int_{-\infty}^{x} e^{-\frac{y^2}{2}}\, dy, \tag{15}$$

$$d_1 = \frac{\ln\left(\frac{S}{X}\right) + \left(r + \frac{\sigma^2}{2}\right) T}{\sigma\sqrt{T}}, \tag{16}$$

$$d_2 = \frac{\ln\left(\frac{S}{X}\right) + \left(r - \frac{\sigma^2}{2}\right) T}{\sigma\sqrt{T}}. \tag{17}$$

Here $N(d)$ is the standard normal cumulative distribution function.

Then, we set the current value of the call option $C = c(0, S)$ as a function of strike price X

$$C(X) = S \cdot N(d_1) - X \cdot e^{-rT} N(d_2). \tag{18}$$

On the other hand, the current value of a put option can be derived as

$$P(X) = -S \cdot N(-d_1) + X \cdot e^{-rT} N(-d_2), \tag{19}$$

similarly.

By combining the call (18) and put (19) options, we obtain the put-call parity formula as

$$C(X) = P(X) + S - X \cdot e^{-rT}. \tag{20}$$

By differentiating the equation (18) with respect to X, we obtain

$$\frac{dC}{dX} = -e^{-rT} N(d_2). \tag{21}$$

In [2], the option pricing theory is related to demand forecasting as follows:

- The payoff in option pricing theory is replaced by the loss function of the Newsboy problem. Then, the option prices are corresponding to the current price of the loss in demand forecasting.
- Strike prices X in option pricing correspond to optimal wholesale u^* in demand forecasting.

Hypotheses in the correspondence between the option pricing theory (OP) and demand forecasting (DF) are as follows:

(H1) OP: The option price follows the Ito process.
DF: The time series of demand quantity follows the Ito process.
(H2) OP: The option can be exercised only at the end of the period (European option).
DF: Profit margin and losses of the commodities are calculable at the pull date.
(H3) OP: It is possible to sell short.
DF: A deposit is charged before the reservation would be accepted.
(H4) OP: The dealing cost is not necessary.
DF: Extra costs other than the wholesale price are unnecessary.
(H5) OP: Riskless arbitrage does not exist.
DF: Disposal and opportunity losses always exist.
(H6) OP: Dealings are done continuously.
DF: Commodities are sold continuously.
(H7) OP: A risk-free interest rate is constant and equal at any expiration period.
DF: The rate of natural increase of demand is constant and the same in any forecasting period.

By letting opportunity and disposal losses correspond to call and put options, respectively, the expected value of the loss function can be represented as

$$E\left(l\left(\cdot, X\right)\right) = \alpha \cdot P(X) + \beta \cdot C(X). \tag{22}$$

By substituting (20) into (22), we obtain

$$E\left(l\left(\cdot, X\right)\right) = \gamma \cdot C(X) - \alpha \cdot S + \alpha \cdot X \cdot e^{-rT}, \tag{23}$$

where we regard r as the natural demand increasing rate corresponding to the risk-free interest rate in the option pricing theory. By differentiating the equation (23) with respect to X, setting it to zero, and substituting the equation (21) and the equation (1), we obtain

$$N\left(d_2\right) = \frac{\alpha}{\gamma/(1+r_c)}. \tag{24}$$

By transforming the equation (24) using the equations (15), (16) and (17), and replacing strike price X with optimal wholesale quantity u^*, we obtain

$$u^* = \exp\left\{\left(r - \frac{1}{2}\sigma^2\right)T - \sigma N^{-1}\left(\frac{\alpha}{\gamma/(1+r_c)}\right)\sqrt{T}\right\}S. \tag{25}$$

This is the optimal wholesale quantity, by which the loss at T is expected to be minimal.

4 Validation Data

4.1 Original Data

For this study, the quantity of wholesale and retail sales of rice balls and sandwiches, and the number of customers that passed the checkout counter at each time period were provided by the Okayama University Co-operative, which manages a retail shop located on the Okayama University campus. Details of the data are as follows:

(a) "Tuna and mayonnaise rice balls" are adopted among the various brands of rice balls and sandwiches because that commodity was sold throughout the target period duration.
(b) Shop hours are 8:00 to 23:00.
(c) The wholesale price of the commodity $\alpha = 70$ (JPY), the profit margin is $\beta = 30$ (JPY) and the retail price $\gamma = 105$ (JPY).
(d) The commodities remaining unsold are disposed at the end of shop hours.
(e) In case 1, we prepare the sequential dataset "Monday, Tuesday, Wednesday, Thursday, Friday, Monday, Tuesday, Wednesday, Thursday, Friday, ..., Monday, Tuesday, Wednesday, Thursday, Friday" excluding Saturdays, Sundays, and holidays.
In case 2, for comparison, we prepare five sequential datasets consisting of the same days of the week, "Monday, ..., Monday", "Tuesday, ..., Tuesday", "Wednesday, ... Wednesday", "Thursday, ..., Thursday", "Friday, ..., Friday", and excluding Saturdays, Sundays, and holidays.
(f) The target period was 30 weeks (147 days) during 21 May 2007–28 March 2008.

Table 1. Number of customers in each time period

	Time period (h)						
Date (d)	8	9	10	···	20	21	22
21/05/07	11	64	163	···	106	71	2
22/05/07	15	63	205	···	92	56	1
⋮	⋮	⋮	⋮	⋱	⋮	⋮	⋮
27/03/08	0	1	43	···	50	66	8
28/03/08	0	2	43	···	75	78	8

Table 2. Wholesale/retail quantities and final sales time

Date (d)	Retail sale (v_r)	Wholesale (u)	Final sales time (T_s)
21/05/07	25	25	21:09
22/05/07	20	20	12:21
⋮	⋮	⋮	⋮
27/03/08	26	26	17:26
28/03/08	28	30	17:20

(g) Table 1 presents an example of the number of customers $A(d, h)$ that passed the checkout counter during each time period of each day. Therein, d is the date and h is time period from 8 to 22.
Table 2 portrays an example of data for quantity of retail sales v_r, the quantity of wholesale u, and the final sales time T_s on each day.

(h) If a part of the commodity remains unsold, regard the quantity of sales v_r itself as the quantity of the demand v. If the commodity is sold out and the final sales time is earlier than the end of the shop hours, the amount of the demand v_d is estimated as

$$v_d = \frac{v_r \times \sum_{h=8}^{22} A(d,h)}{\sum_{h=8}^{T_z} A(d,h)}, \tag{26}$$

where T_z is the integer portion of T_s.
This procedure is used for inferring the likely quantity of the net demand on the day using the fraction of customers who have purchasing intentions for "Tuna and mayonnaise rice ball" among all customers.

4.2 Calculation and Loss Evaluation

We compare three techniques, the Black–Scholes model, simple linear extrapolation, and the decision by a real expert buyer who makes decisions of wholesale quantity every day at an actual retail shop.

Black–Scholes Model

Step 1. We construct the datasets of D days long $s(d,z)$, $1 \le d \le D$, $1 \le z \le Z$. Parameter D denotes the segment length. The total number of evaluation data is $Z = 147 - D$, where 147 is the total number of days of the data included in the target period of this study.
Step 2. Compute $P_r(s(d,\cdot),k)$ from (10) and volatility σ as follows:

$$\sigma(s(d,\cdot)) = \sqrt{\frac{1}{D-2}\sum_{k=1}^{D-2}\left(P_r\left(s\left(d,\cdot\right),k\right)\right)^2 - \left(\sum_{k=1}^{D-2} P_r\left(s\left(d,\cdot\right),k\right)\right)^2}. \tag{27}$$

Step 3. In the expression (25), $T = 1$ and the data of $(D-1)$th day is used as S. The forecast value is rounded off and the integer is taken.

Linear Extrapolation

Step 1. Compute a and b that minimize

$$\sum_{d=1}^{D-1} |v_r - (ad+b)|^2.$$

Step 2. Compute $v_D = aD+b$ using a and b obtained in Step 1. v_D is rounded off and the integer is taken.

Expert Buyer

The wholesale quantity u decided by the expert buyer working in the actual retail shop is used.

5 Numerical Experiments and Result

5.1 Aspects of the Techniques

In fact, Figure 2 portrays an example of the forecasted wholesale quantity using the three techniques to view their aspects. In this example, D is set at 6 for B-S model and linear extrapolation and r is set at 0.2 for B-S model. The days from the 47th to 56th are the sales campaign period, during which the price of a certain commodity is set at lower price for sales promotion.

It is apparent that the Black–Scholes model and linear extrapolation lag a few days after the change of the real demand and can not react to a rapid change in the demand at the sales campaign period. This is not so strange because the real expert buyer knows when the sales campaign starts. Furthermore, the forecasted value by B-S model drops down to zero because the volatility σ in (25) becomes very large. Such a sales campaign period is beyond

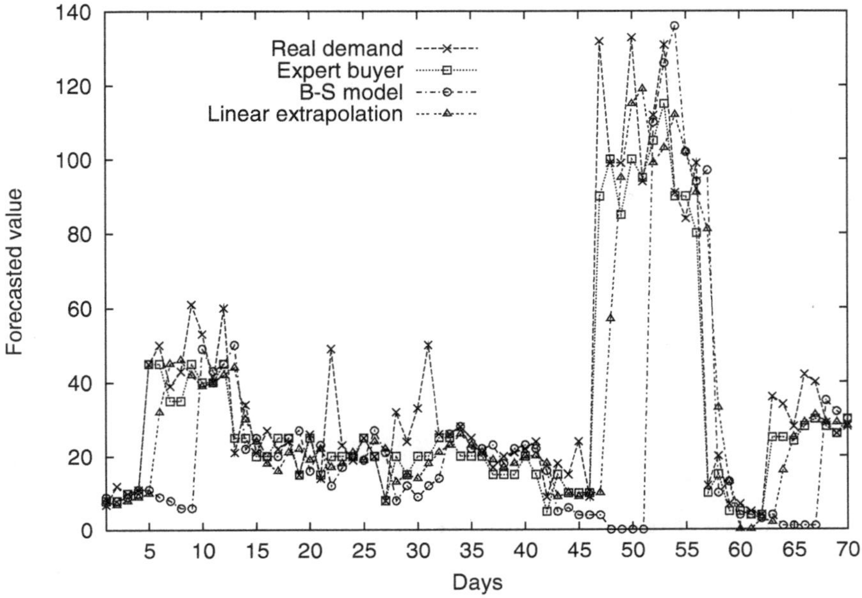

Fig. 2. Forecasted values.

the scope of the demand forecasting model examined in this study. Therefore, we exclude the sales campaign periods from our numerical experiments. Consequently, the total days of data become 119.

Respectively, Figures 3 and 4 show time histories of disposal and opportunity losses over an interval excluding sales campaign period. Results show that the occurrence frequency of disposal losses is lower than that of opportunity losses and that the expert buyer makes decisions that bring about disposal losses only rarely.

5.2 Selection of Parameters

Black–Scholes Model

In the Black–Scholes model, unknown parameters are the natural demand increase rate r and segment length D. To fix these parameters, an iterative procedure is adopted as follows:

Step 1. Provide segment length D between 5 and 99.

Step 2. For each D, the optimal r is sought by testing various values between 0.01 and 1.00. In this test, the demand on $(D-1)$th day is forecasted using the demand on $(D-2)$th day. Finally, the optimal r, which minimizes the loss on $(D-1)$th day, is obtained.

Step 3. Using r fixed at Step 2 and the real demand of the $(D-1)$th day, compute the loss $l(v, u)$ on the Dth day.

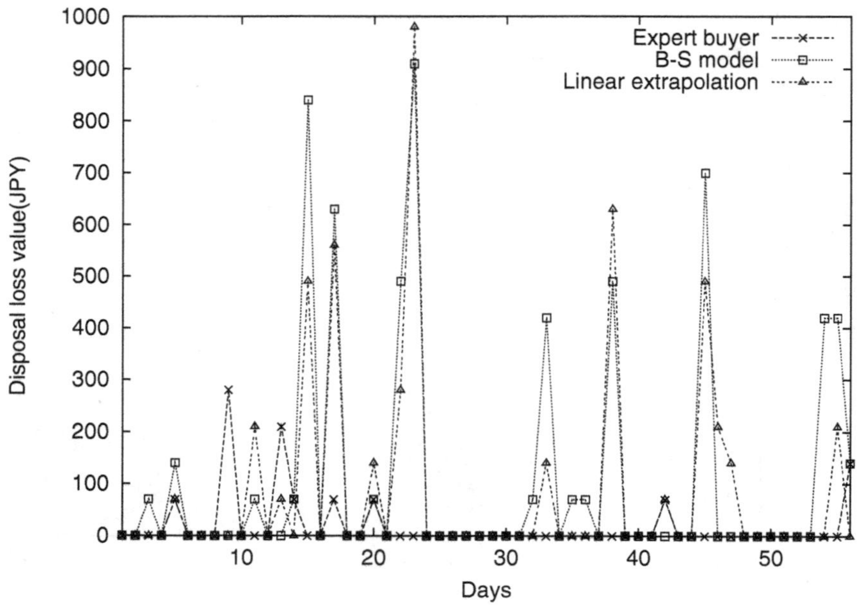

Fig. 3. Disposal loss $l_d(v, u)$.

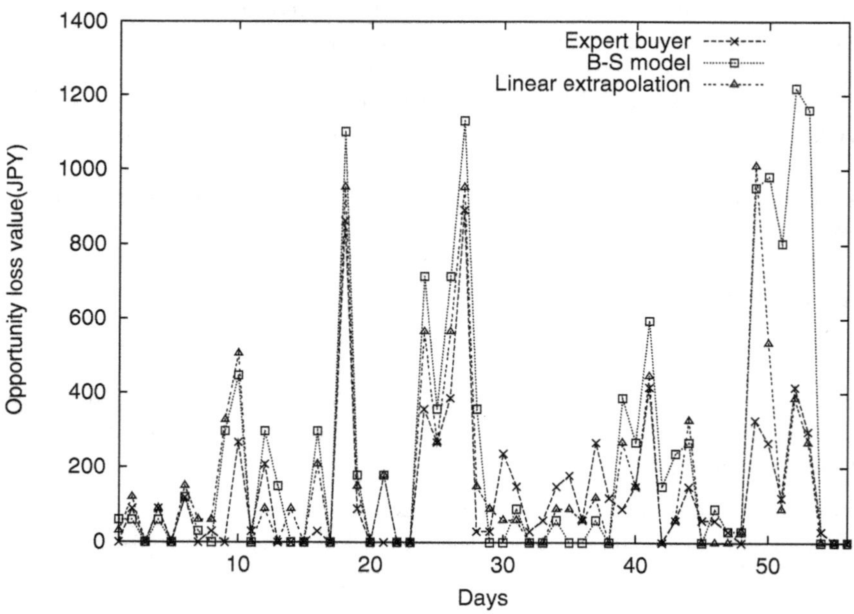

Fig. 4. Opportunity loss $l_o(v, u)$.

Step 4. Compute the averaged loss of the last m days of the total data, where m is the average number of business days in one month:

$$l_m = \frac{\sum_{z=Z-m+1}^{Z} l(v_z, u_z)}{m}. \tag{28}$$

Linear Extrapolation

In linear extrapolation, the parameter is segment length D only, which is provided same as Step 1 of the previous subsection. The average loss is also computed over the same m days.

5.3 Result of Case 1

Actually, Figure 5 shows the dependence of the total loss on segment length D in the case that all days through the weeks are used sequentially. Because the expert buyer does not possess the parameter of the segment length, the loss takes a constant value in this figure. The following have been read from Figure 5:

1. In linear extrapolation, the total loss $l(v, u)$ grows as D increases and approaches a constant value. In forecasting using linear extrapolation, smaller D seems to be appropriate to reflect tendency changes adequately.

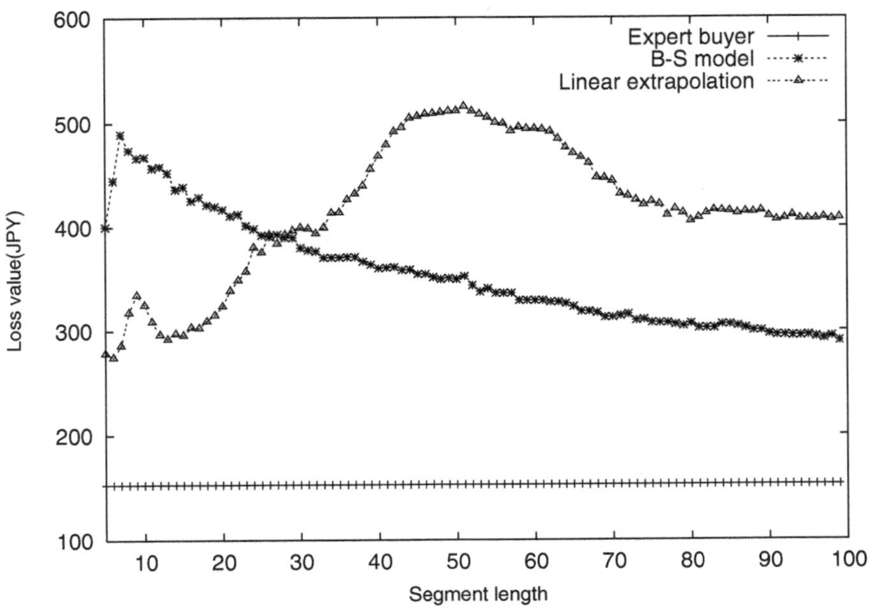

Fig. 5. Dependence on the segment length for the total loss.

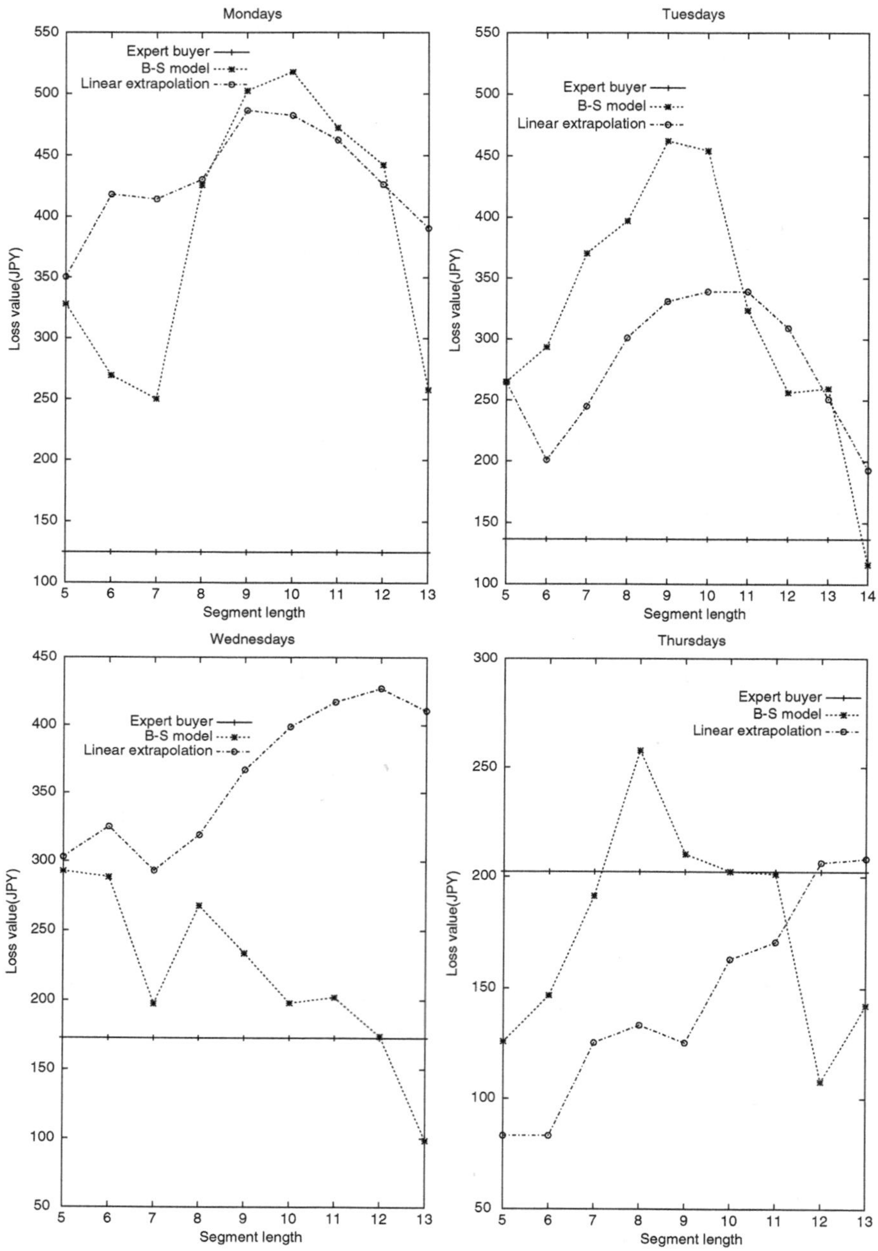

Fig. 6. Dependence of total loss on the segment length from monday to thursday.

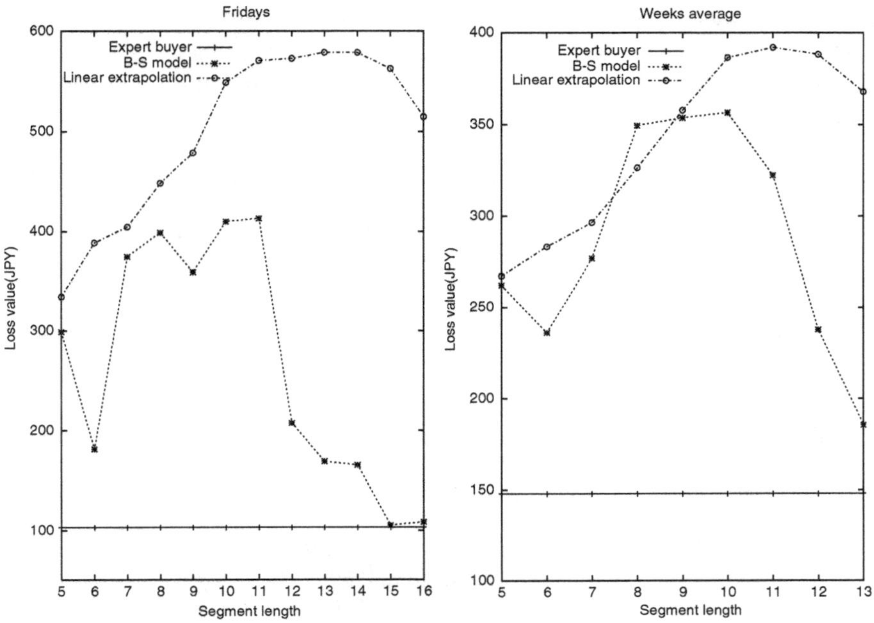

Fig. 7. Dependence of total loss on the segment length of friday and weeks average.

2. For the B-S model, the total loss $l(v, u)$ becomes smaller as D increases, which indicates that the B-S model, as a method based on stochastic processes, seems to need longer D to use more appropriate σ.

5.4 Result of Case 2

Figures 6 and 7 show the total losses obtained using the data divided into each day of the week. By this treatment, the performance of B-S model is considerably improved. For Tuesdays, Wednesdays, and Thursdays, it can do better than the expert buyer by taking the sufficient segment length. The reason might be that the target retail shop is located on a university campus and the trend of customers must be strongly dependent on the day of the week.

6 Conclusion

This study evaluated the performance of a demand forecasting strategy based on stochastic processes using real-world data. Results show that the B-S model proposed by Kawarada and Hachiya can reduce the total loss by taking the appropriate segment length. Unfortunately, regarding the data used for this study, the proposed methods were not always superior to the predictions by the expert buyer. The reason might be that the expert buyer incorporates much information into forecasts and tries to make the disposal loss as low as possible. Nevertheless, the methods described herein can be very useful in retail shops that have no expert buyers.

As a technical problem, the method of presuming the quantity of demand using final sales time data is expected to be improved in the future.

Acknowledgement. The authors thank the Okayama University Co-operative, which kindly provided sales data for their retail shop.

References

1. F. Black and M. Scholes. The pricing of options and corporate liabilities. *Journal of Political Economy*, 81(3):637–654, 1973.
2. H. Kawarada and H. Hachiya. Optimal stock control strategy based on stochastic process. *Journal of Logistics and Informatics*, 1(1):39–47, 2004.
3. S. G. Makridakis and S. C. Wheelwright. *Forecasting methods for management.* John Wiley & Sons, New York, 5th edition, 1989.
4. T. Masui, S. Yurimoto, and N. Katayama. *OR in Logistics.* Makisyoten, Tokyo, 1998.
5. S. Munakata and K. Saito. A new demand forecasting method for newly-launched consumer products based on estimated market parameters. *Hitachi Tohoku Software Technical Report*, 11:34–39, 2005.
6. F. Sekine. Supply chain management in Kao. *MH Journal*, 238(summer issue): 10–14, 2004.

Analytic Bounds for Diagonal Elements of Functions of Matrices

Gérard Meurant

30 rue de Sergent Bauchat, 75012 Paris, France, gerard.meurant@gmail.com

Summary. This paper is concerned with computing analytic lower and upper bounds for diagonal elements of $f(A)$ where A is a real symmetric matrix and f is a smooth function. The mathematical tools to be used are Riemann–Stieltjes integrals, orthogonal polynomials, Gauss quadrature and the Lanczos algorithm.

1 Introduction

In this paper we are concerned with computing analytic lower and upper bounds for diagonal elements of $f(A)$ where A is a real symmetric matrix and f is a smooth function. Typical examples are $f(x) = 1/x$ or $f(x) = \exp(x)$. This leads to compute bounds for $(e^i)^T f(A)e^i$ where e^i is the ith column of the identity matrix. This problem was considered in Golub and Meurant [3] for $f(A) = A^{-1}$. This idea developed in this paper is to write the quadratic form as a Riemann–Stieltjes integral and to approximate this integral by a 2-point Gauss or Gauss–Radau quadrature rule. When the derivatives of the function f have a constant sign over the interval of integration this will lead to lower and upper bounds for the quadratic form. The nodes and weights of the quadrature formulas are obtained by doing two iterations of the Lanczos algorithm. This can be done analytically and it provides formulas for these lower and upper bounds. Of course, except in very special cases, the bounds are not sharp since doing more Lanczos iterations (that is to say increasing the number of points of the quadrature rule) will improve the bounds.

The contents of the paper are the following. In Section 2 we review how the elements of $f(A)$ can be written as a Riemann–Stieltjes integral and how the nodes and weights of the Gauss quadrature rules are obtained. Section 3 describes how results were obtained in [3] for the inverse of the matrix A. Section 4 generalizes these results to the case of a general function f with an application to the exponential function.

W. Fitzgibbon et al. (eds.), *Applied and Numerical Partial Differential Equations*, Computational Methods in Applied Sciences 15, DOI 10.1007/978-90-481-3239-3_12,

2 Quadratic Forms and Gauss Quadrature Rules

Since the matrix A is assumed to be symmetric we have $A = Q\Lambda Q^T$ where Q is the orthonormal matrix whose columns are the normalized eigenvectors of A and Λ is a diagonal matrix whose diagonal elements are the eigenvalues λ_i of A which we order as

$$\lambda_1 \leq \lambda_2 \leq \cdots \leq \lambda_n.$$

By definition of a function of a symmetric matrix, we have

$$f(A) = Qf(\Lambda)Q^T.$$

Therefore, if u is a given vector, we have

$$\begin{aligned} u^T f(A)u &= u^T Q f(\Lambda) Q^T u, \\ &= \beta^T f(\Lambda)\beta, \\ &= \sum_{i=1}^{n} f(\lambda_i)\beta_i^2. \end{aligned}$$

This last sum can be considered as a Riemann–Stieltjes integral,

$$I[f] = u^T f(A)u = \int_a^b f(\lambda)\, d\alpha(\lambda), \tag{1}$$

where the measure α is piecewise constant and defined by

$$\alpha(\lambda) = \begin{cases} 0 & \text{if } \lambda < a = \lambda_1, \\ \sum_{j=1}^{i} \beta_j^2 & \text{if } \lambda_i \leq \lambda < \lambda_{i+1}, \\ \sum_{j=1}^{n} \beta_j^2 & \text{if } b = \lambda_n \leq \lambda. \end{cases}$$

We remark that α is an increasing positive function. The integral in the equation (1) can be approximated by Gauss quadrature rules. A general rule is written as

$$I[f] = \int_a^b f(\lambda)\, d\alpha(\lambda) = \sum_{j=1}^{N} w_j f(t_j) + \sum_{k=1}^{M} v_k f(z_k) + R[f], \tag{2}$$

where the weights $[w_j]_{j=1}^N$, $[v_k]_{k=1}^M$ and the nodes $[t_j]_{j=1}^N$ are unknowns and the nodes $[z_k]_{k=1}^M$ are prescribed, see [1,2,5]. We are interested in the case where we have two nodes. That is the 2-point Gauss rule (with no prescribed node) for which $N = 2$ and $M = 0$ and the Gauss–Radau rule (with one prescribed node) for which $N = 1$, $M = 1$. The prescribed node is either a or b, the ends of the integration interval.

The remainder term is written as

$$R[f] = \frac{f^{(2N+M)}(\eta)}{(2N+M)!} \int_a^b \prod_{k=1}^{M} (\lambda - z_k) \left[\prod_{j=1}^{N} (\lambda - t_j) \right]^2 d\alpha(\lambda), \quad a < \eta < b. \quad (3)$$

Therefore, if the sign of the derivative is constant on the integration interval, we know the sign of the remainder.

How are the nodes and weights obtained? The nodes are the eigenvalues of the tridiagonal matrix J constructed with the coefficients of the three-term recurrence relation satisfied by the orthogonal polynomial associated with the measure α. The weights are the squares of the first elements of the normalized eigenvectors, see [5]. It can be also shown (see [3]) that

$$I[f] = (e^1)^T f(J) e^1 + R[f].$$

The matrix J is obtained by running the Lanczos algorithm (see, for instance, [4]) with the matrix A and a starting vector $u/\|u\|$. In the cases we are interested in J is a 2×2 matrix. The last diagonal element of this matrix has to be modified for the Gauss–Radau rule to obtain the prescribed eigenvalue.

3 Analytic Bounds for the Diagonal Elements of the Inverse

We consider obtaining analytical bounds for the entries of the inverse of A by doing two iterations of the Lanczos algorithm. This is obtained from the general framework of Section 2 by considering the function

$$f(\lambda) = \frac{1}{\lambda}, \quad 0 < a \le \lambda \le b,$$

for which the derivatives are

$$f^{(2k+1)}(\lambda) = -(2k+1)!\, \lambda^{-(2k+2)}, \quad f^{(2k)}(\lambda) = (2k)!\, \lambda^{-(2k+1)}.$$

Therefore, the even derivatives are positive on $[a, b]$ when $a > 0$ and the odd derivatives are negative which implies that the Gauss rule gives a lower bound and the Gauss–Radau rule gives lower and upper bounds depending on the prescribed node. We obtain the following bounds, see [3].

Theorem 1. *Let A be a symmetric positive definite matrix with elements $a_{i,j}$. Let*

$$s_i^2 = \sum_{j \ne i} a_{ji}^2, \quad i = 1, \dots, n.$$

We have the following bounds for the diagonal entries of the inverse given respectively by the Gauss and Gauss–Radau quadrature rules:

$$\frac{\sum_{k\neq i}\sum_{l\neq i} a_{k,i}a_{k,l}a_{l,i}}{a_{i,i}\sum_{k\neq i}\sum_{l\neq i} a_{k,i}a_{k,l}a_{l,i} - \left(\sum_{k\neq i} a_{k,i}^2\right)^2} \leq (A^{-1})_{i,i},$$

$$\frac{a_{i,i} - b + \frac{s_i^2}{b}}{a_{i,i}^2 - a_{i,i}b + s_i^2} \leq (A^{-1})_{i,i} \leq \frac{a_{i,i} - a + \frac{s_i^2}{a}}{a_{i,i}^2 - a_{i,i}a + s_i^2}.$$

Proof. We choose the initial vector $v^1 = e^i$ and we apply the Lanczos algorithm computing the elements of the symmetric tridiagonal matrix J and the Lanczos vectors v^i. The first step of the Lanczos algorithm (see [4]) gives

$$\alpha_1 = (e^i)^T A e^i = a_{ii},$$
$$\eta_1 v^2 = (A - \alpha_1 I)e^i.$$

Let s_i be defined by

$$s_i^2 = \sum_{j\neq i} a_{ji}^2,$$

and

$$d^i = (a_{1,i}, \ldots, a_{i-1,i}, 0, a_{i+1,i}, \ldots, a_{n,i})^T.$$

Then

$$\eta_1 = s_i, \quad v^2 = \frac{1}{s_i} d^i.$$

From this, we have

$$\alpha_2 = (Av^2, v^2) = \frac{1}{s_i^2}\sum_{k\neq i}\sum_{l\neq i} a_{k,i}a_{k,l}a_{l,i}.$$

We can now compute the Gauss rule and obtain a lower bound on the diagonal element by considering the matrix

$$J_2 = \begin{pmatrix} \alpha_1 & \eta_1 \\ \eta_1 & \alpha_2 \end{pmatrix},$$

and its inverse

$$J_2^{-1} = \frac{1}{\alpha_1\alpha_2 - \eta_1^2}\begin{pmatrix} \alpha_2 & -\eta_1 \\ -\eta_1 & \alpha_1 \end{pmatrix}.$$

The lower bound is given by $(e^1)^T J_2^{-1} e^1$, the $(1,1)$ entry of the inverse

$$(e^1)^T J_2^{-1} e^1 = \frac{\alpha_2}{\alpha_1\alpha_2 - \eta_1^2} = \frac{\sum_{k\neq i}\sum_{l\neq i} a_{k,i}a_{k,l}a_{l,i}}{a_{i,i}\sum_{k\neq i}\sum_{l\neq i} a_{k,i}a_{k,l}a_{l,i} - \left(\sum_{k\neq i} a_{k,i}^2\right)^2}.$$

Note that this bound does not depend on the extreme eigenvalues of A. To obtain an upper bound we consider the Gauss–Radau rule. Then, we have to modify the $(2,2)$ element of the Lanczos matrix as

$$\tilde{J}_2 = \begin{pmatrix} \alpha_1 & \eta_1 \\ \eta_1 & \xi \end{pmatrix},$$

to obtain the prescribed node. The eigenvalues λ of $\tilde{J}_2$ are the roots of $(\alpha_1 - \lambda)(\xi - \lambda) - \eta_1^2 = 0$, which gives the relation

$$\xi = \lambda + \frac{\eta_1^2}{\alpha_1 - \lambda}.$$

To obtain an upper bound, we impose to have an eigenvalue equal to the lower bound of the eigenvalues of A, $\lambda = a$. The solution is

$$\xi = \xi_a = a + \frac{\eta_1^2}{\alpha_1 - a},$$

from which we can compute the $(1,1)$ element of the inverse of $\tilde{J}_2$,

$$(e^1)^T \tilde{J}_2^{-1} e^1 = \frac{\xi}{\alpha_1 \xi - \eta_1^2}.$$

Using b as a prescribed node gives a lower bound.

Of course, as we said before these bounds are not sharp since they can be improved by doing more Lanczos iterations, except if the Lanczos algorithm converges in two iterations. Using more iterations can eventually be done analytically by using a symbolic calculation software.

4 Analytic Bounds for Elements of Other Functions

If we would like to obtain analytical bounds of diagonal elements for other functions, we see from the derivation of Section 3 for the inverse that all we have to do is to compute $f(J)$ (and, in fact, only the (1,1) element) for a symmetric matrix J of order 2 whose coefficients are known or the eigenvalues and eigenvectors if $f(J)$ is not available. Let

$$J = \begin{pmatrix} \alpha & \eta \\ \eta & \xi \end{pmatrix}.$$

If we are interested in the exponential function we have to compute $\exp(J)$. This can be done using a symbolic mathematics package. For instance, in the Matlab symbolic toolbox (from MathWorks) there is a function giving the exponential of a symbolic matrix. The result is the following.

Theorem 2. *Let* $\alpha = a_{i,i}$, $\eta = s_i$ *using the notations of the previous section. The element* ξ *is either* α_2 *or* ξ_a *(or* ξ_b*) and let*

$$\delta = (\alpha - \xi)^2 + 4\eta^2,$$
$$\gamma = \exp\left(\frac{1}{2}(\alpha + \xi - \sqrt{\delta})\right),$$
$$\omega = \exp\left(\frac{1}{2}(\alpha + \xi + \sqrt{\delta})\right).$$

Then, the $(1,1)$ *element of the exponential of* J *is*

$$\frac{1}{2}\left[\gamma + \omega + \frac{\omega - \gamma}{\sqrt{\delta}}(\alpha - \xi)\right].$$

Although these expressions are quite complicated, if we substitute the values of the parameters as functions of the elements of A we obtain analytically a lower bound of $[\exp(A)]_{i,i}$ from the Gauss rule and an upper bound (with ξ_a) from the Gauss–Radau rule.

For other functions which are not available in symbolic packages we can compute analytically the eigenvalues and eigenvectors of J. In fact, we just need the first components of the eigenvectors. The eigenvalues are

$$\lambda_+ = \frac{1}{2}(\alpha + \xi + \sqrt{\delta}), \quad \lambda_- = \frac{1}{2}(\alpha + \xi - \sqrt{\delta}).$$

The matrix of the unnormalized eigenvectors is

$$Q = \begin{pmatrix} \theta & \mu \\ 1 & 1 \end{pmatrix},$$

where

$$\theta = -\frac{1}{2\eta}(\alpha - \xi + \sqrt{\delta}), \quad \mu = -\frac{1}{2\eta}(\alpha - \xi - \sqrt{\delta}).$$

The first components of the normalized eigenvectors are $\theta/\sqrt{1+\theta^2}$ and $\mu/\sqrt{1+\mu^2}$. Then we have to compute the $(1,1)$ element of $\tilde{Q}f(\Lambda)\tilde{Q}^T$ where Λ is the diagonal matrix of the eigenvalues λ_+ and λ_- and $\tilde{Q}$ is the matrix of the normalized eigenvectors. We need the values $\theta^2/(1+\theta^2)$ and $\mu^2/(1+\mu^2)$.

Lemma 1. *We have*

$$\frac{\theta^2}{1+\theta^2} = \frac{\alpha - \xi + \sqrt{\delta}}{2\sqrt{\delta}}, \quad \frac{\mu^2}{1+\mu^2} = -\frac{\alpha - \xi - \sqrt{\delta}}{2\sqrt{\delta}}.$$

From this lemma we obtain the $(1,1)$ element of $f(J)$.

Theorem 3. *Using the notations of Theorem 2, the* $(1,1)$ *element of* $f(J)$ *is*

$$\frac{1}{2\sqrt{\delta}}\left[(\alpha - \xi)(f(\lambda_+) - f(\lambda_-)) + \sqrt{\delta}(f(\lambda_+) + f(\lambda_-))\right].$$

Proof. Clearly the $(1, 1)$ element is

$$\frac{\theta^2}{1+\theta^2} f(\lambda_+) + \frac{\mu^2}{1+\mu^2} f(\lambda_-).$$

Using the expressions of Lemma 1 and simplifying, we obtain the result.

We see that if f is the exponential function we recover the results of Theorem 2. From the last theorem we can obtain analytic bounds for the (i, i) element of $f(A)$ for any function for which the integral exists and for which we can compute $f(\lambda_+)$ and $f(\lambda_-)$. Wether we obtain lower and upper bounds depend on the sign of the third and fourth derivatives of f.

Bounds for off diagonal elements of $f(A)$ can be obtained by using the nonsymmetric Lanczos algorithm, see [3].

References

1. P. J. Davis and P. Rabinowitz. *Methods of numerical integration.* Academic Press, Orlando, FL, second edition, 1984.
2. W. Gautschi. *Orthogonal polynomials: computation and approximation.* Oxford University Press, New York, 2004.
3. G. H. Golub and G. Meurant. Matrices, moments and quadrature. In D. F. Griffiths and G. A. Watson, editors, *Numerical Analysis 1993 (Dundee, 1993)*, volume 303 of *Pitman Res. Notes Math. Ser.*, pages 105–156, Harlow, 1994. Longman Sci. Tech.
4. G. H. Golub and C. F. Van Loan. *Matrix computations.* Johns Hopkins University Press, Baltimore, MD, third edition, 1996.
5. G. H. Golub and J. H. Welsch. Calculation of Gauss quadrature rules. *Math. Comp.*, 23:221–230, 1969.

Numerical Methods for Ferromagnetic Plates

Michel Flück, Thomas Hofer, Ales Janka, and Jacques Rappaz

École Polytechnique Fédérale de Lausanne, CH-1015 Lausanne, Switzerland,
michel.flueck@epfl.ch, t.hofer@epfl.ch, jacques.rappaz@epfl.ch

1 Introduction

We present two numerical methods for the simulation of ferromagnetic phenomenons in a metallic plate, with or without holes. First we briefly recall the physical model we use for describing the ferromagnetic phenomenon. This model is based on the use of a scalar potential while other models rather use a vector potential as in [1] or [2]. Next we present the discretization methods we use. We then apply these methods on the simple test-case of a thin ferromagnetic plate placed in front of a rectilineal electric conductor. We show the various obtained results: magnetic field on a line perpendicular to the plate and relative permeability on a given line in the plate. Finally we illustrate our results with an industrial device.

2 Modeling of Ferromagnetism

Let $\Lambda \subset \mathbb{R}^3$ be a domain with boundary $\partial\Lambda$ occupied by a ferromagnetic material with relative magnetic permeability $\mu_r(\|\mathbf{H}\|) \geq 1$ depending on the Euclidean norm of the magnetic field $\mathbf{H}$, denoted by $\|\mathbf{H}\|$. In the following, we suppose that Λ is a bounded open, possibly non simply connected set, surrounded by known stationary electric currents denoted by $\mathbf{j}_0$. We denote by $\mathbf{n}$ the unit normal on $\partial\Lambda$, external to Λ. Moreover, we assume that all the external currents are not modified by the presence of the ferromagnetic material and no electric current flows in the domain Λ. The goal of this paragraph is to establish a modeling of the screen effect due to the presence of Λ on the magnetic fields.

Without the ferromagnetic material, it is possible to explicit the magnetic induction field $\mathbf{B}_0$ due to $\mathbf{j}_0$ by using Biot–Savart law:

$$\mathbf{B}_0(\mathbf{x}) = \mu_0 \int_{\mathbb{R}^3} \boldsymbol{\nabla}_x G(\mathbf{x},\mathbf{y}) \wedge \mathbf{j}_0(\mathbf{y})\, dy, \quad \forall \mathbf{x} \in \mathbb{R}^3, \tag{1}$$

W. Fitzgibbon et al. (eds.), *Applied and Numerical Partial Differential Equations*, Computational Methods in Applied Sciences 15, DOI 10.1007/978-90-481-3239-3_13,

where μ_0 is the magnetic permeability of the void, $G(\mathbf{x}, \mathbf{y})$ is the Green kernel given by

$$G(\mathbf{x}, \mathbf{y}) = \frac{1}{4\pi} \frac{1}{\|\mathbf{x} - \mathbf{y}\|} \quad \text{with } \mathbf{x}, \mathbf{y} \in \mathbb{R}^3, \mathbf{x} \neq \mathbf{y}, \tag{2}$$

and ∇_x denotes the gradient with respect to the variable $\mathbf{x}$.

Let us remark that if $\mathbf{H}_0$ is the magnetic field corresponding to $\mathbf{B}_0$, we have in the whole space $\mathbb{R}^3$ without ferromagnetic materials

$$\mathbf{B}_0 = \mu_0 \mathbf{H}_0, \tag{3}$$

$$\operatorname{div} \mathbf{B}_0 = 0, \tag{4}$$

$$\operatorname{curl} \mathbf{H}_0 = \mathbf{j}_0. \tag{5}$$

Due to the presence of the ferromagnetic domain Λ, the magnetic field $\mathbf{H}$ and the induction field $\mathbf{B}$ cannot be explicitly given in function of $\mathbf{j}_0$, but they are governed by the following relationships (the Maxwell equations), true in the whole space $\mathbb{R}^3$:

$$\mathbf{B} = \mu_0 \mu_r \mathbf{H}, \tag{6}$$

$$\operatorname{div} \mathbf{B} = 0, \tag{7}$$

$$\operatorname{curl} \mathbf{H} = \mathbf{j}_0. \tag{8}$$

We note that outside the domain Λ we have $\mu_r = 1$. Since the magnetization field $\mathbf{M}$ is defined by $\mathbf{M} = \mu_0 (\mu_r - 1) \mathbf{H}$, we will be able to compute $\mathbf{M}$ if we are able to calculate $\mathbf{H}$. In the following, we are looking for the field $\mathbf{H}$.

2.1 A Scalar Potential Model

By subtracting (5) and (8) we obtain the existence of a continuous function ψ satisfying

$$\mathbf{H}(\mathbf{x}) - \mathbf{H}_0(\mathbf{x}) = -\nabla \psi(\mathbf{x}) \quad \forall \mathbf{x} \in \mathbb{R}^3. \tag{9}$$

By using the equalities (3), (4) and (6), (7) together with (9), we easily verify that

$$-\operatorname{div}(\mu_r \nabla \psi) = -\operatorname{div}(\mu_r - 1) \mathbf{H}_0 \quad \text{in } \mathbb{R}^3. \tag{10}$$

In order to obtain a finite energy, we assume that

$$\psi(\mathbf{x}) = \mathcal{O}\left(\frac{1}{\|\mathbf{x}\|}\right) \quad \text{when } \|\mathbf{x}\| \text{ tends to infinity.} \tag{11}$$

Let Λ' be the exterior open domain $\Lambda' = \mathbb{R}^3 \setminus \overline{\Lambda}$. Since $\mu_r = 1$ in Λ', we obtain

$$\Delta \psi = 0 \quad \text{in } \Lambda', \tag{12}$$

where Δ is the Laplace operator.

In fact, the equation (10) is non-linear and it is necessary to precise what is μ_r, which is a discontinuous function since $\mu_r = 1$ in Λ' and $\mu_r = \mu_r(\|\mathbf{H}\|)$ in Λ. In order to write correctly the model, we define the mapping $\bar{\mu} : \mathbb{R}^3 \times \mathbb{R}^+ \to \mathbb{R}^+$ by

$$\bar{\mu}(\mathbf{x}, s) = \begin{cases} 1 & \text{if } \mathbf{x} \in \Lambda', \ s \in \mathbb{R}^+, \\ \mu_r(s) & \text{if } \mathbf{x} \in \Lambda, \ s \in \mathbb{R}^+, \end{cases} \tag{13}$$

where $\mu_r(s)$ is the relative magnetic permeability of the ferromagnetic material occupying Λ given in function of $s = \|\mathbf{H}\|$. Since $\mathbf{H} = \mathbf{H}_0 - \boldsymbol{\nabla}\psi$, it follows that the model consists to find $\psi : \mathbb{R}^3 \to \mathbb{R}$ satisfying

$$-\operatorname{div}(\bar{\mu}(\cdot, \|\mathbf{H}_0 - \boldsymbol{\nabla}\psi\|)\boldsymbol{\nabla}\psi) = -\operatorname{div}(\bar{\mu}(\cdot, \|\mathbf{H}_0 - \boldsymbol{\nabla}\psi\|) - 1)\mathbf{H}_0 \quad \text{in } \mathbb{R}^3, \tag{14}$$

with

$$\psi(\mathbf{x}) = \mathcal{O}\left(\frac{1}{\|\mathbf{x}\|}\right) \quad \text{when } \|\mathbf{x}\| \to \infty. \tag{15}$$

In order to simplify the notation, we will leave out in the following the argument of the mapping $\bar{\mu}$, knowing that it depends on $\mathbf{x} \in \mathbb{R}^3$ and $\|\mathbf{H}_0 - \boldsymbol{\nabla}\psi\|$, in order to write

$$\begin{cases} -\operatorname{div}(\bar{\mu}\boldsymbol{\nabla}\psi) = -\operatorname{div}(\bar{\mu} - 1)\,\mathbf{H}_0 & \text{in } \mathbb{R}^3, \\ \psi(\mathbf{x}) = \mathcal{O}\left(\frac{1}{\|\mathbf{x}\|}\right) & \text{when } \|\mathbf{x}\| \to \infty. \end{cases} \tag{16}$$

Remark that $\mathbf{H}_0$ need not be known in $\mathbb{R}^3$ but only on Λ because $\bar{\mu} = 1$ outside Λ.

3 Two Formulations of the Scalar Potential Problem

Let us now focus on two weak formulations of this scalar potential model. The main difficulty with the problem (16) is that we seek a function ψ defined in the whole space $\mathbb{R}^3$.

We will use two different ways to overcome this problem: the first one uses an integral formulation on $\partial\Lambda$ to replace the so-called "exterior" problem by a relation valid on the boundary of Λ; the numerical approximation leads in practice to a big non sparse matrix to "invert".

The other way uses a Schwarz decomposition method with overlapping technique. By introducing a ball containing the ferromagnetic object Λ, we solve the problem exterior to that ball by means of the Poisson representation formula.

We have seen that the scalar potential model leads to find a mapping ψ satisfying (16). Since $\operatorname{div}\mathbf{H}_0 = 0$ in $\mathbb{R}^3$, we can write this problem in the form

$$\operatorname{div}(\bar{\mu}(\mathbf{H}_0 - \boldsymbol{\nabla}\psi)) = 0 \quad \text{in } \mathbb{R}^3 \tag{17}$$

with $\psi(\mathbf{x}) = \mathcal{O}(\frac{1}{\|\mathbf{x}\|})$ when $\|\mathbf{x}\| \to \infty$. If $W^1(\mathbb{R}^3)$ is the Beppo Levi space given by

$$W^1(\mathbb{R}^3) = \left\{ v : \mathbb{R}^3 \to \mathbb{R} : \frac{v(\mathbf{x})}{1 + \|\mathbf{x}\|} \in L^2(\mathbb{R}^3), \boldsymbol{\nabla} v \in L^2(\mathbb{R}^3) \right\}, \tag{18}$$

it is proven in [8] that there exists a unique $\psi \in W^1(\mathbb{R}^3)$ satisfying

$$\int_{\mathbb{R}^3} \bar{\mu} \left(\mathbf{H}_0 - \boldsymbol{\nabla}\psi \right) \cdot \boldsymbol{\nabla}\varphi dx = 0, \quad \forall \varphi \in W^1(\mathbb{R}^3). \tag{19}$$

It follows that the problem (16) possesses a unique weak solution $\psi \in W^1(\mathbb{R}^3)$.

We now present two different approaches to compute the scalar potential ψ.

3.1 Boundary Integral Formulation of the Scalar Potential Model

It is known [7] that if v is a harmonic function in Λ and in Λ' satisfying $v(\mathbf{x}) = \mathcal{O}(\|\mathbf{x}\|^{-1})$ when $\|\mathbf{x}\| \to \infty$, and sufficiently regular (say C^1 in $\overline{\Lambda}$ and in $\overline{\Lambda'}$), then we have for $\mathbf{x} \in \partial\Lambda$:

$$\frac{1}{2}(v^E(\mathbf{x}) + v^I(\mathbf{x})) = \\ - \int_{\partial\Lambda} \left\lfloor \frac{\partial v}{\partial \mathbf{n}}(\mathbf{y}) \right\rfloor_{\partial\Lambda} G(\mathbf{x}, \mathbf{y})\, ds(\mathbf{y}) + \int_{\partial\Lambda} \lfloor v(\mathbf{y}) \rfloor_{\partial\Lambda} \frac{\partial G(\mathbf{x}, \mathbf{y})}{\partial \mathbf{n}(\mathbf{y})}\, ds(\mathbf{y}), \tag{20}$$

where v^E is the restriction of v to Λ', v^I is the restriction of v to Λ and $\lfloor v \rfloor_{\partial\Lambda} = v^E - v^I$ is the jump of v through the boundary $\partial\Lambda$ of Λ.

If ψ is the solution of (17), let w be a harmonic function in $\Lambda \cup \Lambda'$ satisfying $w = \psi$ on $\partial\Lambda$, $w(\mathbf{x}) = \mathcal{O}\left(\|\mathbf{x}\|^{-1}\right)$ when $\|\mathbf{x}\| \to \infty$. Clearly, because ψ is harmonic in Λ', we have $w = \psi$ in $\overline{\Lambda}'$. Moreover, by using the relationship (20) with $v = w$ in Λ and $v = \psi$ in Λ', we obtain for $x \in \partial\Lambda$

$$\psi(\mathbf{x}) = - \int_{\partial\Lambda} \left(\frac{\partial \psi^E}{\partial \mathbf{n}}(\mathbf{y}) - \frac{\partial w^I}{\partial \mathbf{n}}(\mathbf{y}) \right) G(\mathbf{x}, \mathbf{y})\, ds(\mathbf{y}), \tag{21}$$

which is equivalent to

$$\int_{\partial\Lambda} \psi\eta\, ds = - \int_{\partial\Lambda} \eta(\mathbf{x})\, ds(\mathbf{x}) \int_{\partial\Lambda} \left(\frac{\partial \psi^E}{\partial \mathbf{n}}(\mathbf{y}) - \frac{\partial w^I}{\partial \mathbf{n}}(\mathbf{y}) \right) G(\mathbf{x}, \mathbf{y})\, ds(\mathbf{y}), \tag{22}$$

for all $\eta \in H^{-1/2}(\partial\Lambda)$.[1]

By using the fact that $\operatorname{div} \mathbf{H}_0 = 0$ in $\mathbb{R}^3$, we have

$$\int_{\mathbb{R}^3} \mathbf{H}_0 \cdot \boldsymbol{\nabla}\varphi\, dx = 0, \quad \forall \varphi \in W^1(\mathbb{R}^3), \tag{23}$$

[1] We note by $H^1(\Lambda)$ the classical Sobolev space of order 1, $H^{1/2}(\partial\Lambda)$ the space of traces on $\partial\Lambda$ of mappings belonging to $H^1(\Lambda)$ and $H^{-1/2}(\partial\Lambda)$ its dual space.

and with the weak formulation (19):

$$\int_{\mathbb{R}^3} \bar{\mu} \nabla \psi \cdot \nabla \varphi \, dx = \int_{\mathbb{R}^3} (\bar{\mu} - 1) \mathbf{H}_0 \cdot \nabla \varphi \, dx, \quad \forall \varphi \in W^1(\mathbb{R}^3). \tag{24}$$

Since $\bar{\mu} = 1$ outside Λ we obtain for all $\varphi \in W^1(\mathbb{R}^3)$:

$$\int_{\Lambda} \bar{\mu} \nabla \psi \cdot \nabla \varphi \, dx + \int_{\Lambda'} \nabla \psi \cdot \nabla \varphi \, dx = \int_{\Lambda} (\bar{\mu} - 1) \mathbf{H}_0 \cdot \nabla \varphi \, dx, \tag{25}$$

and, by integrating by parts ($\mathbf{n}$ is pointing inside Λ' and $\Delta \psi = 0$ in Λ'), for all $\varphi \in W^1(\mathbb{R}^3)$:

$$\int_{\Lambda} \bar{\mu} \nabla \psi \cdot \nabla \varphi \, dx - \int_{\partial \Lambda'} \frac{\partial \psi^E}{\partial \mathbf{n}} \varphi \, ds = \int_{\Lambda} (\bar{\mu} - 1) \mathbf{H}_0 \cdot \nabla \varphi \, dx. \tag{26}$$

Since w is harmonic in Λ, we have $w \in H^1(\Lambda)$ satisfying $w = \psi$ on $\partial \Lambda$ and $\int_{\Lambda} \nabla w \cdot \nabla v \, dx = 0$, for all $v \in H_0^1(\Lambda)$.

By using the definition (13) of $\bar{\mu}$, we can replace $\bar{\mu}$ in (26) by $\bar{\mu} = \mu_r(\|\mathbf{H}_0 - \nabla \psi\|)$ and, by setting $\lambda = \frac{\partial \psi^E}{\partial \mathbf{n}}$ (external Steklov–Poincaré operator) in (22), (26), we obtain the nonlinear problem:

Find $\psi \in H^1(\Lambda)$, $w \in H^1(\Lambda)$ and $\lambda \in H^{-\frac{1}{2}}(\partial \Lambda)$ satisfying $w = \psi$ on $\partial \Lambda$ and for all $\varphi \in H^1(\Lambda)$, $v \in H_0^1(\Lambda)$, $\eta \in H^{-\frac{1}{2}}(\partial \Lambda)$:

$$\int_{\Lambda} \mu_r \nabla \psi \cdot \nabla \varphi \, dx - \int_{\partial \Lambda} \lambda \varphi \, ds = \int_{\Lambda} (\mu_r - 1) \, \mathbf{H}_0 \cdot \nabla \varphi \, dx, \tag{27}$$

$$\int_{\Lambda} \nabla w \cdot \nabla v \, dx = 0, \tag{28}$$

$$\int_{\partial \Lambda} \psi \eta \, ds = - \int_{\partial \Lambda} \eta(\mathbf{x}) \, ds(\mathbf{x}) \int_{\partial \Lambda} \left(\lambda(\mathbf{y}) - \frac{\partial w}{\partial \mathbf{n}}(\mathbf{y}) \right) G(\mathbf{x}, \mathbf{y}) \, ds(\mathbf{y}); \tag{29}$$

here $\mu_r = \mu_r(\|\mathbf{H}_0 - \nabla \psi\|)$.

3.2 A Domain Decomposition Formulation for the Scalar Potential Problem

Let us now introduce another way to reduce the "exterior" problem to a problem expressed in a bounded domain.

Let $\mathcal{B}_R$ be the ball of radius $R > 0$ and $\mathcal{B}_r$ be the ball of radius r with $R > r > 0$, both centered at the origin, and such that $\Lambda \subset \mathcal{B}_r$.

From (16) we obtain

$$\Delta \psi = 0 \qquad \text{in } \mathbb{R}^3 \setminus \overline{\mathcal{B}_r}, \tag{30}$$

$$\psi(\mathbf{x}) = \mathcal{O}\left(\frac{1}{\|\mathbf{x}\|} \right) \qquad \text{when } \|\mathbf{x}\| \to \infty \tag{31}$$

and

$$-\operatorname{div}(\bar{\mu}\boldsymbol{\nabla}\psi) = -\operatorname{div}(\bar{\mu}-1)\,\mathbf{H}_0 \quad \text{in } \mathcal{B}_R. \tag{32}$$

We use the Poisson formula which says that since ψ is a harmonic function outside the ball $\mathcal{B}_r$ and radially decreasing at infinity, then for each point $\mathbf{x}$ outside the ball $\mathcal{B}_r$ we have

$$\psi(\mathbf{x}) = \frac{\|\mathbf{x}\|^2 - r^2}{4\pi r} \int_{\partial\mathcal{B}_r} \frac{\psi(\mathbf{y})}{\|\mathbf{y}-\mathbf{x}\|^3}\, ds(\mathbf{y}). \tag{33}$$

By using this formula for $\mathbf{x} \in \partial\mathcal{B}_R$, we obtain the formulation: find $\psi \in H^1(\mathcal{B}_R)$ satisfying for all $\varphi \in H^1_0(\mathcal{B}_R)$,

$$\int_{\mathcal{B}_R} \bar{\mu}\boldsymbol{\nabla}\psi \cdot \boldsymbol{\nabla}\varphi\, dx = \int_{\Lambda} (\bar{\mu}-1)\,\mathbf{H}_0 \cdot \boldsymbol{\nabla}\varphi\, dx, \tag{34}$$

$$\psi(\mathbf{x}) = \frac{R^2 - r^2}{4\pi r} \int_{\partial\mathcal{B}_r} \frac{\psi(\mathbf{y})}{\|\mathbf{y}-\mathbf{x}\|^3}\, ds(\mathbf{y}), \tag{35}$$

for all points $\mathbf{x} \in \partial\mathcal{B}_R$; here $\bar{\mu} = \bar{\mu}(\cdot, \|\mathbf{H}_0 - \boldsymbol{\nabla}\psi\|)$.

4 Discretization and Numerical Methods

Before we describe the methods we derived from the above formulations, let us introduce some notations of discretization spaces, common for all of them.

Let us assume that the ferromagnetic domain Λ is a polyhedron. Let us also consider polyhedra $\mathcal{B}_{Rh}$ and $\mathcal{B}_{rh}$ approximating the big and the small balls $\mathcal{B}_R$, resp. $\mathcal{B}_r$.

Definition 1 (Interior and boundary mesh). *Let Ω be a polyhedral domain.*

1. *Let us denote $\tau_h(\Omega)$ the* tetrahedral mesh of $\overline{\Omega}$ *with conforming tetrahedra, in the finite-element sense. The tetrahedron $K \in \tau_h(\Omega)$ is understood as a closed tetrahedron.*
2. *The set of all internal and external faces of the tetrahedral mesh $\tau_h(\Omega)$ is denoted $\mathcal{F}_h(\Omega)$.*
3. *Let us denote $\tau_h(\partial\Omega)$ the* trace of the mesh $\tau_h(\Omega)$ on the boundary $\partial\Omega$. *The* boundary mesh $\tau_h(\partial\Omega)$ *is composed of all triangular faces $F \in \mathcal{F}_h(\Omega)$ such that $F \subset \partial\Omega$.*

Definition 2 (Finite element spaces). *Let Ω be a polyhedral domain, $\tau_h(\Omega)$ its tetrahedral mesh and $\tau_h(\partial\Omega)$ the corresponding boundary mesh.*

1. *On the tetrahedral mesh $\tau_h(\Omega)$, we define:*
   a) *The finite element space $\mathcal{P}1_h(\Omega)$ of continuous functions which are piecewise polynomial of degree 1 on $K \in \tau_h(\Omega)$.*

   b) *The finite element space* $\mathcal{P}1_{h0}(\Omega)$ *of functions* $w \in \mathcal{P}1_h(\Omega)$ *such that* $w = 0$ *on* $\partial\Omega$.
   c) *The finite element space* $\mathcal{P}0_h(\Omega)$ *of piecewise constant functions on* $K \in \tau_h(\Omega)$.
2. *On the boundary mesh* $\tau_h(\partial\Omega)$ *we define*
   a) *The finite element space* $\mathcal{P}0_h(\partial\Omega)$ *of piecewise constant functions on the faces* $F \in \tau_h(\partial\Omega)$.
   b) *The finite element space* $\mathcal{P}1_h(\partial\Omega)$ *of continuous functions which are piecewise polynomial of degree 1 on each face* $F \in \tau_h(\partial\Omega)$.

4.1 Boundary Integral Method for the Scalar Potential Model

Let us approximate the spaces $H^1(\Lambda)$, resp. $H^{-\frac{1}{2}}(\partial\Lambda)$ by the space $\mathcal{P}1_h(\Lambda)$ of piecewise linear functions, resp. by the space $\mathcal{P}0_h(\partial\Lambda)$ of piecewise constant functions on the boundary mesh. We can write the discrete formulation corresponding to the problem (27)–(29):

Find $(\psi_h, \lambda_h, w_h) \in \mathcal{P}1_h(\Lambda) \times \mathcal{P}0_h(\partial\Lambda) \times \mathcal{P}1_h(\Lambda)$, $w_h = \psi_h$ on $\partial\Lambda$ such that

$$\int_\Lambda \tilde{\mu} \mathbf{\nabla}\psi_h \cdot \mathbf{\nabla}\varphi_h \, dx - \int_{\partial\Lambda} \lambda_h \varphi_h \, ds = \int_\Lambda (\tilde{\mu} - 1)\mathbf{H}_0 \cdot \mathbf{\nabla}\varphi_h \, dx, \tag{36}$$

$$\int_{\partial\Lambda} \psi_h \cdot \eta_h \, ds + \int_{\partial\Lambda} \eta_h(\mathbf{x}) \, ds(\mathbf{x}) \int_{\partial\Lambda} \left(\lambda_h(\mathbf{y}) - \frac{\partial w_h}{\partial \mathbf{n}}(\mathbf{y}) \right) G(\mathbf{x}, \mathbf{y}) \, ds(\mathbf{y}) = 0, \tag{37}$$

$$\int_\Lambda \mathbf{\nabla} w_h \cdot \mathbf{\nabla} v_h \, dx = 0, \tag{38}$$

for all $(\varphi_h, \eta_h, v_h) \in \mathcal{P}1_h(\Lambda) \times \mathcal{P}0_h(\partial\Lambda) \times \mathcal{P}1_{h0}(\Lambda)$. The function $\tilde{\mu} \in \mathcal{P}0_h(\Lambda)$ is the approximation of μ_r in Λ defined by

$$\tilde{\mu} = \mu_r \left(\| Q_h \mathbf{H}_0 - \mathbf{\nabla}\psi_h \| \right), \tag{39}$$

where $Q_h : L^2(\Lambda)^3 \to \mathcal{P}0_h(\Lambda)^3$ is the L^2-orthogonal projection.

Note that all integrals are done exactly except for the one involving the Green kernel G which must be numerically approximated.

The nonlinear problem (36)–(38) is solved using a standard fixed point method, cf. Algorithm 1. The convergence of this fixed point method is proven in [8].

Algorithm 1 (Boundary integral method for scalar potential)
Let us set $\psi_h^0 \in \mathcal{P}1_h(\Lambda)$, $\psi_h^0 = 0$.
For $k = 1, \ldots, J$ do

1. Evaluate $\tilde{\mu}^k = \mu_r \left(\| Q_h \mathbf{H}_0 - \mathbf{\nabla}\psi_h^{k-1} \| \right)$.
2. Find $(\psi_h^k, \lambda_h^k, w_h^k) \in \mathcal{P}1_h(\Lambda) \times \mathcal{P}0_h(\partial\Lambda) \times \mathcal{P}1_h(\Lambda)$, $w_h^k = \psi_h^k$ on $\partial\Lambda$ such that

$$\int_{\Lambda} \tilde{\mu}^k \nabla\psi_h^k \cdot \nabla\varphi_h \, dx - \int_{\partial\Lambda} \lambda_h^k \varphi_h \, ds = \int_{\Lambda} (\tilde{\mu}^k - 1) Q_h \mathbf{H}_0 \cdot \nabla\varphi_h \, dx,$$

$$\int_{\partial\Lambda} \psi_h^k \cdot \eta_h \, ds + \int_{\partial\Lambda} \eta_h(\mathbf{x}) \, ds(\mathbf{x}) \int_{\partial\Lambda} \left(\lambda_h^k(\mathbf{y}) - \frac{\partial w_h^k}{\partial \mathbf{n}}(\mathbf{y}) \right) G(\mathbf{x},\mathbf{y}) \, ds(\mathbf{y}) = 0,$$

$$\int_{\Lambda} \nabla w_h^k \cdot \nabla v_h \, dx = 0,$$

for all $(\varphi_h, \eta_h, v_h) \in \mathcal{P}1_h(\Lambda) \times \mathcal{P}0_h(\partial\Lambda) \times \mathcal{P}1_{h0}(\Lambda)$

until estimated convergence.

In the applications, the ferromagnetic structures are often thin and their discretizations contain a lot of triangles on their surfaces. The main drawback arising from the formulation (36)–(38) is that the second term of the equation (37) leads in this case to a full matrix with a big order. Consequently, the formulation (36)–(38) imposes an important restriction on the mesh, which is not operational in a lot of applications on the presented form.

4.2 Scalar Potential Problem with Poisson Formula Boundary Condition

Let us now discretize the formulation (34)–(35). We are approximating the functional space $H^1(\mathcal{B}_R)$ by the space $\mathcal{P}1_h(\mathcal{B}_{Rh})$ by assuming that $\partial\Lambda$ and $\partial\mathcal{B}_{rh}$ are made of faces of $\mathcal{F}_h(\mathcal{B}_{Rh})$.

The discrete formulation for (34)–(35) reads: find $\psi_h \in \mathcal{P}1_h(\mathcal{B}_{Rh})$ such that for all $\varphi_h \in \mathcal{P}1_{h0}(\mathcal{B}_{Rh})$ we have

$$\int_{\mathcal{B}_{Rh}} \tilde{\mu} \nabla\psi_h \cdot \nabla\varphi_h \, dx = \int_{\Lambda} (\tilde{\mu} - 1) Q_h \mathbf{H}_0 \cdot \nabla\varphi_h \, dx, \tag{40}$$

$$\psi_h(\mathbf{x}_i) = \frac{R^2 - r^2}{4\pi r} \int_{\partial\mathcal{B}_{rh}} \frac{\psi_h(\mathbf{y})}{\|\mathbf{y} - \mathbf{x}_i\|^3} \, ds(\mathbf{y}), \tag{41}$$

point-wise for all vertices $\mathbf{x}_i$ of the mesh $\tau_h(\partial\mathcal{B}_{Rh})$. Here, $\tilde{\mu} \in \mathcal{P}0_h(\mathcal{B}_{Rh})$ is the piecewise-constant approximation of $\bar{\mu}$ defined as in (39) with $\tilde{\mu} = 1$ outside Λ.

Note that all integrals are done exactly except for the one on $\partial\mathcal{B}_{rh}$. In that case, first recall that the sphere is approximated by a triangular mesh, second on each of these triangles we use a simple Gauss integration scheme.

The problem (40)–(41) is nonlinear. Moreover, the coupling of (40) and (41) is non-local, thus filling the matrix of the underlying linear system with full blocks. This is why we propose in Algorithm 2 a fixed point iteration, combined with a multiplicative Dirichlet–Dirichlet domain decomposition between the meshed interior and the exterior, represented by the Poisson representation formula (41) (see [8] for some aspects of convergence).

Algorithm 2 (Domain decomposition for scalar potential)
Let us set $\psi_h^0 \in \mathcal{P}1_h(\mathcal{B}_{Rh})$, $\psi_h^0 = 0$.
For $k = 1, \ldots, J$ do

1. Define $\psi_h^{k-\frac{1}{2}} \in \mathcal{P}1_h(\partial\mathcal{B}_{Rh})$ such that

$$\psi_h^{k-\frac{1}{2}}(\mathbf{x}_i) = \frac{R^2 - r^2}{4\pi r} \int_{\partial\mathcal{B}_{rh}} \frac{\psi_h^{k-1}(\mathbf{y})}{\|\mathbf{y} - \mathbf{x}_i\|^3}\, ds(\mathbf{y}), \tag{42}$$

point-wise on each vertex $\mathbf{x}_i$ of the mesh $\tau_h(\partial\mathcal{B}_{Rh})$,
2. Evaluate $\tilde{\mu}^k \in \mathcal{P}0_h(\mathcal{B}_{Rh})$ by

$$\tilde{\mu}^k = \begin{cases} \mu_r\left(\|Q_h\mathbf{H}_0 - \boldsymbol{\nabla}\psi_h^{k-1}\|\right) & \text{in } \Lambda, \\ 1 & \text{otherwise,} \end{cases}$$

3. Find $\psi_h^k \in \mathcal{P}1_h(\mathcal{B}_{Rh})$, $\psi_h^k = \psi_h^{k-\frac{1}{2}}$ on $\partial\mathcal{B}_{Rh}$, such that for all $\varphi_h \in \mathcal{P}1_{h0}(\mathcal{B}_{Rh})$ there is

$$\int_{\mathcal{B}_{Rh}} \tilde{\mu}^k \boldsymbol{\nabla}\psi_h^k \cdot \boldsymbol{\nabla}\varphi_h\, dx = \int_{\Lambda} (\tilde{\mu}^k - 1) Q_h\mathbf{H}_0 \cdot \boldsymbol{\nabla}\varphi_h\, dx \tag{43}$$

until estimated convergence.

To solve (43) approximatively, we use several iterations of an algebraic multigrid AMG [9]. Moreover, let us remark that computation of (42) can be easily parallelized.

5 Numerical Results

Our aim here is not to give a comprehensive comparison of results obtained by the above described algorithms. This can be found in [4] where several other methods are described to approximate this ferromagnetic problem. We just show a simple example exhibiting the screen effect produced by ferromagnetic materials.

5.1 A Thin Steel Plate Placed in Front of a Conductor

We consider a simple ferromagnetic rectangular plate which is placed in front of an idealized infinitely long wire with zero section (see Figure 1). A given constant electric current runs in the wire. Omitting the plate, this current would produce an induction field $\mathbf{B}_0$ (suggested in Figure 1) given by Biot–Savart law. In the presence of the plate, the induction field $\mathbf{B}$ will be modified, due to ferromagnetic response of the steel plate.

We want to simulate the screen effect of the ferromagnetic plate, i.e., the attenuation of the induction field "behind" the plate. For this, we will compare $\mathbf{B}$ and $\mathbf{B}_0$ along some "observation lines" (see Figure 1): line 1 is orthogonal to the plate; lines 2 and 3 are located in the plate to measure the magnetization in it.

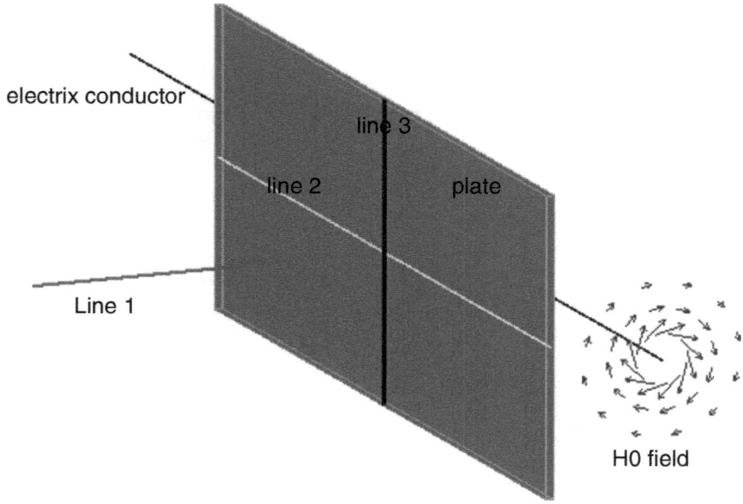

Fig. 1. Geometry of the test-case: the rectangular plate and current support.

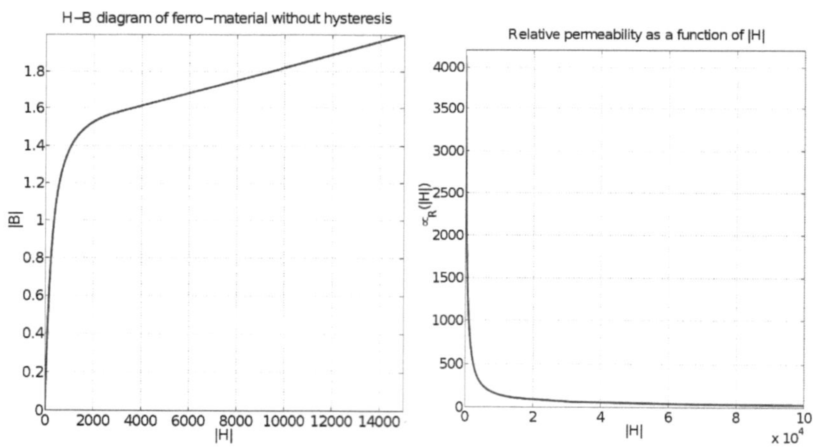

Fig. 2. Material properties of the ferro-material: the H-B diagram without hysteresis (left), relative permeability $\mu_r(\|\mathbf{H}\|)$ as a function of $\|H\|$ (right).

The nonlinear material behaviour is characterized by the $B-H$ diagram, or by the relative magnetic permeability function $\mu_r(\|H\|)$ given in Figure 2.

In order to apply our algorithms, we take the small ball $\mathcal{B}_r$ near the plate and we take $\mathcal{B}_R$ in such a way that the ratio of their radiuses is 1.5. Both balls have their center at the center of the plate.

We then mesh the whole ball $\mathcal{B}_R$ in order to obtain a mesh compatible with the boundary of $\mathcal{B}_r$ and with the plate as well. We also refine twice this mesh to get three meshes, so we can check convergence in space of our approximations.

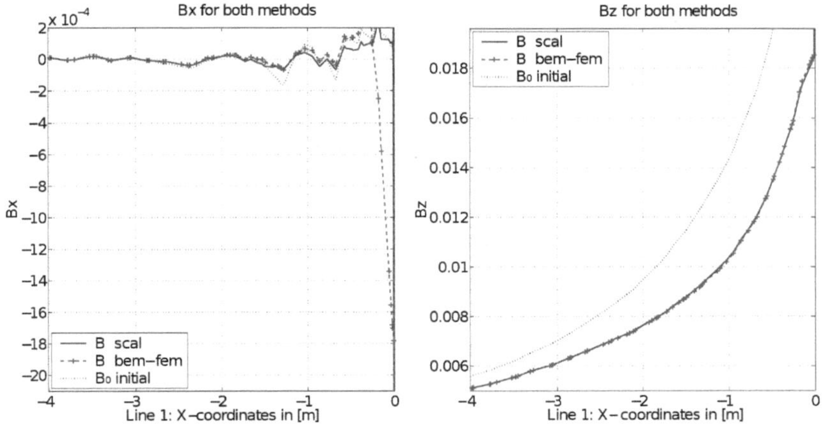

Fig. 3. Magnetic induction $\mathbf{B} = (B_x, B_y, B_z)$ on the observation line, on the screened side of the plate. Left: component B_x, right: component B_z. Here 'scal' stands for Algorithm 2 and 'bem-fem' for Algorithm 1.

5.2 Screen Effect Behind the Plate

Let us compare the two presented algorithms on the described plate case. To demonstrate the screen effect on the plate, we plot in Figure 3 induction $\mathbf{B}_0$ (in dotted line) and $\mathbf{B}$ (in full line) along observation line 1.

The plate is located on the right of each plot, distance to the plate is given with negative values. The left plot represents the B_x component of induction and the right plot represents the B_z component. We first remark both algorithms give very similar results. It is clear from the right plot that component B_z is attenuated while component B_x should be vanishing.

5.3 Magnetization in the Plate

In Figure 4 we see the plots of μ_r on the observation lines 2 and 3 inside the plate. The first plot corresponds to the observation line 2 (horizontal line), the second one corresponds to the observation line 3 (vertical line). Results for both methods are superposed in one plot. We can see that parts of the plate which are the nearest of the conductor are magnetically saturated, while parts of the plate which are the most away from the conductor are not.

5.4 An Industrial Application

Let us finish by presenting an industrial application of our model. On an industrial scale, aluminum is produced by electrolysis of alumine Al_2O_3. This process is realized in big cells 10 m long, 4 m wide and 1 m high in which aluminum as well as the electrolytic bath containing the alumine are liquid.

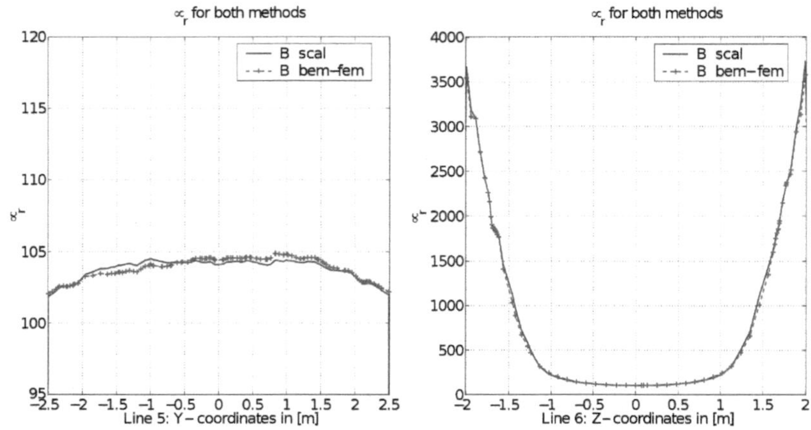

Fig. 4. Relative permeability μ_r on observation line 2 (horizontal line) and line 3 (vertical line) in the plate. As for previous plot 'scal' stands for Algorithm 2 and 'bem-fem' for Algorithm 1.

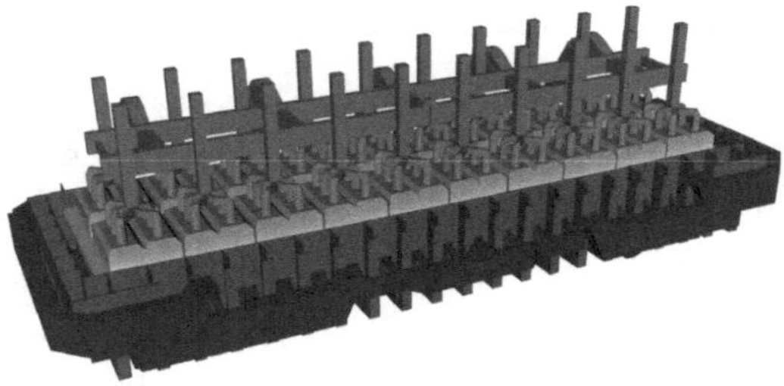

Fig. 5. Contour plot of electric potential for an electrolysis cell: current enters the cell through red parts and quits the cell through blue parts, steel shell is colored in gray.

Strong continuous electric currents run vertically through the cell allowing electrolysis to happen. These currents, coupled to the magnetic induction field induced by currents in the cell as well as by currents going from one cell to the other produce an important Lorentz force. This force causes fluids to move, which can perturb the efficiency of electrolysis. The fact that such cells are built in a 2 cm thick steel shell has an appreciated effect: the ferromagnetic response of this steel shell, in fact, protects in some extent the fluids inside the cell from magnetic induction produced by currents feeding the cell. Thus, modeling ferromagnetic screen effect produced by this steel shell is an important step in the full modelization of an electrolysis cell. Figure 5 shows a typical example of a cell colored by electric potential: current enters the cell through red parts and quits the cell through blue parts, steel shell is colored

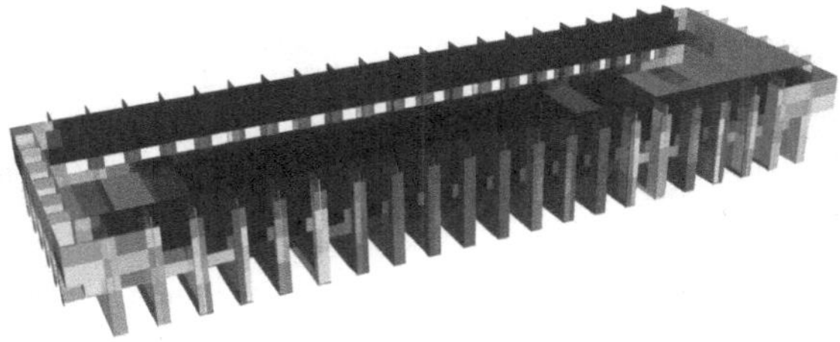

Fig. 6. Same cell as previous with only ferromagnetic parts colored by relative permeability: low permeability is colored in blue, high permeability is colored in red.

in gray. Figure 6 shows only the steel container which has been colored by relative permeability: low permeability is colored in blue, high permeability is colored in red.

6 Conclusion and Remarks

As a known problem for this approximation, both methods produce spurious oscillations of the magnetization vector in the plate. These oscillations are due both to the fact that the plate is thin and to the important jump in permeability when crossing the plate boundary (see [3]). These oscillations can be considerably reduced by simply applying a smoothing operator after computing the induction.

It is clear that the boundary integral formulation leads to a dense matrix to be solved, which is a severe penalty. However, we think it is possible to circumvent this difficulty by using some multipole technique. This method also needs some critical calculations involving a singular kernel which can be found in [5].

The domain decomposition formulation, on the contrary is very fast, since we only solve a Laplacian problem outside the ferromagnetic domain. This is done with best efficiency using an algebraic multigrid algorithm as described in [9]. However, it is not always a trivial problem to mesh the domain between the ferromagnetic material and the sphere $\partial\mathcal{B}_R$, knowing that this mesh will also have an influence on the precision of the approximation.

For simplicity, we used the fixed point method to solve the non-linear problem involved in our formulations, but it is of course possible to use Newton process as well.

Acknowledgement. The authors are grateful to the Alcan-Péchiney Company for supporting this work.

References

1. O. Bíró and K. Preis. On the use of magnetic vector potential in the finite element analysis of three-dimensional eddy currents. *IEEE Trans. Magn.*, 25(4):3145–3159, 1989.
2. O. Bíró, K. Preis, and K. R. Richter. On the use of the magnetic vector potential in the nodal and edge finite element analysis of 3D magnetostatic problems. *IEEE Trans. Magn.*, 32(3):651–654, 1996.
3. J. Descloux, M. Flueck, and M. V. Romerio. A problem of magnetostatics related to thin plates. *RAIRO Modél. Math. Anal. Numér.*, 32(7):859–876, 1998.
4. M. Flück, T. Hofer, A. Janka, and J. Rappaz. Numerical methods for ferromagnetic plates. Research report 08.2007, Institute of Analysis and Scientific Computing (IACS), EPFL, 2007.
5. A. Masserey, J. Rappaz, R. Rozsnyo, and M. Swierkosz. Numerical integration of the three-dimensional Green kernel for an electromagnetic problem. *J. Comput. Phys.*, 205(1):48–71, 2005.
6. A. Masud and T. J. R. Hughes. A stabilized mixed finite element method for Darcy flow. *Comput. Methods Appl. Mech. Engrg.*, 191(39–40):4341–4370, 2002.
7. J.-C. Nédélec. *Acoustic and electromagnetic equations. Integral representations for harmonic problems*, volume 144 of *Applied Mathematical Sciences*. Springer, New York, 2001.
8. J. Rappaz. About the ferromagnetic effects. Some mathematical results. Research report, Institute of Analysis and Scientific Computing (IACS), EPFL. To appear.
9. P. Vaněk, M. Brezina, and J. Mandel. Convergence of algebraic multigrid based on smoothed aggregation. *Numer. Math.*, 88(3):559–579, 2001.

Two-Sided Estimates of the Solution Set for the Reaction–Diffusion Problem with Uncertain Data

Olli Mali[1] and Sergey Repin[2]

[1] Department of Mathematical Information Technology, P.O. Box 35 (Agora), FI-40014 University of Jyväskylä, Finland, olli.mali@jyu.fi

[2] V. A. Steklov Institute of Mathematics in St. Petersburg, Fontanka 27, 191023, St. Petersburg, Russia, repin@pdmi.ras.ru

Summary. We consider linear reaction–diffusion problems with mixed Dirichlét–Neumann–Robin conditions. The diffusion matrix, reaction coefficient, and the coefficient in the Robin boundary condition are defined with an uncertainty which allow bounded variations around some given mean values. A solution to such a problem cannot be exactly determined (it is a function in the set of "possible solutions" formed by generalized solutions related to possible data). The problem is to find parameters of this set. In this paper, we show that computable lower and upper bounds of the diameter (or radius) of the set can be expressed throughout problem data and parameters that regulate the indeterminacy range. Our method is based on using a posteriori error majorants and minorants of the functional type (see [5,6]), which explicitly depend on the coefficients and allow to obtain the corresponding lower and upper bounds by solving the respective extremal problems generated by indeterminacy of coefficients.

1 Introduction

This paper is concerned with boundary-value problems for partial differential equations of elliptic type coefficients of which contain an indeterminacy. Such a situation is quite typical in real-life problems where parameters of mathematical model cannot be determined exactly and instead one knows only that the coefficients belong to a certain set of "admissible" data Λ. In view of this fact, instead of a single exact solution "u", we have to consider a "set of solutions" (we denote it by $\mathcal{S}(\Lambda)$). As a result, the error control problem comes in a more complicated form in which approximation errors must be evaluated together with errors arose due to indeterminant data (various approaches that can be used for such an analysis are exposed in, e.g. [1, 2, 8]).

W. Fitzgibbon et al. (eds.), *Applied and Numerical Partial Differential Equations*, Computational Methods in Applied Sciences 15, DOI 10.1007/978-90-481-3239-3_14,

In this paper, we establish explicit relations between the sets Λ and $\mathcal{S}(\Lambda)$ for the reaction–diffusion problem with mixed Dirichlét–Robin boundary conditions (we call it $\mathcal{P}$) defined by the system

$$-\operatorname{div}(A\nabla u) + \rho u = f \quad \text{in } \Omega \tag{1}$$

$$u = 0 \quad \text{on } \Gamma_1 \tag{2}$$

$$n \cdot A\nabla u = F \quad \text{on } \Gamma_2 \tag{3}$$

$$\alpha u + n \cdot A\nabla u = G \quad \text{on } \Gamma_3. \tag{4}$$

Here, $\Omega \in \mathbb{R}^d$ is a bounded and connected domain with Lipschitz continuous boundary $\Gamma_1 \cup \Gamma_2 \cup \Gamma_3$ and $f \neq 0$. We assume that exact A, ρ, and α are unknown. Instead, we know that $A \in \Lambda_A$, $\rho \in \Lambda_\rho$, and $\alpha \in \Lambda_\alpha$, where

$$\begin{aligned}
\Lambda_A &:= \{A \in L_2(\Omega, \mathbb{M}^{d\times d}) \mid A = A_0 + \delta_1 \Psi,\ \|\Psi\|_{L_\infty(\Omega,\mathbb{M}^{d\times d})} \leq 1\}\\
\Lambda_\rho &:= \{\rho \in L_2(\Omega) \mid \rho = \rho_0 + \delta_2 \psi_\rho,\ \|\psi_\rho\|_{L_\infty(\Omega)} \leq 1\}\\
\Lambda_\alpha &:= \{\alpha \in L_2(\Gamma_3) \mid \alpha = \alpha_0 + \delta_3 \psi_\alpha,\ \|\psi_\alpha\|_{L_\infty(\Gamma_3)} \leq 1\}.
\end{aligned}$$

In other words, we assume that the sets of admissible data are formed by some (limited) variations of some known "mean" data (which are denoted by subindex 0). The parameters δ_i, $i = 1, 2, 3$, represent the magnitude of these variations. Thus, in the case considered,

$$\Lambda := \Lambda_A \times \Lambda_\rho \times \Lambda_\alpha.$$

We note that such a presentation of the data arises in many engineering problems where data are given in a form `mean±error`. The solution associated to non-perturbed data A_0, ρ_0, and α_0 is denoted by u_0.

Our goal is to give computable estimates of the radius of $\mathcal{S}(\Lambda)$ (we denote this quantity by $r_{\mathcal{S}}$). The value of $r_{\mathcal{S}}$ has a large significance for practical applications because it shows an accuracy limit defined by the problem statement. Attempts to find approximate solutions having approximation errors lesser then $r_{\mathcal{S}}$ have no practical sense.

The generalized statement of Problem ($\mathcal{P}$) is as follows: Find $u \in V_0$ such that

$$a(u, w) = l(w) \quad \forall w \in V_0, \tag{5}$$

where space V_0 and functionals $a : V_0 \times V_0 \to R$ and $l : V_0 \to R$ are defined by the relations

$$\begin{aligned}
V_0 &:= \{w \in H^1(\Omega) \mid w_{|\Gamma_1} = 0\},\\
a(u, w) &:= \int_\Omega A\nabla u \cdot \nabla w \, dx + \int_\Omega \rho u w \, dx + \int_{\Gamma_3} \alpha u w \, ds,\\
l(w) &:= \int_\Omega f w \, dx + \int_{\Gamma_2} F w \, ds + \int_{\Gamma_3} G w \, ds.
\end{aligned}$$

We assume that

$$\begin{aligned} \underline{c}_1|\xi|^2 \leq A_0\xi\cdot\xi \leq \overline{c}_1|\xi|^2 \quad \forall \xi \in \mathbb{R}^d \qquad & \text{on } \Omega, \\ \underline{c}_2 \leq \rho_0 \leq \overline{c}_2 \qquad & \text{on } \Omega, \\ \underline{c}_3 \leq \alpha_0 \leq \overline{c}_3 \qquad & \text{on } \Gamma_3, \end{aligned}$$

where $\underline{c}_i > 0$. In view of the above-stated conditions, the "mean" problem is evidently elliptic and has a unique solution u_0. The condition

$$0 \leq \delta_i < \underline{c}_i, \quad i = 1, 2, 3,$$

guarantees that the perturbed problem remains elliptic and possesses a unique solution u.

For each $A, \rho, \alpha \in \Lambda$, the corresponding problem of $\mathcal{P}(\Lambda)$ is natural to analyze using the (energy) norm

$$|\!|\!|v|\!|\!|^2_{A,\rho,\alpha} := a(v,v) = \int_\Omega A\nabla v\cdot\nabla v\,dx + \int_\Omega \rho v^2\,dx + \int_{\Gamma_3} \alpha v^2\,ds. \tag{6}$$

For the sake of simplicity we will also use an abridged notation $|\!|\!|v|\!|\!|$ for the norm $|\!|\!|v|\!|\!|_{A,\rho,\alpha}$. For the "mean" problem, we use the norm $|\!|\!|v|\!|\!|_{A_0,\rho_0,\alpha_0}$, which is also denoted by $|\!|\!|v|\!|\!|_0$. It is easy to see that the norms $|\!|\!|v|\!|\!|_0$ and $|\!|\!|v|\!|\!|$ are equivalent and satisfy the relation

$$\underline{\mathrm{C}}|\!|\!|v|\!|\!|^2 \leq |\!|\!|v|\!|\!|^2_0 \leq \overline{\mathrm{C}}|\!|\!|v|\!|\!|^2, \tag{7}$$

where

$$\overline{\mathrm{C}} := \max_{i\in\{1,2,3\}} \frac{\overline{c}_i}{\underline{c}_i - \delta} \quad \text{and} \quad \underline{\mathrm{C}} := \min_{i\in\{1,2,3\}} \frac{\underline{c}_i}{\overline{c}_i + \delta_i} \tag{8}$$

These constants $\overline{\mathrm{C}}$ and $\underline{\mathrm{C}}$ depend only on the problem data and indeterminacy range. They play an important role in our analysis.

Now, we can define the quantity we are interested in

$$r_{\mathcal{S}} := \sup_{\tilde{u}\in\mathcal{S}} |\!|\!|u_0 - \tilde{u}|\!|\!|_0. \tag{9}$$

A normalized counterpart of $r_{\mathcal{S}}$ is defined by the relation

$$\hat{r}_{\mathcal{S}} := \sup_{\tilde{u}\in\mathcal{S}} \frac{|\!|\!|u_0 - \tilde{u}|\!|\!|_0}{|\!|\!|u_0|\!|\!|_0}.$$

2 Lower Bound of $r_{\mathcal{S}}$

Problem $\mathcal{P}$ has a variational statement and the solution u can be considered as a minimizer of the functional

$$J(v) := \frac{1}{2}a(v,v) - l(v)$$

on the set V_0. Using this statement, we can easily derive computable lower bounds of the difference between u and an arbitrary function $v \in V_0$ in terms of the energy norm (see [5] where such bounds have been derived for a wide class of variational problems).

First, we use the identity

$$\|u - v\|^2 = a(u - v, u - v) = 2(J(v) - J(u)), \tag{10}$$

which for quadratic functionals is established in [4]. Let w be an arbitrary function in V_0. Then,

$$J(v) - J(u) \geq J(v) - J(v + w)$$

and by (10) we conclude that

$$\|u - v\|^2 \geq -a(w + 2v, w) + 2l(w) \quad \forall w \in V_0. \tag{11}$$

We note that for $w = u - v$ the estimate (11) holds as equality, so that there is no "gap" between the left- and right-hand sides of (11). The right-hand side of (11) is explicitly computable. It provides the so-called functional error minorant, which we denote by $\mathcal{M}_{\ominus}^{A,\rho,\alpha}(v, w)$ (if no confusion may arise, we also use a simplified notation $\mathcal{M}_{\ominus}(v, w)$). This functional serves as the main tool when deriving the lower bound of $r_{\mathcal{S}}$.

Theorem 1. *Assume that all the assumptions of Section 1 hold. Then*

$$r_{\mathcal{S}}^2 \geq \underline{\mathrm{C}} \sup_{w \in V_0} \mathrm{M}_{\ominus}^{rs}(u_0, w), \tag{12}$$

where w is an arbitrary function in V_0 and

$$\begin{aligned} \mathrm{M}_{\ominus}^{rs}(u_0, w) := -\|w\|_0^2 + \delta_1 \int_{\Omega} |\nabla w + 2\nabla u_0|\,|\nabla w|\, dx \\ + \delta_2 \int_{\Omega} |(w + 2u_0)w|\, dx + \delta_3 \int_{\Gamma_3} |(w + 2u_0)w|\, ds. \end{aligned} \tag{13}$$

Proof. We have

$$r_{\mathcal{S}} = \sup_{\tilde{u} \in \mathcal{S}} \|u_0 - \tilde{u}\|_0 \geq \underline{C} \sup_{\tilde{u} \in \mathcal{S}} \|u_0 - \tilde{u}\|. \tag{14}$$

On the other hand,

$$\begin{aligned} \sup_{\tilde{u} \in \mathcal{S}} \|u_0 - \tilde{u}\|^2 &= \sup_{\tilde{u} \in \mathcal{S}} \left\{ \sup_{w \in V_0} \mathcal{M}_{\ominus}(u_0, w) \right\} \\ &= \sup_{w \in V_0} \left\{ \sup_{A \in \Lambda_A, \rho \in \Lambda_\rho, \alpha \in \Lambda_\alpha} \mathcal{M}_{\ominus}^{A,\rho,\alpha}(u_0, w) \right\} \end{aligned}$$

and we conclude that

$$r_S^2 \geq \underline{C} \sup_{w \in V_0} \left\{ \sup_{A \in \Lambda_A, \rho \in \Lambda_\rho, \alpha \in \Lambda_\alpha} \mathcal{M}_{\ominus}^{A,\rho,\alpha}(u_0, w) \right\}. \tag{15}$$

Now our goal is to estimate the right-hand side of (15) from below. For this purpose, we exploit the structure of the minorant, which allows to explicitly evaluate effects caused by indeterminacy of the coefficients.

It is easy to see that the minorant can be represented as follows:

$$\begin{aligned} \mathcal{M}_{\ominus}^{A,\rho,\alpha}(u_0, w) = & - \int_\Omega (A_0 + \delta_1 \Psi)(\nabla w + 2\nabla u_0) \cdot \nabla w \, dx \\ & - \int_\Omega (\rho_0 + \delta_2 \psi_\rho)(w + 2u_0) w \, dx \\ & - \int_{\Gamma_3} (\alpha_0 + \delta_3 \psi_\alpha)(w + 2u_0) w \, ds + 2l(w). \end{aligned} \tag{16}$$

Note that

$$\int_\Omega (A_0 \nabla u_0 \cdot \nabla w \, dx + \rho_0 u_0 w) \, dx + \int_{\Gamma_3} \alpha_0 u_0 w \, ds = l(w).$$

Hence,

$$\begin{aligned} \mathcal{M}_{\ominus}^{A,\rho,\alpha}(u_0, w) = & - \int_\Omega A_0 \nabla w \cdot \nabla w \, dx - \int_\Omega \rho_0 w^2 \, dx - \int_{\Gamma_3} \alpha_0 w^2 \, ds \\ & - \delta_1 \int_\Omega \Psi(\nabla w + 2\nabla u_0) \cdot \nabla w \, dx - \delta_2 \int_\Omega \psi_\rho (w + 2u_0) w \, dx \\ & - \delta_3 \int_{\Gamma_3} \psi_\alpha (w + 2u_0) w \, ds \end{aligned} \tag{17}$$

and we obtain

$$\begin{aligned} \sup_{A \in \Lambda_A, \rho \in \Lambda_\rho, \alpha \in \Lambda_\alpha} \mathcal{M}_{\ominus}^{A,\rho,\alpha}(u_0, w) = & - \| w \|_0^2 \\ & + \delta_1 \sup_{|\Psi| \leq 1} \int_\Omega \Psi(\nabla w + 2\nabla u_0) \cdot \nabla w \, dx + \delta_2 \sup_{|\psi_\rho| \leq 1} \int_\Omega \psi_\rho (w + 2u_0) w \, dx \\ & + \delta_3 \sup_{|\psi_\alpha| \leq 1} \int_{\Gamma_3} \psi_\alpha (w + 2u_0) w \, ds. \end{aligned} \tag{18}$$

The integrand of the first integral in the right-hand side of (18) can be presented as $\Psi : \tau$, where

$$\tau = \nabla w \otimes (\nabla w + 2\nabla u_0)$$

and $\otimes$ stands for the diad product. For the first supremum we have

$$\sup_{|\Psi|\leq 1}\left\{\int_\Omega \Psi : \tau\, dx\right\} = \int_\Omega |\tau|\, dx. \tag{19}$$

Analogously, we find that

$$\sup_{|\psi_\rho|\leq 1}\int_\Omega \psi_\rho(w+2u_0)w\, dx \leq \int_\Omega |(w+2u_0)w|\, dx, \tag{20}$$

$$\sup_{|\psi_\alpha|\leq 1}\int_\Omega \psi_\rho(w+2u_0)w\, dx \leq \int_{\Gamma_3} |(w+2u_0)w|\, ds. \tag{21}$$

By (18)–(21), we arrive at the relation

$$\sup_{A,\rho,\alpha} \mathcal{M}_\ominus^{A,\rho,\alpha}(u_0,w) = -\|w\|_0^2 + \delta_1\int_\Omega |(\nabla w + 2\nabla u_0)\otimes\nabla w|\, dx$$
$$+\,\delta_2\int_\Omega |(w+2u_0)w|\, dx + \delta_3\int_{\Gamma_3} |(w+2u_0)w|\, ds, \tag{22}$$

which together with (15) leads to (12).

Theorem 1 gives a general form of the lower bound of $r_{\mathcal{S}}$. Also, it creates a basis for practical computation of this quantity. Indeed, let $V_{0h}\subset V_0$ be a finite dimensional space. Then

$$r_{\mathcal{S}}^2 \geq \underline{\mathrm{C}} \sup_{w\in V_{0h}} \mathrm{M}_\ominus^{r_S}(u_0,w). \tag{23}$$

It is worth noting that the wider set V_{0h} we take the better lower bound of the radius we compute. However, as it is shown below, a meaningful lower bound can be deduced even analytically.

Corollary 1. *Under assumptions of Theorem 1,*

$$r_{\mathcal{S}}^2 \geq \underline{C} r_{\mathcal{S}\ominus}^2 \quad \textit{and} \quad \hat{r}_{\mathcal{S}}^2 \geq \underline{C}\hat{r}_{\mathcal{S}\ominus}^2, \tag{24}$$

where

$$r_{\mathcal{S}\ominus}^2 = \frac{\|u_0\|_\delta^4}{\|u_0\|_0^2 - \|u_0\|_\delta^2} \geq \frac{\Theta^2}{1-\Theta}\|u_0\|_0^2, \tag{25}$$

where

$$\|u_0\|_\delta^2 := \delta_1\|\nabla u_0\|_\Omega^2 + \delta_2\|u_0\|_\Omega^2 + \delta_3\|u_0\|_{\Gamma_3}^2$$

and

$$\Theta := \min_{i\in\{1,2,3\}} \frac{\delta_i}{\overline{c}_i}.$$

For the normalized radius, we have

$$\hat{r}_{\mathcal{S}\ominus}^2 = \frac{\Theta^2}{1-\Theta}. \tag{26}$$

Proof. Use (12) and set

$$w = \lambda u_0, \tag{27}$$

where $\lambda \in \mathbb{R}$. Then we observe that

$$r_{\mathcal{S}}^2 \geq \underline{C}\left(-\lambda^2 \|u_0\|_0^2 + \lambda(\lambda+2)\|u_0\|_\delta^2\right). \tag{28}$$

The right-hand side of (28) is a quadratic function with respect to λ. It attains its maximal value if

$$\lambda \|u_0\|_0^2 = (\lambda+1)\|u_0\|_\delta^2,$$

that is, if

$$\lambda = \frac{\|u_0\|_\delta^2}{\|u_0\|_0^2 - \|u_0\|_\delta^2}.$$

Substituting this λ, we arrive at (25). Note that

$$\begin{aligned}
\|u_0\|_0^2 &= \int_\Omega (A_0 \nabla u_0 \cdot \nabla u_0 + \rho_0 u_0^2\, dx + \int_{\Gamma_3} \alpha_0 u_0\, ds \\
&\geq \int_\Omega (\underline{c}_1 \nabla u_0 \cdot \nabla u_0 + \underline{c}_2 u_0^2)\, dx + \int_{\Gamma_3} \underline{c}_3 u_0\, ds \\
&> \delta_1 \|\nabla u_0\|_\Omega^2 + \delta_2 \|u_0\|_\Omega^2 + \delta_3 \|u_0\|_{\Gamma_3}^2 = \|u_0\|_\delta^2,
\end{aligned} \tag{29}$$

so that λ (and the respective lower bound) is positive. Moreover,

$$\begin{aligned}
\|u_0\|_\delta^2 \geq \frac{\delta_1}{\overline{c}_1} \int_\Omega A_0 \nabla u_0 \cdot \nabla u_0\, dx + \frac{\delta_2}{\overline{c}_2} \int_\Omega \rho_0 u_0^2\, dx + \frac{\delta_3}{\overline{c}_3} \int_{\Gamma_3} \alpha_0 u_0^2\, ds \\
\geq \Theta \|u_0\|_0^2.
\end{aligned} \tag{30}$$

Also,

$$\begin{aligned}
&\|u_0\|_0^2 - \|u_0\|_\delta^2 \\
&\quad = \int_\Omega (A_0 - \delta_1 I)\nabla v \cdot \nabla v\, dx + \int_\Omega (\rho_0 - \delta_2) v^2\, dx + \int_{\Gamma_3} (\alpha_0 - \delta_3) v^2\, ds \\
&\geq \left(1 - \frac{\delta_1}{\overline{c}_1}\right) \int_\Omega A_0 \nabla v \cdot \nabla v\, dx + \left(1 - \frac{\delta_2}{\overline{c}_2}\right) \int_\Omega \rho_0 v^2\, dx + \left(1 - \frac{\delta_3}{\overline{c}_3}\right) \int_{\Gamma_3} \alpha_0 v^2\, ds \\
&\geq \max_{i=1,2,3} \left(1 - \frac{\delta_i}{\overline{c}_i}\right) \|u_0\|_0^2 = (1-\Theta)\|u_0\|_0^2.
\end{aligned} \tag{31}$$

By (30) and (31), we arrive at the relation

$$r_{\mathcal{S}\Theta}^2 \geq \frac{\Theta^2}{1-\Theta} \|u_0\|_0^2, \tag{32}$$

which implies (25) and (26).

3 Upper Bound of r_S

A computable upper bound of r_S can be derived with the help of a posteriori error majorant of the functional type, which are derived by purely functional methods without attracting any information on the mesh and method used. For a wide class of problems they were derived in [5–7] by variational techniques and in [7, 9, 10] by transformations of integral identities (see also the papers cited therein). Below we derive functional error majorant for our class of problems using the latter method based on transformations of the respective integral identity. After that, we use it's properties to derive the desired upper bound in Section 3.2.

3.1 Error Majorant

Let $v \in V_0$ be an admissible approximation of the exact solution u (generated by A, ϱ, and α). From (5) it follows that for any $w \in V_0$

$$\begin{aligned} a(u-v,w) = \int_\Omega fw\,dx + \int_{\Gamma_2} Fw\,ds + \int_{\Gamma_3} Gw\,ds \\ - \int_\Omega A\nabla v\cdot\nabla w\,dx - \int_\Omega \rho vw\,dx - \int_{\Gamma_3}\alpha vw\,ds \\ + \int_\Omega \left(\mathrm{div}(y)w + y\cdot\nabla w\right)\,dx - \int_{\Gamma_2\cup\Gamma_3}(y\cdot\nu)\,w\,ds, \quad (33) \end{aligned}$$

where ν denotes unit outward normal to Γ and

$$y \in H^+(\mathrm{div},\Omega) := \{y \in H(\mathrm{div},\Omega) \mid y\cdot\nu \in L^2(\Gamma_2\cup\Gamma_3)\}.$$

We note that the last line is zero for all $y \in H(\Omega,\mathrm{div})$ (in view of the integration-by-parts formula). We regroup the terms and rewrite the relation as follows:

$$a(u-v,w) = I_1 + I_2 + I_3 + I_4, \quad (34)$$

where

$$\begin{aligned} I_1 &:= \int_\Omega r_1(v,y)w\,dx := \int_\Omega (f-\rho v+\mathrm{div}(y))w\,dx, \\ I_2 &:= \int_\Omega r_2(v,y)w\,dx := \int_{\Gamma_3}(G-\alpha v - y\cdot\nu)w\,ds, \\ I_3 &:= \int_{\Gamma_2}(F-y\cdot\nu)w\,ds, \\ I_4 &:= \int_\Omega (y-A\nabla v)\cdot\nabla w\,dx. \end{aligned}$$

Now we can estimate each term separately by the Friedrich and trace inequalities (which holds due to our assumptions concerning Ω). We have

$$\begin{aligned} \|w\|^2_{\Omega} &\le C_1(\Omega)\|\nabla w\|^2_{\Omega} && \forall w \in V_0, \\ \|w\|^2_{2,\Gamma_2} &\le C_2(\Omega,\Gamma_2)\|w\|^2_{\Omega} && \forall w \in V_0, \\ \|w\|^2_{2,\Gamma_3} &\le C_3(\Omega,\Gamma_3)\|w\|^2_{\Omega} && \forall w \in V_0. \end{aligned}$$

When estimating the integrands of I_1 and I_2, we introduce additional functions μ_1 and μ_2, which have values in $[0,1]$:

$$\begin{aligned} I_1 &= \int_\Omega \frac{\mu_1}{\sqrt{\rho}} r_1(v,y)\sqrt{\rho} w\,dx + \int_\Omega (1-\mu_1) r_1(v,y) w\,dx \\ &\le \left\| \frac{\mu_1}{\sqrt{\rho}} r_1(v,y)\right\|_\Omega \|\sqrt{\rho} w\|_\Omega + \sigma_1 \|(1-\mu_1) r_1(v,y)\|_\Omega \left(\int_\Omega A\nabla w\cdot\nabla w\,dx\right)^{1/2} \end{aligned}$$

and

$$\begin{aligned} I_2 = \int_{\Gamma_3} \frac{\mu_2}{\sqrt{\alpha}} r_2(v,y)\sqrt{\alpha} w\,dx &+ \int_{\Gamma_3} (1-\mu_2) r_2(v,y) w\,dx \\ &\le \left\| \frac{\mu_2}{\sqrt{\alpha}} r_2(v,y)\right\|_{\Gamma_3} \|\sqrt{\alpha} w\|_{\Gamma_3} \\ &\quad + \|(1-\mu_2) r_2(v,y)\|_{\Gamma_3}\sigma_3 \left(\int_\Omega A\nabla w\cdot\nabla w\,dx\right)^{1/2}, \end{aligned}$$

and

$$I_3 \le \|F - y\cdot\nu\,\|_{\Gamma_2}\sigma_2 \left(\int_\Omega A\nabla w\cdot\nabla w\,dx\right)^{1/2}, \tag{35}$$

where

$$\sigma_1 = \sqrt{\frac{C_1(\Omega)}{\underline{c}_1}}, \quad \sigma_2 = \sqrt{\frac{C_1(\Omega)C_2(\Omega,\Gamma_2)}{\underline{c}_1}}, \quad \sigma_3 = \sqrt{\frac{C_1(\Omega)C_3(\Omega,\Gamma_3)}{\underline{c}_1}}.$$

We note that a similar approach was used in [9] for the reaction–diffusion problem and in [10] for the generalized Stokes problem. In these publications it was shown that splitting of the residual term (performed with the help of a single function μ) allows to obtain error majorants that are insensitive with respect to small values of the lower term coefficient and at the same time sharp (i.e. have no irremovable gap between the left- and right-hand sides). In our case, we have two lower terms, so that we need two functions μ_1 and μ_2 to split the respective residual terms.

The term I_4 is estimated as follows:

$$I_4 \le D(\nabla v, y)^{\frac{1}{2}} \left(\int_\Omega A\nabla w\cdot\nabla w\,dx\right)^{1/2}, \tag{36}$$

where

$$D(\nabla v, y) = \int_\Omega (y - A\nabla v) \cdot (\nabla v - A^{-1} y)\, dx. \tag{37}$$

We collect all the terms and obtain

$$\begin{aligned} a(u - v, w) \leq \Big(& D(\nabla v, y)^{1/2} + \sigma_1 \|(1-\mu_1) r_1(v,y)\|_\Omega \\ & + \sigma_3 \|(1-\mu_2) r_2(v,y)\|_{\Gamma_3} + \sigma_2 \|F - y \cdot \nu\,\|_{\Gamma_2} \Big) \left(\int_\Omega A\nabla w \cdot \nabla w\, dx \right)^{1/2} \\ & + \left\| \frac{\mu_1}{\sqrt{\rho}} r_1(v,y) \right\|_\Omega \|\sqrt{\rho} w\|_\Omega + \left\| \frac{\mu_2}{\sqrt{\alpha}} r_2(v,y) \right\|_{\Gamma_3} \|\sqrt{\alpha} w\|_{\Gamma_3}. \end{aligned} \tag{38}$$

Set $w = u - v$ and use the Cauchy–Schwartz inequality

$$\sum_{i=1}^{d} x_i y_i \leq \sqrt{\sum_{i=1}^{d} x_i^2} \sqrt{\sum_{i=1}^{d} y_i^2}. \tag{39}$$

Then, we arrive at the estimate

$$\begin{aligned} \|u - v\|^2 \leq \Big(& D(\nabla v, y)^{1/2} + \sigma_1 \|(1-\mu_1) r_1(v,y)\|_\Omega \\ & + \sigma_3 \|(1-\mu_2) r_2(v,y)\|_{\Gamma_3} + \sigma_2 \|F - y \cdot \nu\,\|_{\Gamma_2} \Big)^2 \\ & + \left\| \frac{\mu_1}{\sqrt{\rho}} r_1(v,y) \right\|_\Omega^2 + \left\| \frac{\mu_2}{\sqrt{\alpha}} r_2(v,y) \right\|_{\Gamma_3}^2. \end{aligned} \tag{40}$$

It is worth remarking that the estimate (40) provides a guaranteed upper bound of the error for *any* conforming approximation of the problem (1)–(4). The estimate has a form typical for all functional a posteriori estimates: it is presented by the sum of residuals of the basic relations with multipliers that depend on the constants in the respective functional (embedding) inequalities for the domain and boundary parts.

However, for our subsequent goals, it is desirable to have the majorant in a form that involve only quadratic terms. Such a form can be easily derived from (40) if we square both parts and apply the algebraic inequality (39) to the first term (with multipliers $\gamma_i > 0$, $i = 1, 2, 3, 4$). Then, we obtain

$$\begin{aligned} \|u - v\|^2 \leq \kappa\Big(& \gamma_1 D(\nabla v, y) + \gamma_2 \|(1-\mu_1) r_1(v,y)\|_\Omega^2 \\ & + \gamma_3 \|(1-\mu_2) r_2(v,y)\|_{\Gamma_3}^2 + \gamma_4 \|F - y \cdot \nu\,\|_{\Gamma_2}^2 \Big) \\ & + \left\| \frac{\mu_1}{\sqrt{\rho}} r_1(v,y) \right\|_\Omega^2 + \left\| \frac{\mu_2}{\sqrt{\alpha}} r_2(v,y) \right\|_{\Gamma_3}^2, \end{aligned} \tag{41}$$

where

$$\kappa := \frac{1}{\gamma_1} + \frac{\sigma_1^2}{\gamma_2} + \frac{\sigma_2^2}{\gamma_3} + \frac{\sigma_3^2}{\gamma_4}.$$

We note that (41) coincides with (40) if

$$\gamma_1 = \bar{\gamma}_1 := D(\nabla v, y)^{-1/2}, \tag{42}$$

$$\gamma_2 = \bar{\gamma}_2 := \frac{\sigma_1}{\|(1-\mu_1)r_1(v,y)\|_\Omega}, \tag{43}$$

$$\gamma_3 = \bar{\gamma}_3 := \frac{\sigma_2}{\|(1-\mu_2)r_2(v,y)\|_{\Gamma_3}}, \tag{44}$$

$$\gamma_4 = \bar{\gamma}_4 := \frac{\sigma_3}{\|F - y\cdot\nu\,\|_{\Gamma_2}}. \tag{45}$$

Certainly, the estimate (41) looks more complicated with respect to (40). However, it has an important advantage: the weight functions μ_1 and μ_2 enter it as quadratic integrands, so that we can easily find their optimal form adapted to a particular v and the respective error distribution.

In the simplest case, we take $\mu_1 = \mu_2 = 0$, which yields the estimate

$$\|u-v\|^2 \le \kappa \times \Big(\gamma_1 D(\nabla v, y) + \gamma_2\|r_1(v,y)\|_\Omega^2 + \gamma_3\|r_2(v,y)\|_{\Gamma_3}^2 + \gamma_4\|F - y\cdot\nu\,\|_{\Gamma_2}^2\Big). \tag{46}$$

Another estimate arises if we set $\mu_1 = \mu_2 = 1$. In this case, the terms with factors σ_1 and σ_3 in (40) are equal to zero, so that subsequent relations do not contain the terms with multipliers formed by γ_2 and γ_3. Hence, we arrive at the estimate

$$\|u-v\|^2 \le \left(\frac{1}{\gamma_1} + \frac{\sigma_3^2}{\gamma_4}\right) \times \left(\gamma_1 D(\nabla v, y) + \gamma_4\|F - y\cdot\nu\,\|_{\Gamma_2}^2\right) + \left\|\frac{1}{\sqrt{\rho}}r_1(v,y)\right\|_\Omega^2 + \left\|\frac{1}{\sqrt{\alpha}}r_2(v,y)\right\|_{\Gamma_3}^2. \tag{47}$$

The estimate (47) involves "free" parameters γ_1 and γ_4 and a "free" vector-valued function y (which can be thought of as an image of the true flux). There exists a combination of these free parameters that makes the left-hand side of the estimate equal to the right-hand one. Indeed, set $y = A\nabla u$. Then

$$\begin{aligned} r_1(v,y) &= \rho(u-v) && \text{in } \Omega,\\ r_2(v,y) &= \alpha(u-v) && \text{on } \Gamma_3,\\ F - y\cdot\nu &= 0 && \text{on } \Gamma_2 \end{aligned}$$

and we find that for γ_4 tending to infinity and for any $\gamma_1 > 0$ the right-hand side of (47) coincides with the energy norm of the error. However, the estimate (47) has a drawback: it is sensitive with respect to ρ and α and may essentially overestimate the error if ρ or α are small. On the other hand, the right-hand side of (46) is stable with respect to small values of ρ and α. Regrettably, it does not possess the "exactness" (in the above-discussed sense) because it may have a "gap" between the left- and right-hand sides for any y.

An upper bound of the error that combines positive features of (46) and (47) can be derived (as in [9,10]) if a certain optimization procedure is used in order to select optimal functions μ_1 and μ_2. If y is given, then optimal μ_1 and μ_2 can be found analytically. It is not difficult to see that μ_1 must minimize the quantity

$$\int_\Omega \left(\kappa\gamma_2(1-\mu_1)^2 + \frac{\mu_1^2}{\rho} \right) r_1(v,y)^2\,dx. \tag{48}$$

The quantity attains its minimum with

$$\mu_1(x) = \mu_1^{opt}(x) := \frac{\kappa\gamma_2}{\kappa\gamma_2 + \rho(x)^{-1}} \quad \text{in } \Omega. \tag{49}$$

Similarly, we find that the integrals associated with Γ_3 attain minimum if

$$\mu_2(x) = \mu_2^{opt}(x) := \frac{\kappa\gamma_3}{\kappa\gamma_3 + \alpha(x)^{-1}} \quad \text{on } \Gamma_3. \tag{50}$$

Substituting these values to (41) results in the estimate

$$\begin{aligned} |\!|\!| u - v |\!|\!|^2 \leq \kappa \Bigg(& \gamma_1 D(\nabla v, y) + \gamma_2 \left\| \frac{\sqrt{\kappa^2\gamma_2^2\rho + 1}}{\kappa\gamma_2\rho + 1} r_1(v,y) \right\|_\Omega^2 \\ & + \gamma_3 \left\| \frac{\sqrt{\kappa^2\gamma_3^2\alpha + 1}}{\kappa\gamma_3\alpha + 1} r_2(v,y) \right\|_{\Gamma_3}^2 + \gamma_4 \| F - y\cdot\nu \|_{\Gamma_2}^2 \Bigg). \end{aligned} \tag{51}$$

Remark 1. For practical computations, it may be easier to use (41) and directly minimize its right-hand side with respect to μ_i, γ_i, and y using the following iteration procedure:

Step 1. Keep γ_i and μ_j fixed in (41) and minimize resulting quadratic functional of y in suitable finite subspace. This task can be reduced to solving a system of linear equations.
Step 2. Compute γ_i^{opt}.
Step 3. Compute μ_j^{opt} and repeat from Step 1.

We denote the right-hand side of (41) by $\mathcal{M}_\oplus(v, y, \gamma, \mu_1, \mu_2)$. This error majorant provides a guaranteed upper bound of the error, i.e.

$$|\!|\!| u - v |\!|\!|^2 \leq \mathcal{M}_\oplus(v, y, \gamma, \mu_1, \mu_2). \tag{52}$$

It is exact (in the above-discussed sense). Indeed, for $y = A\nabla u$ and $\mu_1 = \mu_2 = 1$ we obtain

$$\inf_{\gamma_i > 0} \mathcal{M}_\oplus(v, A\nabla u, \gamma, 1, 1) = |\!|\!| u - v |\!|\!|^2. \tag{53}$$

Also, it directly follows from the structure of (51) that the right-hand side is insensitive to small values of ρ and α.

3.2 Upper Bound

In this section, we derive an upper bound of $r_{\mathcal{S}}$. For this purpose, we use the majorant $\mathcal{M}_{\oplus}(v, y, \gamma, \mu_1, \mu_2)$. Since the majorant nonlinearly depends on A, ρ, and α, taking supremum over the respective indeterminacy sets imposes a more complicated task than that for the minorant. For a class of diffusion problems this task was solved in [5, 8]. Below, we deduce a simpler estimate, which can be easily exploited in practical computations and serves as a natural counterpart for the lower bound derived in Corollary 1.

Proposition 1. *Assume that all the assumptions of Section 1 hold. Then*

$$r_{\mathcal{S}}^2 \leq \overline{\mathrm{C}} r_{\mathcal{S}\oplus}^2 \quad \text{and} \quad \hat{r}_{\mathcal{S}}^2 \leq \overline{\mathrm{C}} \hat{r}_{\mathcal{S}\oplus}^2, \tag{54}$$

where

$$r_{\mathcal{S}\oplus}^2 = \frac{\delta_1^2}{\underline{c}_1 - \delta_1} \|\nabla u_0\|_{\Omega}^2 + \frac{\delta_2^2}{\underline{c}_2 - \delta_2} \|u_0\|_{\Omega}^2 + \frac{\delta_3^2}{\underline{c}_3 - \delta_3} \|u_0\|_{\Gamma_3}^2 \tag{55}$$

and

$$\hat{r}_{\mathcal{S}\oplus}^2 = \max_{i \in \{1,2,3\}} \frac{\delta_i^2}{\underline{c}_i(\underline{c}_i - \delta_i)}. \tag{56}$$

Proof. By properties of the majorant, we have

$$\begin{aligned}
\sup_{\tilde{u} \in \mathcal{S}} \|u_0 - \tilde{u}\|^2 &= \sup_{\tilde{u} \in \mathcal{S}} \left\{ \inf_{y, \mu_i, \gamma_j} \mathcal{M}_{\oplus}^{A,\rho,\alpha}(u_0, y, \gamma, \mu_1, \mu_2) \right\} \\
&\leq \inf_{y, \mu_i, \gamma_j} \left\{ \sup_{A,\rho,\alpha} \mathcal{M}_{\oplus}^{A,\rho,\alpha}(u_0, y, \gamma, \mu_1, \mu_2) \right\}.
\end{aligned}$$

Applying (7), we obtain

$$r_{\mathcal{S}}^2 \leq \overline{\mathrm{C}} \inf_{y, \mu_i, \gamma_j} \left\{ \sup_{A,\rho,\alpha} \mathcal{M}_{\oplus}^{A,\rho,\alpha}(u_0, y, \gamma, \mu_1, \mu_2) \right\}. \tag{57}$$

Our task is to explicitly estimate the term in brackets. For this purpose, we estimate from above the last two terms of the majorant and represent it in the form

$$\begin{aligned}
&\mathcal{M}_{\oplus}^{A,\rho,\alpha}(u_0, y, \gamma, \mu_1, \mu_2) \\
&\leq \kappa \Bigg(\gamma_1 D(\nabla v, y) + \left\| \sqrt{\gamma_2 \kappa (1 - \mu_1)^2 + \frac{\mu_1^2}{\kappa(\underline{c}_2 - \delta_2)}}\, r_1(v, y) \right\|_{\Omega}^2 \\
&+ \left\| \sqrt{\gamma_3 \kappa (1 - \mu_2) + \frac{\mu_2}{\kappa(\underline{c}_3 - \delta_3)}}\, r_2(v, y) \right\|_{\Gamma_3}^2 + \gamma_4 \|F - y \cdot \nu\|_{\Gamma_2}^2 \Bigg).
\end{aligned} \tag{58}$$

Now, we find upper bounds with respect to $A \in \Lambda_A$, $\rho \in \Lambda_\rho$, and $\alpha \in \Lambda_\alpha$ separately.

First, we consider the term D generated by A and A^{-1}:

$$\sup_{A \in \Lambda_A} D(\nabla u_0, y) = \sup_{|\Psi|<1} \int_\Omega (A_0 + \delta_1 \Psi)^{-1} |(A_0 + \delta \Psi)\nabla u_0 - y|^2 \, dx$$

$$\leq \frac{1}{\underline{c}_1 - \delta_1} \sup_{|\Psi|<1} \left\{ \|A_0 \nabla u_0 - y\|^2 + 2\delta_1 \int_\Omega \Psi \nabla u_0 \cdot (A_0 \nabla u_0 - y) \, dx + \delta_1^2 \|\Psi \nabla u_0\|^2 \right\}$$

$$\leq \frac{1}{\underline{c}_1 - \delta_1} \left(\|A_0 \nabla u_0 - y\|_\Omega^2 + 2\delta_1 \int_\Omega |\nabla u_0| \, |A_0 \nabla u_0 - y| \, dx + \delta_1^2 \|\nabla u_0\|_\Omega^2 \right). \quad (59)$$

For the term related to the error in equilibrium equation, we have

$$\sup_{\rho \in \Lambda_\rho} \|r_1^\rho(u_0, y)\|_\Omega^2 = \sup_{|\psi_2|<1} \int_\Omega (f - (\rho_0 + \delta_2 \psi_2) u_0 + \operatorname{div} y)^2 \, dx$$

$$= \sup_{|\psi_2|<1} \int_\Omega \left(\operatorname{div} y - \operatorname{div}(A_0 \nabla u_0) - \delta_2 \psi_2 u_0 \right)^2 \, dx$$

$$\leq \| \operatorname{div}(y - A_0 \nabla u_0)\|_\Omega^2 + 2\delta_2 \int_\Omega | \operatorname{div}(y - A_0 \nabla u_0)| \, |u_0| \, dx + \delta_2 \|u_0\|^2. \quad (60)$$

Similarly, for the term related to the error in the Robin boundary condition we have

$$\sup_{\alpha \in \Lambda_\alpha} \|r_2^\alpha(u_0, y)\|_{\Gamma_3}^2 \leq \left\| \frac{\partial(y - A_0 \nabla u_0)}{\partial \nu} \right\|_{\Gamma_3}^2 + 2\delta_3 \int_{\Gamma_3} \left| \frac{\partial(y - A_0 \nabla u_0)}{\partial \nu} \right| |u_0| \, ds + \delta_3^2 \|u_0\|_{\Gamma_3}^2. \quad (61)$$

It is clear that for $y = y_0 := A_0 \nabla u_0$, the estimates (59)–(61) attain minimal values. In addition, we set in (58) $\mu_1 = \mu_2 = 1$ and find that

$$\mathcal{M}_\oplus^{A,\rho,\alpha}(u_0, A_0 \nabla u_0, \gamma, 1, 1) \leq \kappa \left(\frac{\delta_1^2 \gamma_1}{\underline{c}_1 - \delta_1} \|\nabla u_0\|_\Omega^2 + \frac{\delta_2^2}{\underline{c}_2 - \delta_2} \|u_0\|_\Omega^2 + \frac{\delta_3^2}{\underline{c}_3 - \delta_3} \|u_0\|_{\Gamma_3}^2 \right). \quad (62)$$

Now we tend γ_2, γ_3 and γ_4 (which are contained in κ) to infinity. Then, (62) and (57) imply (55). An upper bound for the normalized radius follows from the relation

$$\mathcal{M}_\oplus^{A,\rho,\alpha}(u_0, A_0 \nabla u_0, \gamma, 1, 1) \leq \frac{\delta_1^2}{\underline{c}_1(\underline{c}_1 - \delta_1)} \int_\Omega A \nabla u_0 \cdot \nabla u_0 \, dx + \frac{\delta_2^2}{\underline{c}_2(\underline{c}_2 - \delta_2)} \|\sqrt{\rho_0} u_0\|_\Omega^2 + \frac{\delta_3^2}{\underline{c}_3(\underline{c}_3 - \delta_3)} \|\sqrt{\alpha_0} u_0\|_{\Gamma_3}^2$$

$$\leq \max_{i \in \{1,2,3\}} \frac{\delta_i^2}{\underline{c}_i(\underline{c}_i - \delta_i)} |\!|\!| u_0 |\!|\!|^2,$$

which leads to (56).

Remark 2. The normalized lower bound in Corollary 1 is less than the upper bound established in Proposition 1. Indeed, using

$$\left(1 - \min_i \frac{\delta_i}{\overline{c}_i}\right)^{-1} = \left(\max_i \left(1 - \frac{\delta_i}{\overline{c}_i}\right)\right)^{-1} = \left(\min_i \frac{\overline{c}_i}{\overline{c}_i - \delta_i}\right)$$

to $\hat{r}^2_{\mathcal{S}\ominus}$, we arrive at the following relation between bounds:

$$\begin{aligned}\frac{\overline{C}\hat{r}^2_{\mathcal{S}\oplus}}{\underline{C}\hat{r}^2_{\mathcal{S}\ominus}} &= \frac{\left(\max_i \frac{\overline{c}_i}{\underline{c}_i - \delta_i}\right)\left(\max_i \frac{\delta_i^2}{\underline{c}_i(\underline{c}_i - \delta_i)}\right)}{\left(\min_i \frac{\underline{c}_i}{\overline{c}_i + \delta_i}\right)\left(\min_i \frac{\delta_i^2}{\overline{c}_i^2}\right)\left(\min_i \frac{\overline{c}_i}{\overline{c}_i - \delta_i}\right)} \\ &= \left(\max_i \frac{\overline{c}_i}{\underline{c}_i - \delta_i}\right)\left(\max_i \frac{\delta_i^2}{\underline{c}_i(\underline{c}_i - \delta_i)}\right)\left(\max_i \frac{\overline{c}_i + \delta_i}{\overline{c}_i}\right)\left(\max_i \frac{\overline{c}_i^2}{\delta_i^2}\right)\left(\max_i \frac{\overline{c}_i - \delta_i}{\overline{c}_i}\right).\end{aligned}$$

Maximums can be estimated from below by setting $i = 1$ everywhere. The expression simplifies to

$$\frac{\overline{C}\hat{r}^2_{\mathcal{S}\oplus}}{\underline{C}\hat{r}^2_{\mathcal{S}\ominus}} \geq \frac{\overline{c}_1}{\underline{c}_1}\left(\frac{\overline{c}_1 + \delta_1}{\underline{c}_1 - \delta_1}\right) \geq 1. \tag{63}$$

Acknowledgement. The work was supported by grant 116895 of the Academy of Finland and grant 08-01-00655-a of the Russian Foundation for Basic Researches.

References

1. I. Hlaváček, J. Chleboun, and I. Babuška. *Uncertain input data problems and the worst scenario method.* Elsevier, Amsterdam, 2004.
2. O. Mali and S. Repin. Estimates of the indeterminacy set for elliptic boundary-value problems with uncertain data. In K. Runesson and P. Diez, editors, *Adaptive Modeling and Simulation 2007*, pages 158–161, Barcelona, 2007. CIMNE.
3. O. Mali and S. Repin. Estimates of the indeterminacy set for elliptic boundary-value problems with uncertain data. *J. Math. Sci. (N. Y.)*, 150(1):1869–1874, 2008.
4. S. G. Mikhlin. *Variational methods in mathematical physics.* Macmillan, New York, 1964.
5. P. Neittaanmäki and S. Repin. *Reliable methods for computer simulation. Error control and a posteriori estimates.* Elsevier, Amsterdam, 2004.
6. S. Repin. A posteriori error estimation for variational problems with uniformly convex functionals. *Math. Comp.*, 69(230):481–500, 2000.
7. S. Repin. Two-sided estimates of deviation from exact solutions of uniformly elliptic equations. In *Proc. of the St. Petersburg Mathematical Society, Vol. IX*, pages 143–171, 2001. Translation in Amer. Math. Soc. Transl. Ser. 2, 209, Amer. Math. Soc., Providence, RI, 2003.
8. S. Repin. A posteriori error estimates taking into account indeterminacy of the problem data. *Russian J. Numer. Anal. Math. Modelling*, 18(6):507–519, 2003.

9. S. Repin and S. Sauter. Functional a posteriori estimates for the reaction-diffusion problem. *C. R. Math. Acad. Sci. Paris*, 343(5):349–354, 2006.
10. S. Repin and R. Stenberg. A posteriori error estimates for the generalized Stokes problem. *J. Math. Sci. (N. Y.)*, 142(1):1828–1843, 2007.

Guaranteed Error Bounds for Conforming Approximations of a Maxwell Type Problem

Pekka Neittaanmäki[1] and Sergey Repin[2]

[1] Department of Mathematical Information Technology, University of Jyväskylä, P.O. Box 35 (Agora), FI-40014 University of Jyväskylä, Finland, pekka.neittaanmaki@mit.jyu.fi
[2] Department of Applied Mathematics, St. Petersburg State Technical University, 195251, St. Petersburg, Russia, repin@pdmi.ras.ru

Summary. This paper is concerned with computable error estimates for approximations to a boundary-value problem

$$\operatorname{curl} \mu^{-1} \operatorname{curl} u + \kappa^2 u = j \quad \text{in } \Omega,$$

where $\mu > 0$ and κ are bounded functions. We derive a posteriori error estimates valid for any conforming approximations of the considered problems. For this purpose, we apply a new approach that is based on certain transformations of the basic integral identity. The consistency of the derived a posteriori error estimates is proved and the corresponding computational strategies are discussed.

Key words: A posteriori estimates, the Maxwell equation, guaranteed bounds of approximation errors

1 Introduction

Boundary-value problems related to the Maxwell equation are interesting from the mathematical viewpoint and arise in numerous applications. Existence and regularity properties of solutions and viable methods of approximation are well investigated and presented in the literature. Approximation methods for the Maxwell equation were investigated in, e.g. [5,6,8,11]. A posteriori estimates were obtained in [1] in the framework of the residual approach and in [2] with the help of equilibrated approach. A posteriori estimates for nonconforming approximations of $H(\text{curl})$-elliptic partial differential equations were studied in [7].

In this paper, we derive consistent a posteriori estimates by a different method, which is based upon purely functional analysis of the problem in question and do not attract specific properties of approximations or exact solutions. Earlier, such type of methods were applied to many other classes

W. Fitzgibbon et al. (eds.), *Applied and Numerical Partial Differential Equations*, Computational Methods in Applied Sciences 15, DOI 10.1007/978-90-481-3239-3_15,

of boundary-value problems (see [10, 12, 13, 16] and the references therein). The so-called functional error majorants derived by this techniques are able to estimate the error for any conforming approximation of the exact solution. We show that for the Maxwell type problem (1) such estimates follow from the corresponding generalized statement (integral identity), which defines a weak solution to the problem. The integral identity can be transformed in various ways. The more sophisticated methods of transforming (3) we apply the better estimates of the difference between an approximate solution v and the exact one u we obtain.

The outline of the paper is as follows. Section 2 contains definitions and the generalized statement of the primal problem. In Section 3, we derive a posteriori error estimate of the first type using the simple modus operandi. For problems with $\kappa > 0$ the respective estimate is presented in Proposition 1. This estimate consists of two parts related to errors in the duality relations and in the differential equation and contains no geometrical constants. An important property of this estimate is that it gives a guaranteed upper bound of the error, which is as close to the exact error as it is required provided that the parameters of the majorant are properly selected. However, as the estimates derived for the reaction-diffusion problem the estimate looses the efficiency for small κ. In Section 4, we derive another upper bound of the error, which is insensitive with respect to small values of the coefficients. This estimate contains global constants that depend only on Ω. Regrettably, we cannot prove that error majorants established in Propositions 2 and 3 are equal to the corresponding error norms if the "free" function y is properly selected. Thus, the estimates exposed in Sections 3 and 4 has certain drawbacks that may affect practical efficiency of error estimation. A way out is presented in Section 5, which is devoted to establishing a more general error majorant. The latter encompasses majorants derived in the previous sections as special cases. The majorants defined in Propositions 4 and 5 are also insensitive with respect to small values of the coefficients and as the estimate obtained in Section 3 have no gap between its right- and left-hand sides (so that a computable upper bound of the error can be as close to the exact error as it is required).

2 Notation and Basic Relations

We consider the simplest version of the Maxwell equation

$$\operatorname{curl} \mu^{-1} \operatorname{curl} u + \kappa^2 u = j \quad \text{in } \Omega, \tag{1}$$

where Ω is a bounded domain in $\mathbb{R}^d$, j is a given current density, and μ is the permeability of a medium (may be a positive constant or a positive bounded function). The case $\kappa = 0$ corresponds to stationary transverse magnetic (TM) or transverse electric (TE) equations that arise if one of the components of the electromagnetic field is excluded (e.g., see [8, 11]). The equation (1) with positive κ arises in semidiscrete approximations of the evolutionary Maxwell problem.

On Γ the condition

$$n \times u = 0 \tag{2}$$

is stated. Here, n denotes the unit outward normal to Γ. By $V(\Omega)$ we denote the space $H(\Omega, \mathrm{curl})$, which is a Hilbert space endowed with the norm

$$\|w\|_{\mathrm{curl}} := \left(\|w\|^2 + \|\,\mathrm{curl}\, w\|^2\right)^{1/2}.$$

Here and later on, the symbol $:=$ means 'equals by definition' and $\|\cdot\|$ stands for L^2-norm of scalar- and vector-valued functions.

The generalized solution $u \in V_0$ is defined by the integral relation

$$\int_\Omega \mu^{-1} \,\mathrm{curl}\, u \cdot \mathrm{curl}\, w + \kappa^2 u \cdot w \, dx = \int_\Omega j \cdot w \, dx, \tag{3}$$

where $u \cdot w$ means scalar product of vector-valued functions u and w and

$$V_0 := \{w \in V \mid w \times n = 0 \text{ on } \partial\Omega\}.$$

Henceforth, we assume that j satisfies the condition

$$\int_\Omega j \cdot \nabla\phi \, dx = 0 \quad \forall \phi \in \overset{\circ}{H}{}^1(\Omega) \tag{4}$$

and

$$0 < \mu_\ominus \leq \mu(x) \leq \mu_\oplus, \tag{5}$$
$$0 < \kappa_\ominus \leq \kappa(x) \leq \kappa_\oplus. \tag{6}$$

By scaling arguments, we can set $\mu_\oplus = 1$ without a loss of generality.

Our goal is to derive computable estimates of the difference $u - v$ where $v \in V_0$ is a function viewed as an approximation of u. Estimates are obtained for the weighted energy norm defined by the relation

$$|[w]|^2_{(\gamma,\delta)} := \int_\Omega (\gamma |\,\mathrm{curl}\, w|^2 + \delta |w|^2) \, dx.$$

The derivation method is based on transformations of the integral relation (3). It does not use specific properties of the exact solution or its approximation v (e.g., Galerkin orthogonality). Therefore, the estimates are valid for conforming approximations of all types regardless of the numerical method applied for their construction. These estimates belong to the class of functional a posteriori error estimates that has been derived for some other elliptic and parabolic problems (see [10, 13, 16] and the references therein).

3 A Posteriori Error Estimates of the First Type

3.1 Upper Bound of the Error

Proposition 1. *Assume that $\kappa > 0$ and $v \in V_0$ is an approximation of u. For any $y \in H(\Omega, \mathrm{curl})$ the following estimate holds:*

$$|[u-v]|^2_{(\mu^{-1},\kappa^2)} \leq \mathcal{M}_1^2(v,y) := \left\| \frac{1}{\kappa}\mathrm{r}(v,y) \right\|^2 + \|\mu^{1/2}\mathrm{d}(v,y)\|^2, \qquad (7)$$

where

$$\mathrm{r}(v,y) := j - \mathrm{curl}\, y - \kappa^2 v,$$
$$\mathrm{d}(v,y) := y - \mu^{-1}\,\mathrm{curl}\, v.$$

Proof. From (3) it follows that

$$\int_\Omega \left(\mu^{-1}\,\mathrm{curl}(u-v)\cdot \mathrm{curl}\, w + \kappa^2(u-v)\cdot w\right) dx$$
$$= \int_\Omega (j\cdot w - \mu^{-1}\,\mathrm{curl}\, v \cdot \mathrm{curl}\, w - \kappa^2 v \cdot w)\, dx. \qquad (8)$$

Take $y \in H(\Omega, \mathrm{curl})$ and use the identity

$$(\mathrm{curl}\, y)\cdot w = \mathrm{div}(y\times w) + y\cdot \mathrm{curl}\, w. \qquad (9)$$

Since

$$\int_\Omega \mathrm{div}(y\times w)\, dx = \int_{\partial\Omega} n\cdot(y\times w)\, ds = \int_{\partial\Omega} y\cdot(w\times n)\, ds = 0,$$

we find that

$$\int_\Omega (\mathrm{curl}\, y\cdot w - y\cdot \mathrm{curl}\, w)\, dx = 0 \quad \forall w \in V_0. \qquad (10)$$

By (8) and (10) we obtain

$$\int_\Omega \left(\mu^{-1}\,\mathrm{curl}(u-v)\cdot \mathrm{curl}\, w + \kappa^2(u-v)\cdot w\right) dx$$
$$= \int_\Omega (j - \mathrm{curl}\, y - \kappa^2 v)\cdot w\, dx + \int_\Omega (y - \mu^{-1}\,\mathrm{curl}\, v)\cdot \mathrm{curl}\, w\, dx. \qquad (11)$$

Set $w = u - v$ and estimate two integrals in the right-hand side by the Hölder inequality. We have

$$|[u-v]|^2_{(\mu^{-1},\kappa^2)} \leq \left\| \frac{1}{\kappa}\mathrm{r}(v,y) \right\| \|\kappa(u-v)\| + \|\mu^{1/2}\mathrm{d}(v,y)\| \|\mu^{-1/2}\,\mathrm{curl}(u-v)\|,$$

which implies (7).

The estimate (7) shows that the distance between u and v measured in terms of the weighted norm $|[u-v]|_{(\mu^{-1},\kappa^2)}$ is bounded from above by the sum of two residuals $\mathrm{r}(v,y)$ and $\mathrm{d}(v,y)$ that are associated with the decomposition of (1), which has the form

$$\operatorname{curl} p + \kappa^2 u - j = 0,$$
$$p = \mu^{-1} \operatorname{curl} u.$$

We note that the estimate (7) has no gap between its left- and right-hand sides. Indeed, if we set $y = \mu^{-1} \operatorname{curl} u$ then

$$\|\mu^{1/2}\mathrm{d}(v,y)\| = \|\mu^{-1/2} \operatorname{curl}(u-v)\|$$

and

$$\left\|\frac{1}{\kappa}\mathrm{r}(v,y)\right\| = \|\kappa(u-v)\|$$

so that (7) holds as the equality. However, for small κ the estimate becomes sensitive with respect to $\mathrm{r}(v,y)$ and may loose practical efficiency if the value of this residual is not much smaller than $\mathrm{r}(v,y)$.

Remark 1. If $\kappa > 0$ only in $\Omega_+ \subset \Omega$, then (11) can be transformed as follows:

$$\int_\Omega \left(\mu^{-1} \operatorname{curl}(u-v) \cdot \operatorname{curl} w + \kappa^2 (u-v) \cdot w\right) dx$$
$$= \int_{\Omega_+} (j - \operatorname{curl} y - \kappa^2 v) \cdot w \, dx + \int_\Omega (y - \mu^{-1} \operatorname{curl} v) \cdot \operatorname{curl} w \, dx,$$

which implies the estimate

$$|[u-v]|^2_{(\mu^{-1},\kappa^2)} \leq \left\|\frac{1}{\kappa}\mathrm{r}(v,y)\right\|^2_{\Omega_+} + \|\mu^{1/2}\mathrm{d}(v,y)\|^2.$$

3.2 Lower Bound of the Error

A lower bound of the error norm is derived by the following arguments. First, we note that

$$\sup_{w \in V_0} \int_\Omega \Big(\mu^{-1} \operatorname{curl}(u-v) \cdot \operatorname{curl} w$$
$$+ \kappa^2 w \cdot (u-v) - \frac{1}{2}(\mu^{-1} \operatorname{curl} w \cdot \operatorname{curl} w + \kappa^2 w \cdot w)\Big) \, dx$$
$$\leq \sup_{\substack{\tau \in L^2(\Omega,\mathbb{R}^d) \\ w \in L^2(\Omega,\mathbb{R}^d)}} \int_\Omega \Big(\mu^{-1} \operatorname{curl}(u-v) \cdot \tau - \frac{1}{2}\mu^{-1} \tau \cdot \tau$$
$$+ \kappa^2 w \cdot (u-v) - \frac{1}{2}\kappa^2 w \cdot w\Big) \, dx = \frac{1}{2}|[u-v]|^2_{(\mu^{-1},\kappa^2)}.$$

On the other hand,

$$\sup_{w\in V_0}\int_\Omega\Big(\mu^{-1}\operatorname{curl}(u-v)\cdot\operatorname{curl}w$$
$$+\kappa^2 w\cdot(u-v)-\frac{1}{2}(\mu^{-1}\operatorname{curl}w\cdot\operatorname{curl}w+\kappa^2 w\cdot w)\Big)\,dx$$
$$\geq\int_\Omega\Big(\mu^{-1}\operatorname{curl}(u-v)\cdot\operatorname{curl}(u-v)$$
$$+\kappa^2(u-v)\cdot(u-v)-\frac{1}{2}(\mu^{-1}|\operatorname{curl}(u-v)|^2+\kappa^2|u-v|^2)\Big)\,dx$$
$$=\frac{1}{2}|[u-v]|^2_{(\mu^{-1},\kappa^2)}.$$

Thus, we conclude that

$$\frac{1}{2}|[u-v]|^2_{(\mu^{-1},\kappa^2)}=\sup_{w\in V_0}\int_\Omega\Big(\mu^{-1}\operatorname{curl}(u-v)\cdot\operatorname{curl}w$$
$$+\kappa^2 w\cdot(u-v)-\frac{1}{2}(\mu^{-1}\operatorname{curl}w\cdot\operatorname{curl}w+\kappa^2 w\cdot w)\Big)\,dx.$$

By (3), we obtain

$$|[u-v]|^2_{(\mu^{-1},\kappa^2)}\geq\mathcal{M}^2_\ominus(v,w),\tag{12}$$

where

$$\mathcal{M}^2_\ominus(v,w):=\int_\Omega\big(2j\cdot w-\mu^{-1}|\operatorname{curl}w|^2-\kappa^2|w|^2$$
$$-2\mu^{-1}\operatorname{curl}v\cdot\operatorname{curl}w-2\kappa^2 v\cdot w\big)\,dx.$$

For any $w\in V_0$ the quantity $\mathcal{M}^2_\ominus(v,w)$ provides a lower bound of the error. Certainly, the sharpest bound is given by the quantity

$$M^2_\ominus(v):=\sup_{w\in V_0}\mathcal{M}^2_\ominus(v,w).$$

It is not difficult to prove that this quantity coincides with the squared error (to prove that it suffices to set $w=u-v$).

3.3 Practical Implementation

Practically computable upper (lower) bounds can be determined if minimization of the majorant (maximization of the minorant) is performed over a finite-dimensional subspace $V_k\subset V$, $\dim V_k=k$ ($V_{0m}\subset V_0$, $\dim V_{0m}=m$). Then, finding the quantities

$$M_{k\oplus}(v):=\inf_{y\in V_k}\mathcal{M}^2_\oplus(v,y),\tag{13}$$

$$M_{m\ominus}(v) := \sup_{w \in V_{0m}} \mathcal{M}^2_{\ominus}(v, w) \tag{14}$$

requires solving quadratic type minimization (maximization) problems what can be done by standard methods.

We note that conforming approximations in V (and in V_0) are usually constructed by the Nédélec elements (see [9]), which are also natural to use for the construction of V_k (and V_{0m}). If V_k and V_{0m} are limit dense in V and V_0, respectively (for $k, m \to +\infty$), then it is easy to prove that

$$M_{k\oplus}(v) \to |[u - v]|_{(\mu^{-1},\kappa^2)} \quad \text{and} \quad M_{m\ominus}(v) \to |[u - v]|_{(\mu^{-1},\kappa^2)}.$$

The ratio

$$i_{km} := \frac{M_{k\oplus}(v)}{M_{m\ominus}(v)}$$

is, indeed, computable. It shows the efficiency of the error estimation.

4 A Posteriori Error Estimate of the Second Type

In this section, we derive a posteriori estimates of a more general type assuming that κ is a positive constant. By the Helmholtz decomposition of a vector-valued function, we represent the exact solution u

$$u = u_0 + \nabla\psi,$$

where u_0 is a solenoidal vector-valued function and $\psi \in \overset{\circ}{H}{}^1(\Omega)$. Since $\operatorname{curl}\nabla\psi = 0$, we rewrite (3) as follows:

$$\int_\Omega \mu^{-1} \operatorname{curl} u_0 \cdot \operatorname{curl} w + \kappa^2(u_0 + \nabla\psi) \cdot w \, dx = \int_\Omega j \cdot w \, dx. \tag{15}$$

Next, we make the same decomposition for the trial function and set $w = w_0 + \nabla\phi$. Recall that

$$\int_\Omega j \cdot \nabla\phi \, dx = \int_\Omega u_0 \cdot \nabla\phi \, dx = \int_\Omega w_0 \cdot \nabla\psi \, dx = 0.$$

We observe that

$$\int_\Omega (\mu^{-1} \operatorname{curl} u_0 \cdot \operatorname{curl} w_0 + \kappa^2 u_0 \cdot w_0 + \kappa^2 \nabla\psi \cdot \nabla\phi) \, dx = \int_\Omega j \cdot w_0 \, dx. \tag{16}$$

In (16), we set $w_0 = 0$ and $\phi = \psi$. We find that $\|\nabla\psi\| = 0$. Hence, u is a divergence-free function.

We use this fact to rearrange (11) in a different way. We have

$$\int_\Omega \mathrm{r}(v, y) \cdot w \, dx = \int_\Omega \mathrm{r}(v, y) \cdot (w_0 + \nabla\phi) \, dx \leq \|\mathrm{r}(v, y)\| \left(\|w_0\| + \|\nabla\phi\|\right). \tag{17}$$

Note that ϕ satisfies the relation

$$\int_\Omega \nabla\phi \cdot \nabla\widetilde{\phi}\, dx = \int_\Omega w \cdot \nabla\widetilde{\phi}\, dx = -\int_\Omega (\operatorname{div} w)\widetilde{\phi}\, dx \quad \forall \widetilde{\phi} \in \overset{\circ}{H}{}^1(\Omega), \tag{18}$$

which implies the estimate

$$\|\nabla\phi\| \le C_1(\Omega)\|\operatorname{div} w\|, \tag{19}$$

where $C_1(\Omega)$ is the constant in the Friedrich inequality for the domain Ω. For solenoidal fields we also have the estimate (see, e.g. [4, 8, 18])

$$\|w_0\| \le C_2(\Omega)\|\operatorname{curl} w_0\| = C_2(\Omega)\|\operatorname{curl} w\|. \tag{20}$$

Hence,

$$\int_\Omega \mathrm{r}(v,y) \cdot w\, dx \le \|\mathrm{r}(v,y)\| \left(C_1(\Omega)\|\operatorname{div} w\| + C_2(\Omega)\|\operatorname{curl} w\|\right) \tag{21}$$

and we arrive at the estimate

$$\begin{aligned}\int_\Omega (\mu^{-1}|\operatorname{curl}(u-v)|^2 + \kappa^2|u-v|^2)\, dx \\ \le (\|\mathrm{d}(v,y)\| + C_2(\Omega)\mathrm{r}(v,y))\, \|\operatorname{curl}(u-v)\| \\ + C_1(\Omega)\|\mathrm{r}(v,y)\|\|\operatorname{div}(u-v)\| \le \frac{\alpha}{4}\left(\|\mathrm{d}(v,y)\| + C_2(\Omega)\mathrm{r}(v,y)\|\right)^2 \\ + \frac{1}{\alpha}\|\operatorname{curl}(u-v)\|^2 + C_1(\Omega)\|\mathrm{r}(v,y)\|\|\operatorname{div} v\|,\end{aligned} \tag{22}$$

where $\alpha \ge \mu$.

Hence, we arrive at the following result:

Proposition 2. *If κ is a positive constant and $v \in V_0 \cap H(\Omega, \operatorname{div})$ then for any $y \in H(\Omega, \operatorname{curl})$*

$$\begin{aligned}|[u-v]|^2_{((\frac{1}{\mu}-\frac{1}{\alpha}),\kappa^2)} \le \frac{\alpha}{4}\left(\|\mathrm{d}(v,y)\| + C_2(\Omega)\|\mathrm{r}(v,y)\|\right)^2 \\ + C_1(\Omega)\|\mathrm{r}(v,y)\|\|\operatorname{div} v\|.\end{aligned} \tag{23}$$

If $\operatorname{div} v = 0$, *then the estimate is simplified and has the form*

$$|[u-v]|_{((\frac{1}{\mu}-\frac{1}{\alpha}),\kappa^2)} \le \frac{\sqrt{\alpha}}{2}\left(\|\mathrm{d}(v,y)\| + C_2(\Omega)\|\mathrm{r}(v,y)\|\right). \tag{24}$$

We can use a somewhat different way and estimate the first term in the right-hand side of (11) as follows:

$$\int_\Omega \mathrm{r}(v,y) \cdot w\, dx \le \|\mathrm{r}(v,y)\| \left(C_1(\Omega)\|\operatorname{div} w\| + C_2(\Omega)\mu_\oplus^{1/2}\|\mu^{-1/2}\operatorname{curl} w\|\right). \tag{25}$$

Set $w = u - v$ and note that $\operatorname{div}(u - v) = \operatorname{div} v$. Then we obtain

$$
\begin{aligned}
|[u-v]|^2_{(\mu^{-1},\kappa^2)} &\leq C_1(\Omega)\|\operatorname{div} v\|\|\mathrm{r}(v,y)\| \\
&+ \left(C_2(\Omega)\mu_\oplus^{1/2}\|\mathrm{r}(v,y)\| + \|\mu^{1/2}\mathrm{d}(v,y)\|\right)\|\mu^{-1/2}\operatorname{curl}(u-v)\| \\
&\leq C_1(\Omega)\|\operatorname{div} v\|\|\mathrm{r}(v,y)\| \\
&+ \left(C_2(\Omega)\mu_\oplus^{1/2}\|\mathrm{r}(v,y)\| + \|\mu^{1/2}\mathrm{d}(v,y)\|\right)|[u-v]|_{(\mu^{-1},\kappa^2)} \qquad (26)
\end{aligned}
$$

and arrive at the following result.

Proposition 3. *If κ is a positive constant and $v \in V_0 \cap H(\Omega, \operatorname{div})$ then for any $y \in H(\Omega, \operatorname{curl})$*

$$
|[u-v]|_{(\mu^{-1},\kappa^2)} \leq \mathcal{M}_2(v,y) := \frac{R_2}{2} + \sqrt{R_1 + \frac{R_2^2}{4}}, \qquad (27)
$$

where

$$
R_1 = C_1(\Omega)\|\operatorname{div} v\|\|\mathrm{r}(v,y)\|
$$

and

$$
R_2 := C_2(\Omega)\mu_\oplus^{1/2}\|\mathrm{r}(v,y)\| + \|\mu^{1/2}\mathrm{d}(v,y)\|.
$$

If, in addition, $\operatorname{div} v = 0$, *then*

$$
|[u-v]|_{(\mu^{-1},\kappa^2)} \leq R_2. \qquad (28)
$$

Remark 2. If $\kappa = 0$, then (28) has the form

$$
\|\mu^{-1}\operatorname{curl}(u-v)\| \leq R_2. \qquad (29)
$$

For $\kappa = 0$, this estimate was earlier derived in [14,15].

The estimates (23) and (27) are insensitive with respect to small values of κ (what differs them from (7)). However, we made a certain overestimation of the right-hand side in the last transformation of (22). Therefore, we cannot guarantee that this upper bound has no gap (substitution of $y = \mu^{-1}\operatorname{curl} u$ does not make the respective right-hand sides equal to the error).

5 A Posteriori Estimate of the Third Type

5.1 An Advanced Form of the Error Majorant

To derive upper bounds that possess all positive features of the estimates of the first and second types we use a more sophisticated method.

Proposition 4. *Let v and y satisfy the assumptions of Proposition 2. Then*

$$|[u-v]|^2_{\gamma,\delta} \le \mathcal{M}^2_3(\lambda,\alpha_1,\alpha_2,v,y), \tag{30}$$

where

$$\mathcal{M}^2_3(\lambda,\alpha_1,\alpha_2,v,y) := R_1(\lambda,v,y) + \frac{\alpha_1}{4}R^2_2(\lambda,v,y) + \frac{\alpha_2}{4}R^2_3(\lambda,v,y),$$

α_1 and α_2 are arbitrary numbers in $[1,+\infty)$,

$$\gamma = \left(1-\frac{1}{\alpha_1}\right)\mu^{-1}, \quad \delta = \left(1-\frac{1}{\alpha_2}\right)\kappa^2,$$
$$\lambda \in I_{[0,1]} := \{\lambda \in L^\infty(\Omega) \mid \lambda(x) \in [0,1] \text{ for a.e. } x \in \Omega\},$$

and R_i, $i=1,2,3$, are defined by (33)–(35).

Proof. With the help of λ we decompose the integral identity (11) as follows (in [17], a similar method was used for the decomposition of the reaction-diffusion equation):

$$\begin{aligned}\int_\Omega &\left(\mu^{-1}\operatorname{curl}(u-v)\cdot\operatorname{curl} w + \kappa^2(u-v)\cdot w\right)dx\\ &= \int_\Omega \lambda(j-\operatorname{curl} y-\kappa^2 v)\cdot w\,dx + \int_\Omega (1-\lambda)(j-\operatorname{curl} y-\kappa^2 v)\cdot w\,dx\\ &\qquad + \int_\Omega (y-\mu^{-1}\operatorname{curl} v)\cdot\operatorname{curl} w\,dx,\end{aligned} \tag{31}$$

where $\lambda \in I_{[0,1]}$. Since

$$\int_\Omega \lambda \mathrm{r}(v,y)\cdot(u-v)\,dx \le \left\|\frac{\lambda}{\kappa}\mathrm{r}(v,y)\right\| \|\kappa(u-v)\|$$

and

$$\begin{aligned}\int_\Omega &(1-\lambda)\mathrm{r}(v,y)\cdot(u-v)\,dx\\ &\le \|(1-\lambda)\mathrm{r}(v,y)\|\left(C_1(\Omega)\|\operatorname{div} v\| + C_2(\Omega)\mu_\oplus^{1/2}\|\mu^{-1/2}\operatorname{curl}(u-v)\|\right),\end{aligned}$$

we obtain

$$\begin{aligned}\int_\Omega &\left(\mu^{-1}|\operatorname{curl}(u-v)|^2 + \kappa^2|u-v|^2\right)dx\\ &\le R_1 + R_2\|\mu^{-1/2}\operatorname{curl}(u-v)\| + R_3\|\kappa(u-v)\|,\end{aligned} \tag{32}$$

where

$$R_1(\lambda,v,y) = C_1(\Omega)\|(1-\lambda)\mathrm{r}(v,y)\|\|\operatorname{div} v\|, \tag{33}$$

$$R_2(\lambda, v, y) = C_2(\Omega)\mu_\oplus^{1/2}\|(1-\lambda)\mathrm{r}(v,y)\| + \|\mu^{1/2}\mathrm{d}(v,y)\|, \tag{34}$$

$$R_3(\lambda, v, y) = \left\|\frac{\lambda}{\kappa}\mathrm{r}(v,y)\right\|. \tag{35}$$

By applying the Young inequality to the right-hand side of (32), we obtain

$$\int_\Omega \left(1-\frac{1}{\alpha_1}\right)\mu^{-1}|\operatorname{curl}(u-v)|^2\,dx + \int_\Omega \left(1-\frac{1}{\alpha_2}\right)\kappa^2|u-v|^2\,dx \\ \leq R_1 + \frac{\alpha_1}{4}R_2^2 + \frac{\alpha_2}{4}R_3^2, \tag{36}$$

which implies (30).

Corollary 1. *If $\alpha_1 = \alpha_2 = 2$ then (30) comes in the form*

$$|[u-v]|^2_{(\mu^{-1},\kappa^2)} \leq 2R_1(\lambda,v,y) + R_2^2(\lambda,v,y) + R_3^2(\lambda,v,y). \tag{37}$$

The proposition below shows that the estimate (30) possesses the same principal property as (7): it has no gap between the left- and right-hand sides.

Proposition 5. *If $\alpha_1 = \alpha_2 = 2$, then*

$$|[u-v]|^2_{(\mu^{-1},\kappa^2)} = M_\oplus(v), \tag{38}$$

where

$$M_\oplus(v) := \inf_{\substack{\lambda\in I_{[0,1]},\\ y\in H(\Omega,\mathrm{curl})}} \left\{2R_1(\lambda,v,y) + R_2^2(\lambda,v,y) + R_3^2(\lambda,v,y)\right\}. \tag{39}$$

Proof. Obviously,

$$M_\oplus \leq 2R_1(1,v,p) + R_2^2(1,v,p) + R_3^2(1,v,p),$$

where $p = \mu^{-1}\operatorname{curl} u$. Note that

$$R_1(1,v,p) = 0,$$
$$R_2(1,v,p) = \|\mu^{-1/2}\operatorname{curl}(u-v)\|,$$
$$R_3(1,v,p) = \|\kappa(u-v)\|.$$

Therefore,

$$M_\oplus = \|\mu^{-1/2}\operatorname{curl}(u-v)\|^2 + \|\kappa(u-v)\|^2 = |[u-v]|^2_{(\mu^{-1},\kappa^2)}.$$

5.2 Optimal Form of λ

Now our goal is to derive the sharpest upper bound by defining the function $\lambda(x)$ in an "optimal" way. For this purpose, we first reform (36) by introducing positive parameters α_3 and α_4 and noting that

$$\begin{aligned} R_1(\lambda, v, y) &\leq \frac{\alpha_3}{2} C_1^2(\Omega) \|(1-\lambda)\mathrm{r}(v,y)\|^2 + \frac{1}{2\alpha_3}\|\operatorname{div} v\|^2, \\ R_2^2(\lambda, v, y) &\leq (1+\alpha_4) C_2^2(\Omega)\mu_\oplus \|(1-\lambda)\mathrm{r}(v,y)\|^2 \\ &\quad + \left(1+\frac{1}{\alpha_4}\right)\|\mu^{1/2}\mathrm{d}(v,y)\|^2. \end{aligned}$$

Therefore, (36) implies

$$\begin{aligned} &\int_\Omega \left(1-\frac{1}{\alpha_1}\right)\mu^{-1}|\operatorname{curl}(u-v)|^2\,dx + \int_\Omega \left(1-\frac{1}{\alpha_2}\right)\kappa^2|u-v|^2\,dx \\ &\leq \int_\Omega \left((1-\lambda)^2 P + \lambda^2 Q\right)\mathrm{r}^2(v,y)\,dx + \left(1+\frac{1}{\alpha_4}\right)\frac{\alpha_1}{4}\|\mu^{1/2}\mathrm{d}(v,y)\|^2 + \frac{1}{2\alpha_3}\|\operatorname{div} v\|^2, \end{aligned} \tag{40}$$

where

$$\begin{aligned} P &= \frac{\alpha_3}{2} C_1^2(\Omega) + (1+\alpha_4)\frac{\alpha_1}{4} C_2^2(\Omega)\mu_\oplus, \\ Q &= \frac{\alpha_2}{4\kappa^2}. \end{aligned}$$

Optimal λ is defined by the relation

$$\lambda = \frac{P}{P+Q} \in [0,1]$$

and the respective estimate reads

$$\begin{aligned} &\int_\Omega \left(1-\frac{1}{\alpha_1}\right)\mu^{-1}|\operatorname{curl}(u-v)|^2\,dx + \int_\Omega \left(1-\frac{1}{\alpha_2}\right)\kappa^2|u-v|^2\,dx \\ &\leq \int_\Omega \frac{PQ}{P+Q}\mathrm{r}^2(v,y)\,dx + \left(1+\frac{1}{\alpha_4}\right)\frac{\alpha_1}{4}\|\mu^{1/2}\mathrm{d}(v,y)\|^2 + \frac{1}{2\alpha_3}\|\operatorname{div} v\|^2. \end{aligned} \tag{41}$$

Remark 3. Note that

$$\frac{PQ}{P+Q} \leq \min\{P, Q\}.$$

Therefore, the estimate (41) is insensitive to small values of κ^2.

Acknowledgement. This research was supported by grant 116895 of the Academy of Finland and grant 08-01-00655-a of the Russian Foundation for Basic Researches.

References

1. R. Beck, R. Hiptmair, R. H. W. Hoppe, and B. Wohlmuth. Residual based a posteriori error estimators for eddy current computation. *M2AN Math. Model. Numer. Anal.*, 34(1):159–182, 2000.
2. D. Braess and Schöberl. Equilibrated residual error estimator for Maxvell's equation. To appear.
3. G. Duvaut and J.-L. Lions. *Les inéquations en mécanique et en physique*. Dunod, Paris, 1972.
4. V. Girault and P.-A. Raviart. *Finite element methods for Navier–Stokes equations. Theory and algorithms.* Springer, Berlin, 1986.
5. G. Haase, M. Kuhn, and U. Langer. Parallel multigrid 3D Maxwell solvers. *Parallel Comput.*, 27(6):761–775, 2001.
6. R. Hiptmair. Multigrid method for Maxwell's equations. *SIAM J. Numer. Anal.*, 36(1):204–225, 1999.
7. P. Houston, I. Perugia, and D. Schötzau. An a posteriori error indicator for discontinuous Galerkin discretizations of H(curl)-elliptic partial differential equations. *IMA J. Numer. Anal.*, 27(1):122–150, 2007.
8. P. Monk. *Finite element methods for Maxwell's equations.* Oxford University Press, New York, 2003.
9. J.-C. Nédélec. A new family of mixed finite elements in R^3. *Numer. Math.*, 50(1):57–81, 1986.
10. P. Neittaanmäki and S. Repin. *Reliable methods for computer simulation. Error control and a posteriori estimates.* Elsevier, Amsterdam, 2004.
11. O. Pironneau. Computer solutions of Maxwell's equations in homogeneous media. *Internat. J. Numer. Methods Fluids*, 43(8):823–838, 2003. ECCOMAS Computational Fluid Dynamics Conference, Part III (Swansea, 2001).
12. S. Repin. A posteriori error estimation for nonlinear variational problems by duality theory. *Zap. Nauchn. Sem. S.-Peterburg. Otdel. Mat. Inst. Steklov. (POMI)*, 243:201–214, 1997. Translation in J. Math. Sci. (New York) 99(1):927–935, 2000.
13. S. Repin. A posteriori error estimation for variational problems with uniformly convex functionals. *Math. Comp.*, 69(230):481–500, 2000.
14. S. Repin. On the derivation of functional a posteriori estimates from integral identities. In W. Fitzgibbon, R. Hoppe, J. Periaux, O. Pironneau, and Y. Vassilevski, editors, *Advances in Numerical Mathematics. Proceedings of International Conference on the Occasion of the 60th birthday of Yu. A. Kuznetsov (Moscow, 2005)*, pages 217–242, Moscow and Houston, TX, 2006. Institute of Numerical Mathematics of Russian Academy of Sciences and Department of Mathematics, University of Houston.
15. S. Repin. Functional a posteriori estimates for Maxwell's equation. *J. Math. Sci. (N. Y.)*, 142(1):1821–1827, 2007.
16. S. Repin. A posteriori error estimation methods for partial differential equations. In M. Kraus and U. Langer, editors, *Lectures on Advanced Computational Methods in Mechanics*, pages 161–226. Walter de Gruyter, Berlin, 2007.
17. S. Repin and S. Sauter. Functional a posteriori estimates for the reaction-diffusion problem. *C. R. Math. Acad. Sci. Paris*, 343(5):349–354, 2006.
18. J. Saranen. On an inequality of Friedrichs. *Math. Scand.*, 51(2):310–322, 1982.

A Componentwise Splitting Method for Pricing American Options Under the Bates Model

Jari Toivanen

Institute for Computational and Mathematical Engineering, Stanford University, Stanford, CA 94305, USA, toivanen@stanford.edu

Summary. A linear complementarity problem (LCP) is formulated for the price of American options under the Bates model which combines the Heston stochastic volatility model and the Merton jump-diffusion model. A finite difference discretization is described for the partial derivatives and a simple quadrature is used for the integral term due to jumps. A componentwise splitting method is generalized for the Bates model. It is leads to solution of sequence of one-dimensional LCPs which can be solved very efficiently using the Brennan and Schwartz algorithm. The numerical experiments demonstrate the componentwise splitting method to be essentially as accurate as the PSOR method, but order of magnitude faster. Furthermore, pricing under the Bates model is less than twice more expensive computationally than under the Heston model in the experiments.

1 Introduction

During the last couple of decades, the trading of options has grown to tremendous scale. The most basic options give either the right to sell (put) or buy (call) the underlying asset with the strike price. European options can be exercised only at the expiry time while American options can be exercised any time before the expiry. Usually American options need to be priced numerically due to the early exercise possibility. One approach is to formulate a linear complementarity problem (LCP) or variational inequality with a partial (integro-)differential operator for the price and then solve it numerically after discretization. Since the books by Glowinski, Lions, and Trémolières [17] and by Glowinski [14], these problems have been extensively studied.

For pricing options, a model is needed for the behavior of the value of the underlying asset. Many such models of varying complexity have been developed. More complicated models reproduce more realistic paths for the value and match between the market price and model prices of options is better, but they also make pricing more challenging. In the Black–Scholes model [5], the value is a geometric Brownian motion. The Merton model [26] adds log-normally distributed jumps to the Black–Scholes model while in the Kou

W. Fitzgibbon et al. (eds.), *Applied and Numerical Partial Differential Equations*, Computational Methods in Applied Sciences 15, DOI 10.1007/978-90-481-3239-3_16,

model [23], the jumps are log-doubly-exponentially distributed. The Heston model [19] makes the volatility also stochastic in the Black–Scholes model. The Bates model [4] which is also sometimes called as the Heston–Merton model adds to the Heston model log-normally distributed jumps. The correlated jump model [12] allows also the volatility in the Bates model to jump.

Many methods have been proposed for solving the resulting LCPs. The Brennan and Schwartz algorithm [6] is a direct method for pricing American options under the Black–Scholes model; see also [21]. Numerical methods pricing under the Heston model have been considered in [8, 20, 22, 27, 35]. The treatment of the jumps in the Merton and Kou models have been studied in [2, 3, 9, 10, 25, 32]. Pricing under the Bates model has been considered in [7] and under the correlated jump model in [13].

In this paper, we consider pricing American call options under the Bates model. We discretize the spatial partial derivatives in the resulting partial integro-differential operator using a seven-point finite difference stencil. The integral term is discretized using a simple quadrature. The Rannacher scheme [29] is employed in the time stepping. We treat the LCP by introducing a generalization for the componentwise splitting method in [20]. The numerical experiments demonstrate that the proposed method is orders of magnitude faster than the PSOR method.

The outline of the paper is the following. The Bates model and an LCP for an American call option is described in Section 2. The discretization of LCPs is constructed in Section 3. The componentwise splitting method is proposed in Section 4. Numerical experiments are presented in Section 5 and conclusions are given in Section 6.

2 Option Pricing Model

In the following, we give coupled stochastic differential equations describing the Bates model. Then, we give an LCP for the price of an American call option when the market prices of the volatility and jump risks are zero.

2.1 Bates Model

The Bates stochastic volatility model with jumps [4] combines the Merton jump model [26] and the Heston stochastic volatility model [19]. It describes the behavior of the asset value x and its variance y by the coupled stochastic differential equations

$$\begin{aligned} dx &= (\mu - \lambda\xi)x dt + \sqrt{y} x dw_1 + (J-1)x dn, \\ dy &= \kappa(\theta - y)dt + \sigma\sqrt{y} dw_2, \end{aligned} \tag{1}$$

where μ is the growth rate of the asset value, κ is the rate of reversion to the mean level of y, θ is the mean level of y, and σ is the volatility of the variance y.

The two Wiener processes w_1 are w_2 have the correlation ρ. The Poisson arrival process n has the rate λ. The jump size J is taken from a distribution

$$f(J) = \frac{1}{\sqrt{2\pi}\delta J} \exp\left(-\frac{[\ln J - (\gamma - \delta^2/2)]^2}{2\delta^2}\right), \tag{2}$$

where γ and δ define the mean and variance of the jump. The mean jump ξ is given by $\xi = \exp(\gamma) - 1$.

2.2 Linear Complementarity Problem for American Options

We define a partial integro-differential operator L acting on a price function u as

$$\begin{aligned} Lu = u_\tau - \frac{1}{2}yx^2u_{xx} - \rho\sigma yxu_{xy} - \frac{1}{2}\sigma^2yu_{yy} - (r - q - \lambda\xi)xu_x \\ - \kappa(\theta - y)u_y + (r + \lambda)u - \lambda\int_0^\infty u(Jx, y, \tau)f(J)dJ, \end{aligned} \tag{3}$$

where $\tau = T - t$ is the time to expiry and q is the dividend yield. For computations, the unbounded domain is truncated to be

$$(x, y, \tau) \in (0, X) \times (0, Y) \times (0, T] \tag{4}$$

with sufficiently large X and Y.

The initial value for u is defined by the payoff function $g(x, y)$ which gives the value of option at the expiry. In the following, we consider only call options. A similar approach can be also applied for put options. The payoff function for a call option with the strike price K is

$$g(x, y) = \max\{x - K, 0\}, \quad x \in (0, X),\ y \in (0, Y). \tag{5}$$

The price u of an American option satisfies an LCP

$$\begin{cases} Lu \geq 0, \quad u \geq g, \\ (Lu)\,(u - g) = 0. \end{cases} \tag{6}$$

We pose the boundary conditions

$$\begin{aligned} u(0, y, \tau) = g(0, y), \quad u(X, y, \tau) = g(X, y), \quad y \in (0, Y), \\ u_y(x, Y, \tau) = 0, \quad x \in (0, X). \end{aligned} \tag{7}$$

Beyond the boundary $x = X$, the price u is approximated to be the same as the payoff g, that is, $u(x, y, \tau) = g(x, y)$ for $x \geq X$. On the boundary $y = 0$, the LCP (6) holds and no additional boundary condition needs to be posed.

3 Discretization

We approximate the price u on a space–time grid defined by the grid points (x_i, y_j, τ_k), $0 \le i \le m$, $0 \le j \le n$, $0 \le k \le l$.

3.1 Discretization of Spatial Differential Operator

We use a uniform space grid with the grid steps in the x-direction and y-direction being $\Delta x = X/m$ and $\Delta y = Y/n$, respectively. Figure 1 shows a coarse space grid. A semidiscrete approximation for the price u is given by the time-dependent grid point values

$$u_{i,j}(\tau) \approx u(x_i, y_j, \tau) = u(i\Delta x, j\Delta y, \tau), \quad 0 \le i \le m, \ 0 \le j \le n. \tag{8}$$

We need to discretize the spatial partial derivatives in L given by

$$a_{11}u_{xx} + a_{12}u_{xy} + a_{22}u_{yy} + b_1 u_x + b_2 u_y + cu, \tag{9}$$

where

$$\begin{aligned} a_{11} &= -\frac{1}{2}yx^2, & a_{12} &= -\rho\sigma yx, & a_{22} &= -\frac{1}{2}\sigma^2 y, \\ b_1 &= -(r - q - \lambda\xi)x, & b_2 &= -\kappa(\theta - y), & c &= r + \lambda. \end{aligned} \tag{10}$$

The spatial partial derivatives are discretized using finite differences. For the non cross-derivatives, we use the standard central difference approximations

$$\begin{aligned} u_{xx}(x_i, y_j, \tau) &\approx \frac{1}{(\Delta x)^2}\left(2u(x_i, y_j, \tau) - u(x_i - \Delta x, y_j, \tau) - u(x_i + \Delta x, y_j, \tau)\right), \\ u_{yy}(x_i, y_j, \tau) &\approx \frac{1}{(\Delta y)^2}\left(2u(x_i, y_j, \tau) - u(x_i - \Delta x, y_j, \tau) - u(x_i + \Delta x, y_j, \tau)\right), \\ u_x(x_i, y_j, \tau) &\approx \frac{1}{2\Delta x}\left(u(x_i + \Delta x, y_j, \tau) - u(x_i - \Delta x, y_j, \tau)\right), \\ u_y(x_i, y_j, \tau) &\approx \frac{1}{2\Delta y}\left(u(x_i + \Delta x, y_j, \tau) - u(x_i - \Delta x, y_j, \tau)\right). \end{aligned} \tag{11}$$

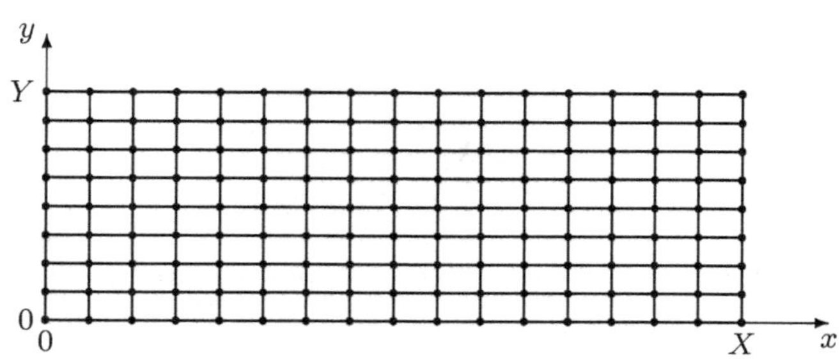

Fig. 1. A coarse 17×9 uniform grid for the computational domain $(0, X) \times (0, Y)$.

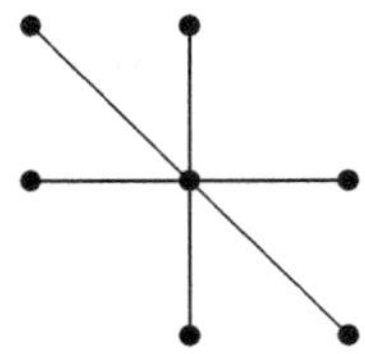

Fig. 2. A seven-point finite difference stencil used with a negative correlation $\rho < 0$ between the Wiener processes for the asset value x and its variance y.

In this paper, we assume that the correlation ρ is negative and we use a seven-point stencil shown in Figure 2. A similar stencil has been described in [7]. For a positive correlation ρ, a suitable seven-point stencil is given in [20, 22]. The cross-derivative u_{xy} is approximated by

$$u_{xy}(x_i, y_j, \tau) \approx \frac{1}{2\Delta x \Delta y}\big(2u(x_i, y_j, \tau) - u(x_i - \Delta x, y_j + \Delta y)$$
$$-u(x_i + \Delta x, y_j - \Delta y) - (\Delta x)^2 u_{xx}(x_i, y_j, \tau) - (\Delta y)^2 u_{yy}(x_i, y_j, \tau)\big). \quad (12)$$

Due to additional derivative terms in (12), we define modified coefficients for u_{xx} and u_{yy} as

$$\tilde{a}_{11} = a_{11} + \frac{1}{2}\frac{\Delta x}{\Delta y} a_{12}, \quad \text{and} \quad \tilde{a}_{22} = a_{22} + \frac{1}{2}\frac{\Delta y}{\Delta x} a_{12}. \quad (13)$$

It is well-known that the central finite differences can lead to positive weights in difference stencil when the convection dominates the diffusion. To avoid positive weights, we add some artificial diffusion according to

$$\hat{a}_{11} = \min\left\{\tilde{a}_{11}, -\frac{1}{2}b_1 \Delta x, \frac{1}{2}b_1 \Delta x\right\} \quad (14)$$

and

$$\hat{a}_{22} = \min\left\{\tilde{a}_{22}, -\frac{1}{2}b_2 \Delta y, \frac{1}{2}b_2 \Delta y\right\}. \quad (15)$$

This is equivalent to using a combination of one-sided and central differences for the convection. The resulting matrix is an M-matrix. Its off-diagonals are nonpositive and the diagonal is positive. It is strictly diagonally dominant when $c = r + \lambda > 0$.

3.2 Discretization of Integral Term

The integral term due to the jumps in (3) needs to computed at each grid point $x = x_i$. We denoted it by

$$I_i = \int_0^\infty u(Jx_i, y, \tau) f(J) dJ. \quad (16)$$

In order to perform the integration, we make a change of variable $J = e^s$ which leads to

$$I_i = \int_{-\infty}^{\infty} u(e^s x_i, y, \tau) p(s) ds, \tag{17}$$

where p is the probability density function of the normal distribution with the mean $\gamma - \delta^2/2$ and the variance δ^2 given by

$$p(s) = \frac{1}{\sqrt{2\pi}\delta} \exp\left(-\frac{[s - (\gamma - \delta^2/2)]^2}{2\delta^2} \right). \tag{18}$$

We decompose I_i into integrals over grid intervals as

$$I_i = \sum_{j=0}^{n-1} I_{i,j} + \int_{\ln x_n - \ln x_i}^{\infty} g(e^s x_i, y) p(s) ds, \tag{19}$$

where

$$I_{i,j} = \int_{\ln x_{j+1} - \ln x_i}^{\ln x_j - \ln x_i} u(e^s x_i, y, \tau) p(s) ds. \tag{20}$$

The price function $u(x, y, \tau)$ needs to be approximated between each grid point pair (x_i, x_{i+1}). For this, we use a piecewise linear interpolation

$$u(x, y, \tau) \approx \frac{x_{i+1} - x}{x_{i+1} - x_i} u(x_i, y, \tau) + \frac{x - x_i}{x_{i+1} - x_i} u(x_{i+1}, y, \tau) \tag{21}$$

for $x \in [x_j, x_{j+1}]$.

By performing the integration, we obtain

$$\begin{aligned} I_{i,j} \approx \frac{e^\gamma}{2} &\left[\operatorname{erf}\left(\frac{s_{i,j+1} - \gamma - \delta^2/2}{\delta\sqrt{2}} \right) - \operatorname{erf}\left(\frac{s_{i,j} - \gamma - \delta^2/2}{\delta\sqrt{2}} \right) \right] \alpha_j x_i \\ + \frac{1}{2} &\left[\operatorname{erf}\left(\frac{s_{i,j+1} - \gamma + \delta^2/2}{\delta\sqrt{2}} \right) - \operatorname{erf}\left(\frac{s_{i,j} - \gamma + \delta^2/2}{\delta\sqrt{2}} \right) \right] \beta_j x_i, \end{aligned} \tag{22}$$

where $\operatorname{erf}(\cdot)$ is the error function, $s_{i,j} = \ln x_j - \ln x_i$,

$$\alpha_j = \frac{u(x_{j+1}, y, \tau) - u(x_j, y, \tau)}{x_{j+1} - x_j}, \quad \text{and} \quad \beta_j = \frac{u(x_j, y, \tau) x_{j+1} - u(x_{j+1}, y, \tau) x_j}{x_{j+1} - x_j}. \tag{23}$$

3.3 Semidiscrete LCP

The space discretization leads to an LCP

$$\begin{cases} \mathbf{u}_\tau + \mathbf{A}\mathbf{u} + \mathbf{a} \geq \mathbf{0}, \quad \mathbf{u} \geq \mathbf{g}, \\ (\mathbf{u}_\tau + \mathbf{A}\mathbf{u} + \mathbf{a})^T (\mathbf{u} - \mathbf{g}) = 0, \end{cases} \tag{24}$$

where $\mathbf{A}$ is $(m+1)(n+1) \times (m+1)(n+1)$ matrix, $\mathbf{a}$ is a vector resulting from the second term in (19), $\mathbf{u}$ and $\mathbf{g}$ are vectors containing the grid point values of the price u and the payoff g, respectively. In the above LCP, the inequalities hold componentwise. The entries in the rows of $\mathbf{A}$ corresponding to the grid points on the boundaries $x = 0$ and $x = X$ are set to zero. The submatrix of $\mathbf{A}$ corresponding to the grid points not on the boundaries $x = 0$ and $x = X$ is an M-matrix. When the numbering of the grid points first goes through the grid points in the x-direction and then in the y-direction, the $(n+1) \times (n+1)$ diagonal blocks of $\mathbf{A}$ are essentially full matrices due to the jump term.

3.4 Time Discretization

We use the Rannacher scheme [29] with nonuniform time steps. It takes a few first time steps with the implicit Euler method and then it uses the Crank–Nicolson method. This leads to better stability properties than using just the Crank–Nicolson method. The solution vector $\mathbf{u}$ is approximated at times

$$\tau_k = \begin{cases} \left(\dfrac{k}{2l}\right)^2 T, & k = 0, 1, 2, 3, \\ \left(\dfrac{k-2}{l-2}\right)^2 T, & k = 4, 5, \dots, l. \end{cases} \tag{25}$$

In order to simplify the following notations, we define time step sizes $\Delta\tau_k = \tau_{k+1} - \tau_k$, $k = 0, 1, \dots, l-1$.

In order to simplify the notations in the following, we denote by $\mathrm{LCP}(\mathbf{B}, \mathbf{u}, \mathbf{b}, \mathbf{g})$ the linear complementarity problem

$$\begin{cases} (\mathbf{Bu} - \mathbf{b}) \geq \mathbf{0}, \quad \mathbf{u} \geq \mathbf{g}, \\ (\mathbf{Bu} - \mathbf{b})^T (\mathbf{u} - \mathbf{g}) = 0. \end{cases} \tag{26}$$

The Rannacher time stepping leads to the solution of the following sequence of LCPs:

$$\mathrm{LCP}(\mathbf{B}^{(k+1)}, \mathbf{u}^{(k+1)}, \mathbf{b}^{(k+1)}, \mathbf{g}), \tag{27}$$

where $\mathbf{u}^{(k)}$ denotes the vector $\mathbf{u}$ at the time τ_k. For the first four time steps $k = 0, 1, 2, 3$, we use the implicit Euler method defined by

$$\mathbf{B}^{(k+1)} = \mathbf{I} + \Delta\tau_k \mathbf{A} \qquad \text{and} \qquad \mathbf{b}^{(k+1)} = \Delta\tau_k \mathbf{u}^{(k)} - \Delta\tau_k \mathbf{a}. \tag{28}$$

The rest of the time steps $k = 4, 5, \dots, l-1$ are performed using the Crank–Nicolson method defined by

$$\mathbf{B}^{(k+1)} = \mathbf{I} + \frac{1}{2}\Delta\tau_k \mathbf{A} \qquad \text{and} \qquad \mathbf{b}^{(k+1)} = \left(\mathbf{I} - \frac{1}{2}\Delta\tau_k \mathbf{A}\right) \mathbf{u}^{(k)} - \Delta\tau_k \mathbf{a}. \tag{29}$$

4 Componentwise Splitting Method

Componentwise splitting methods are inspired by ADI (Alternating Direction Implicit) schemes which were introduced in [11, 28]. Instead of treating a part of operator explicitly, we use fully implicit splittings considered in [15, 16, 24, 33], for example. For the Heston model, the componentwise splitting method were introduced in [20] with a positive correlation ρ. In [7], the splitting method was considered in the case of a negative correlation.

The matrix $\mathbf{A}$ is split into three matrices which correspond to the couplings in the x-direction, y-direction, and diagonal direction. Figure 3 shows the matrix splitting and also the corresponding splitting of the finite difference stencil. The simplest fractional step method based on the implicit Euler method is given in Figure 4. The formal accuracy of this method is $\mathcal{O}(\Delta\tau_{l-1}) = \mathcal{O}\left(\frac{1}{l}\right)$.

We increase the accuracy of the splitting method by performing a Strang symmetrization [30] and use the Crank–Nicolson method; see also [15]. This leads one time step to have the following fractional steps:

$$\text{Step 1. LCP}\left(\mathbf{I} + \frac{\Delta\tau_k}{4}\mathbf{A}_y, \mathbf{u}^{(k+1/5)}, \left(\mathbf{I} - \frac{\Delta\tau_k}{4}\mathbf{A}_y\right)\mathbf{u}^{(k)}, \mathbf{g}\right)$$

$$\text{Step 2. LCP}\left(\mathbf{I} + \frac{\Delta\tau_k}{4}\mathbf{A}_d, \mathbf{u}^{(k+2/5)}, \left(\mathbf{I} - \frac{\Delta\tau_k}{4}\mathbf{A}_y\right)\mathbf{u}^{(k+1/5)}, \mathbf{g}\right)$$

$$\text{Step 3. LCP}\left(\mathbf{I} + \frac{\Delta\tau_k}{2}\mathbf{A}_x, \mathbf{u}^{(k+3/5)}, \left(\mathbf{I} - \frac{\Delta\tau_k}{2}\mathbf{A}_x\right)\mathbf{u}^{(k+2/5)} - \Delta\tau_k\mathbf{a}, \mathbf{g}\right)$$

$$\text{Step 4. LCP}\left(\mathbf{I} + \frac{\Delta\tau_k}{4}\mathbf{A}_d, \mathbf{u}^{(k+4/5)}, \left(\mathbf{I} - \frac{\Delta\tau_k}{4}\mathbf{A}_y\right)\mathbf{u}^{(k+3/5)}, \mathbf{g}\right)$$

$$\text{Step 5. LCP}\left(\mathbf{I} + \frac{\Delta\tau_k}{4}\mathbf{A}_y, \mathbf{u}^{(k+1)}, \left(\mathbf{I} - \frac{\Delta\tau_k}{4}\mathbf{A}_y\right)\mathbf{u}^{(k+4/5)}, \mathbf{g}\right)$$

In order to maintain the good stability of the Rannacher scheme, we use the implicit Euler method instead the Crank–Nicolson method for the first four time steps $k = 0, 1, 2, 3$ in the above symmetrized splitting method.

4.1 Solution of One-Dimensional LCPs

For an American call option, typical early exercise boundaries at different times are shown in Figure 5. The boundary can be described by a relation

$$\mathbf{A} \quad = \quad \mathbf{A}_x \quad + \quad \mathbf{A}_y \quad + \quad \mathbf{A}_d$$

Fig. 3. The matrix splitting of $\mathbf{A}$ and the corresponding splitting of the finite difference stencil.

1. $\mathrm{LCP}(\mathbf{I} + \Delta\tau_k \mathbf{A}_y, \mathbf{u}^{(k+1/3)}, \Delta\tau_k \mathbf{u}^{(k)}, \mathbf{g})$
   Solve the sequence of one-dimensional LCPs:

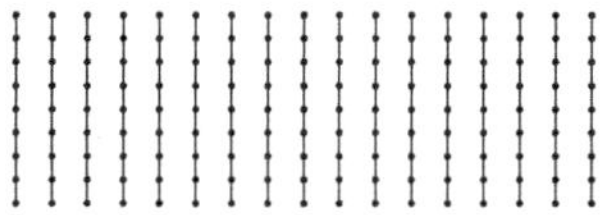

2. $\mathrm{LCP}(\mathbf{I} + \Delta\tau_k \mathbf{A}_d, \mathbf{u}^{(k+2/3)}, \Delta\tau_k \mathbf{u}^{(k+1/3)}, \mathbf{g})$
   Solve the sequence of one-dimensional LCPs:

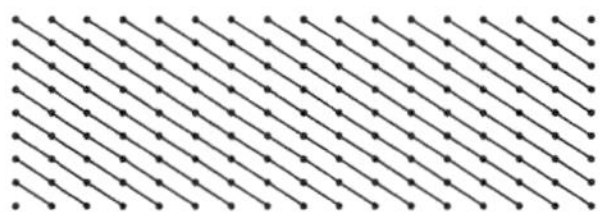

3. $\mathrm{LCP}(\mathbf{I} + \Delta\tau_k \mathbf{A}_x, \mathbf{u}^{(k+1)}, \Delta\tau_k \mathbf{u}^{(k+2/3)} - \Delta\tau_k \mathbf{a}, \mathbf{g})$
   Solve the sequence of one-dimensional LCPs:

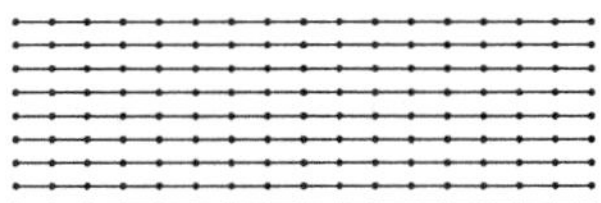

Fig. 4. Three fractional splitting steps for performing the time step from τ_k to τ_{k+1}.

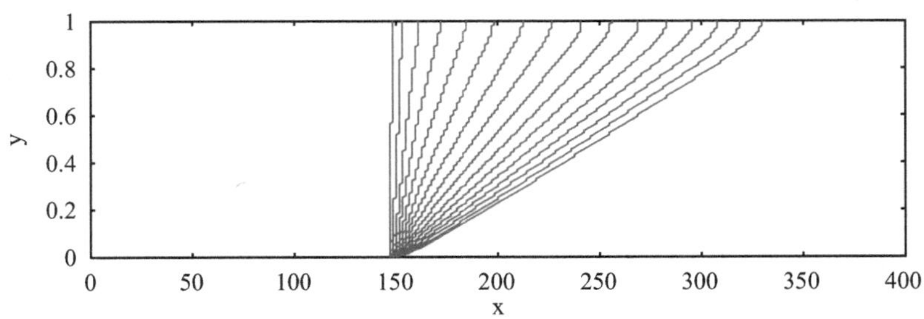

Fig. 5. The time evolution of the early exercise boundaries for an American call option.

$y = h(x, \tau)$, where h an increasing function with respect to x. Thus, a given point (x, y, τ) belongs to

- The hold region if $y > h(x, \tau)$ or
- The early exercise region if $y \leq h(x, \tau)$

Similarly, the early exercise boundary divides each x-directional line, y-directional line, and $(1, -1)$-directional line into two parts. Due to this solution structure and the tridiagonal matrices defining the LCPs in the y-direction and $(1,-1)$-direction, the Brennan and Schwartz algorithm can be used to solve these problems. The LCPs in the x-direction have full matrices due to the integral term. An iterative solution procedure for these problems is described in the end of this section.

Brennan and Schwartz Algorithm

The Brennan and Schwartz algorithm for American put options under the Black–Scholes model was described in [6]. The algorithm can be modified to use a standard **LU**-decomposition [1, 21]. We formulate it for a tridiagonal linear complementarity problem:

$$\mathbf{Tx} = \begin{pmatrix} \mathbf{T}_{1,1} & \mathbf{T}_{1,2} & & & \\ \mathbf{T}_{2,1} & \ddots & \ddots & & \\ & \ddots & \mathbf{T}_{m-1,m-1} & \mathbf{T}_{m-1,m} \\ & & \mathbf{T}_{m,m-1} & \mathbf{T}_{m,m} \end{pmatrix} \begin{pmatrix} \mathbf{x}_1 \\ \mathbf{x}_2 \\ \vdots \\ \mathbf{x}_m \end{pmatrix} \geq \begin{pmatrix} \mathbf{b}_1 \\ \mathbf{b}_2 \\ \vdots \\ \mathbf{b}_m \end{pmatrix} = \mathbf{b}, \tag{30}$$

$$\mathbf{x} \geq \mathbf{g}, \qquad (\mathbf{Tx} - \mathbf{b})^T(\mathbf{x} - \mathbf{g}) = 0. \tag{31}$$

The Brennan and Schwartz algorithm assumes the solution $\mathbf{x}$ to be such that for some integer k it holds that

$$\begin{aligned} \mathbf{x}_i &> g_i, \quad i = 1, \ldots, k, \qquad \text{and} \\ \mathbf{x}_i &= g_i, \quad i = k+1, \ldots, m. \end{aligned} \tag{32}$$

The algorithm with **LU**-decomposition is described as follows:

Brennan and Schwartz algorithm
Computation of **LU**-decomposition and forward substitution:
$\mathbf{U}_{1,1} = \mathbf{T}_{1,1}$
$\mathbf{y}_1 = \mathbf{b}_1$
Do $i = 2, \ldots, m$
  $\mathbf{L}_{i,i-1} = \mathbf{T}_{i,i-1}/\mathbf{U}_{i-1,i-1}$
  $\mathbf{U}_{i-1,i} = \mathbf{T}_{i-1,i}$
  $\mathbf{U}_{i,i} = \mathbf{T}_{i,i} - \mathbf{L}_{i,i-1}\mathbf{U}_{i-1,i}$
  $\mathbf{y}_i = \mathbf{b}_i - \mathbf{L}_{i,i-1}\mathbf{y}_{i-1}$
End Do
Backward substitution with a projection:
$\mathbf{x}_m = \mathbf{y}_m/\mathbf{U}_{m,m}$
$\mathbf{x}_m = \max\{\mathbf{x}_m, \mathbf{g}_m\}$
Do $i = m-1, \ldots, 1$
  $\mathbf{x}_i = (\mathbf{y}_i - \mathbf{U}_{i,i+1}\mathbf{x}_{i+1})/\mathbf{U}_{i,i}$
  $\mathbf{x}_i = \max\{\mathbf{x}_i, \mathbf{g}_i\}$
End Do

In this algorithm the only modification to a standard solution with **LU**-decomposition is the additional projection in the backward substitution.

After a suitable numbering of unknowns the assumption (32) holds for the one-dimensional LCPs in all three directions. The Brennan and Schwartz algorithm can be use directly to solve the one-dimensional LCPs in the y-direction and in the $(1, -1)$-direction.

LCPs with Full Matrices Associated to the **x**-*Direction*

A matrix associated to one-dimensional LCP in the x-direction is denoted by $\mathbf{B}$. It has a regular splitting [34]

$$\mathbf{B} = \mathbf{T} - \mathbf{J}, \tag{33}$$

where $-\mathbf{J}$ is a full matrix resulting from the integral term and $\mathbf{T}$ is the rest which a tridiagonal matrix. We generalize a fixed point iteration described in [31] and used in [2, 10, 32]. The fixed point iteration for LCP$(\mathbf{B}, \mathbf{x}, \mathbf{b}, \mathbf{g})$ reads

$$\text{LCP}(\mathbf{T}, \mathbf{x}^{j+1}, \mathbf{b} + \mathbf{J}\mathbf{x}^{j}, \mathbf{g}), \quad j = 0, 1, \ldots \tag{34}$$

Each iteration requires the solution of an LCP with the tridiagonal $\mathbf{T}$ and the multiplication of a vector by $\mathbf{J}$. The Brennan and Schwartz algorithm can be used to solve the LCPs (34). The iteration converges very rapidly and only a couple of iterations are needed to reach sufficient accuracy for practical purposes.

5 Numerical Experiments

In the numerical experiments for call options, we use the model parameter values:

- The risk free interest rate $r = 0.03$
- The dividend yield $q = 0.05$
- The strike price $K = 100$
- The correlation between the price and variance processes $\rho = -0.5$
- The mean level of the variance $\theta = 0.04$
- The rate of reversion to the mean level $\kappa = 2.0$
- The volatility of the variance $\sigma = 0.25$
- The jump rate $\lambda = 0.2$
- The mean jump $\gamma = -0.5$
- The variance of jump $\delta = 0.4$

The computational domain is $(x, y, \tau) \in [0, 400] \times [0, 1] \times [0, 0.5]$.

Our first experiment compares the PSOR method and the Strang symmetrized componentwise splitting method for call options under the Heston model, that is, $\lambda = 0$. In this case, the LCPs in the x-direction are tridiagonal and they can be solved using the Brennan and Schwartz algorithm without the iteration (34).

Table 1. The numerical results for the Heston model

Method	Grid (m, n, l)	Iteration	Error	Ratio	CPU
PSOR	$(64, 32, 8)$	34.6	0.14470		0.05
	$(128, 64, 16)$	42.3	0.05607	2.58	0.48
	$(256, 128, 32)$	95.3	0.01006	5.58	8.18
	$(512, 256, 64)$	196.6	0.00350	2.87	128.51
	$(1024, 512, 128)$	372.2	0.00066	5.31	1890.76
Componentwise	$(64, 32, 8)$		0.14412		0.01
splitting	$(128, 64, 16)$		0.05621	2.56	0.06
	$(256, 128, 32)$		0.01019	5.51	0.51
	$(512, 256, 64)$		0.00355	2.87	6.36
	$(1024, 512, 128)$		0.00067	5.28	58.27

Table 1 reports the numerical results. It (and also Table 2) has the following columns: Grid (m, n, l) defines the number of grid steps in x, y, and τ to be m, n, and l, respectively. Iteration gives the average number of PSOR iterations on each time step with the relaxation parameter $\omega = 1.5$. With the componentwise splitting method iteration specifies the number of iterations (34) to solve the LCPs in the x-direction at each time step. Error column gives the root mean square relative error given by

$$\text{error} = \left[\frac{1}{5} \sum_{i=1}^{5} \left(\frac{u(\mathbf{x}_i, \theta, T) - U(\mathbf{x}_i, \theta, T)}{U(\mathbf{x}_i, \theta, T)} \right)^2 \right]^{1/2}, \tag{35}$$

where $\mathbf{x} = (80, 90, 100, 110, 120)^T$ and U is the reference price. Ratio is the ratio of the consecutive root mean square relative errors. CPU gives the CPU time in seconds on a 3.8 GHz Intel Xeon PC. The reference prices under the Heston model at $(\mathbf{x}_i, \theta, T)$, $i = 1, 2, \ldots, 5$, are 0.131563, 1.255396, 4.999888, 11.680219, 20.325463 which were computed using the componentwise splitting method on the grid $(4096, 2048, 512)$.

We can observe from Table 1 that the discretizations with both methods appears to be roughly second-order accurate as the ratio is four on average. Furthermore, the splitting increases the error only about 2%. On the coarsest grid, the splitting method is five times faster than the PSOR method, and on the finest grid it is 32 times faster.

In our second experiment, we performed the same comparison under the Bates model. The reference prices computed using the componentwise splitting method on the grid $(4096, 2048, 512)$ are 0.328526, 2.109397, 6.711622, 13.749337, 22.143307. In the componentwise splitting method, the LCPs in the x-direction lead to full matrices and the iteration (34) is employed to solve them. Based on a few experiments, we observed that already after two iterations the accuracy is sufficient. Thus, we use two iterations in our comparison. The multiplication by the matrix $\mathbf{J}$ is the most expensive operation in the iteration. In order to perform it efficiently, we collected all n multiplications

Table 2. The numerical results for the Bates model

Method	Grid (m, n, l)	Iteration	Error	Ratio	CPU
PSOR	$(64, 32, 8)$	39.6	0.10887		0.16
	$(128, 64, 16)$	48.1	0.03803	2.86	2.14
	$(256, 128, 32)$	108.2	0.00670	5.68	138.92
	$(512, 256, 64)$	222.6	0.00209	3.20	8605.09
	$(1024, 512, 128)$	420.5	0.00034	6.13	275191.73
Componentwise splitting	$(64, 32, 8)$	2.0	0.10833		0.01
	$(128, 64, 16)$	2.0	0.03790	2.86	0.09
	$(256, 128, 32)$	2.0	0.00668	5.67	0.81
	$(512, 256, 64)$	2.0	0.00210	3.19	10.18
	$(1024, 512, 128)$	2.0	0.00035	6.07	109.45

corresponding to all x-grid lines together and then performed the resulting matrix–matrix multiplication using the optimized GotoBLAS library [18].

The numerical results under the Bates model are given in Table 2. Absolute errors are comparable to the ones under the Heston model, but as the option prices are higher under the Bates model the relative errors reported in the table are smaller. Again roughly second-order accuracy is observed with both methods. The CPU times with the componentwise splitting method were less than twice the times under the Heston model. The componentwise splitting method is 16 times faster on the coarsest grid, and it is about 2,500 times faster on the finest grid. On finer grids, the PSOR method leads to infeasible CPU times while the times with componentwise splitting method are still reasonable.

6 Conclusions

We described a linear complementarity problem (LCP) for pricing American options under the Bates model and we considered a finite difference discretization. We proposed a componentwise splitting method to solve approximately the LCPs. It leads to a sequence of LCPs with tridiagonal matrices. The Brennan and Schwartz algorithm can solve these LCPs efficiently.

Our numerical experiments showed that the additional splitting error do not essentially increase the discretization error. The componentwise splitting method is orders of magnitude faster than the PSOR method under the Bates model. Pricing under the Bates model was at most two times more expensive than under the Heston model with the componentwise splitting method.

Acknowledgement. The author thanks Prof. Roland Glowinski for introduction and discussions on operator splitting methods. The author is grateful to Dr. Samuli Ikonen for many fruitful discussions on numerical methods for option pricing.

References

1. Y. Achdou and O. Pironneau. *Computational methods for option pricing*, volume 30 of *Frontiers in Applied Mathematics*. SIAM, Philadelphia, PA, 2005.
2. A. Almendral and C. W. Oosterlee. Numerical valuation of options with jumps in the underlying. *Appl. Numer. Math.*, 53(1):1–18, 2005.
3. L. Andersen and J. Andreasen. Jump-diffusion processes: Volatility smile fitting and numerical methods for option pricing. *Rev. Deriv. Res.*, 4(3):231–262, 2000.
4. D. S. Bates. Jumps and stochastic volatility: Exchange rate processes implicit Deutsche mark options. *Review Financial Stud.*, 9(1):69–107, 1996.
5. F. Black and M. Scholes. The pricing of options and corporate liabilities. *J. Polit. Econ.*, 81:637–654, 1973.
6. M. J. Brennan and E. S. Schwartz. The valuation of American put options. *J. Finance*, 32:449–462, 1977.
7. C. Chiarella, B. Kang, G. H. Mayer, and A. Ziogas. The evaluation of American option prices under stochastic volatility and jump-diffusion dynamics using the method of lines. Research Paper 219, Quantitative Finance Research Centre, University of Technology, Sydney, 2008.
8. N. Clarke and K. Parrott. Multigrid for American option pricing with stochastic volatility. *Appl. Math. Finance*, 6:177–195, 1999.
9. R. Cont and E. Voltchkova. A finite difference scheme for option pricing in jump diffusion and exponential Lévy models. *SIAM Numer. Anal.*, 43(4):1596–1626, 2005.
10. Y. d'Halluin, P. A. Forsyth, and K. R. Vetzal. Robust numerical methods for contingent claims under jump diffusion processes. *IMA J. Numer. Anal.*, 25(1):87–112, 2005.
11. J. Douglas, Jr. and H. H. Rachford, Jr. On the numerical solution of heat conduction problems in two and three space variables. *Trans. Amer. Math. Soc.*, 82:421–439, 1956.
12. D. Duffie, J. Pan, and K. Singleton. Transform analysis and asset pricing for affine jump-diffusions. *Econometrica*, 68(6):1343–1376, 2000.
13. L. Feng and V. Linetsky. Pricing options in jump-diffusion models: an extrapolation approach. *Oper. Res.*, 56:304–325, 2008.
14. R. Glowinski. *Numerical methods for nonlinear variational problems*. Springer, New York, 1984.
15. R. Glowinski. Finite element methods for incompressible viscous flow. In P. G. Ciarlet and J.-L. Lions, editors, *Handbook of Numerical Analysis, Vol. IX*, pages 3–1176. North-Holland, Amsterdam, 2003.
16. R. Glowinski and P. Le Tallec. *Augmented Lagrangian and operator splitting methods in nonlinear mechanics*. SIAM, Philadelphia, PA, 1989.
17. R. Glowinski, J.-L. Lions, and R. Trémolières. *Analyse numérique des inéquations variationnelles. Tome 1 & 2*. Dunod, Paris, 1976.
18. K. Goto and R. A. van de Geijn. Anatomy of high-performance matrix multiplication. *ACM Trans. Math. Software*, 34(3):Art. 12, 25 pp., 2008.
19. S. Heston. A closed-form solution for options with stochastic volatility with applications to bond and currency options. *Rev. Financial Stud.*, 6:327–343, 1993.
20. S. Ikonen and J. Toivanen. Componentwise splitting methods for pricing American options under stochastic volatility. *Int. J. Theor. Appl. Finance*, 10:331–361, 2007.

21. S. Ikonen and J. Toivanen. Pricing American options using LU decomposition. *Appl. Math. Sci.*, 1:2529–2551, 2007.
22. S. Ikonen and J. Toivanen. Efficient numerical methods for pricing American options under stochastic volatility. *Numer. Methods Partial Differential Equations*, 24:104–126, 2008.
23. S. G. Kou. A jump-diffusion model for option pricing. *Management Sci.*, 48(8):1086–1101, 2002.
24. G. I. Marchuk. Splitting and alternating direction methods. In P. G. Ciarlet and J.-L. Lions, editors, *Handbook of Numerical Analysis, Vol. I*, pages 197–462. North-Holland, Amsterdam, 1990.
25. A.-M. Matache, C. Schwab, and T. P. Wihler. Fast numerical solution of parabolic integrodifferential equations with applications in finance. *SIAM J. Sci. Comput.*, 27(2):369–393, 2005.
26. R. Merton. Option pricing when underlying stock returns are discontinuous. *J. Financial Econ.*, 3:125–144, 1976.
27. C. W. Oosterlee. On multigrid for linear complementarity problems with application to American-style options. *Electron. Trans. Numer. Anal.*, 15:165–185, 2003.
28. D. W. Peaceman and H. H. Rachford, Jr. The numerical solution of parabolic and elliptic differential equations. *J. Soc. Indust. Appl. Math.*, 3:28–41, 1955.
29. R. Rannacher. Finite element solution of diffusion problems with irregular data. *Numer. Math.*, 43:309–327, 1982.
30. G. Strang. On the construction and comparison of difference schemes. *SIAM J. Numer. Anal.*, 5:506–517, 1968.
31. D. Tavella and C. Randall. *Pricing financial instruments: The finite difference method.* John Wiley & Sons, Chichester, 2000.
32. J. Toivanen. Numerical valuation of European and American options under Kou's jump-diffusion model. *SIAM J. Sci. Comput.*, 30:1949–1970, 2008.
33. N. N. Yanenko. *The method of fractional steps. The solution of problems of mathematical physics in several variables.* Springer, New York, 1971.
34. D. M. Young. *Iterative solution of large linear systems.* Academic Press, New York, 1971.
35. R. Zvan, P. A. Forsyth, and K. R. Vetzal. Penalty methods for American options with stochastic volatility. *J. Comput. Appl. Math.*, 91(2):199–218, 1998.

Exact Controllability of the Time Discrete Wave Equation: A Multiplier Approach

Xu Zhang[1], Chuang Zheng[2], and Enrique Zuazua[3]

[1] Academy of Mathematics and Systems Sciences, Chinese Academy of Sciences, CN-100080 Beijing, China, xuzhang@amss.ac.cn
[2] School of Mathematical Sciences, Beijing Normal University, CN-100875 Beijing, China, chuang.zheng@bnu.edu.cn
[3] Basque Center for Applied Mathematics, Bizkaia Technology Park, Building 500, ES-48160 Derio, Basque Country, Spain, zuazua@bcamath.org, http://www.bcamath.org/zuazua/

Summary. In this paper we summarize our recent results on the exact boundary controllability of a trapezoidal time discrete wave equation in a bounded domain. It is shown that the projection of the solution in an appropriate space in which the high frequencies have been filtered is exactly controllable with uniformly bounded controls (with respect to the time-step). By classical duality arguments, the problem is reduced to a boundary observability inequality for a time-discrete wave equation. Using multiplier techniques the uniform observability property is proved in a class of filtered initial data. The optimality of the filtering parameter is also analyzed.

Key words: Exact controllability, observability, time discretization, wave equation, multiplier technique, filtering.

1 Introduction

Let Ω be an open bounded domain in $\mathbb{R}^d$ ($d \in \mathbb{N}^*$) with C^2 boundary Γ. Let $T > 0$ be a given time duration. We consider the following wave equation with a state $y = y(x,t)$ and a controller $u = u(x,t)$ acting on the nonempty subset Γ_0 of the boundary $\Gamma = \partial\Omega$:

$$\begin{cases} y_{tt} - \Delta y = 0 & \text{in } (0,T) \times \Omega, \\ y = u 1_{\Gamma_0} & \text{on } (0,T) \times \Gamma, \\ y(0) = y_0, \quad y_t(0) = y_1 & \text{in } \Omega. \end{cases} \tag{1}$$

Here 1_{Γ_0} is the characteristic function of the set Γ_0.

This paper is devoted to analyze whether the known controllability results for (1) can be recovered as a consequence of similar results for the time-discrete versions. This kind of problems has been the object of intensive research in

W. Fitzgibbon et al. (eds.), *Applied and Numerical Partial Differential Equations*, Computational Methods in Applied Sciences 15, DOI 10.1007/978-90-481-3239-3_17,

the past few years but mainly in the context of space semi-discretizations. In the present paper we summarize the main results by the authors [14] in the time discrete case. This issue is of interest from a numerical analysis point of view but also in what concerns the link between the control properties of time continuous and time-discrete distributed parameter systems. The topic of numerical approximation of boundary controls for wave equations was initiated by R. Glowinski, J.-L. Lions and coworkers (see, for instance, [3, 14]) and has motivated intensive research (we refer to [17] for a survey).

The exact controllability of (1) requires that the subset Γ_0 of the boundary fulfills some geometric conditions. It holds, in particular, for those subsets that are obtained through the multiplier method. More precisely, fix some $x_0 \in \mathbb{R}^d$, and put

$$\begin{cases} R \triangleq \max\limits_{x\in\Omega} |x - x_0|, \\ \Gamma_0 \triangleq \{x \in \Gamma \mid (x - x_0)\cdot \nu(x) > 0\}, \end{cases} \tag{2}$$

where $\nu(x)$ is the unit outward normal vector of Ω at $x \in \Gamma$. For these subsets Γ_0 the exact controllability property of (1) holds provided $T > 2R$.

To be more precise, the following exact controllability result for (1) is well known (see [6]): *For any* $(y_0, y_1) \in L^2(\Omega) \times H^{-1}(\Omega)$, *there exists a control* $u \in L^2((0,T)\times\Gamma_0)$ *such that the solution* $y = y(t,x)$ *of* (1), *defined by the classical transposition method, satisfies*

$$y(T) = y_t(T) = 0 \quad \text{in } \Omega. \tag{3}$$

By classical duality arguments [6], the above controllability property is equivalent to a (boundary) observability one of the following uncontrolled wave equation:

$$\begin{cases} \varphi_{tt} - \Delta\varphi = 0, & \text{in } (0,T)\times\Omega \\ \varphi = 0 & \text{on } (0,T)\times\Gamma \\ \varphi(T) = \varphi_0, \quad \varphi_t(T) = \varphi_1, & \text{in } \Omega, \end{cases} \tag{4}$$

i.e. to the fact that *solutions of* (4) *satisfy*

$$E(0) \le C \int_0^T \int_{\Gamma_0} \left|\frac{\partial\varphi}{\partial\nu}\right|^2 d\Gamma_0 dt, \quad \forall(\varphi_0,\varphi_1) \in H_0^1(\Omega)\times L^2(\Omega). \tag{5}$$

Here and thereafter, we will use C to denote a generic positive constant (depending only on T, Ω and Γ_0) which may vary from line to line. On the other hand, $E(0)$ stands for the energy $E(t)$ of (4) at $t = 0$, with

$$E(t) = \frac{1}{2}\int_\Omega \left[|\varphi_t(t,x)|^2 + |\nabla\varphi(t,x)|^2\right] dx, \tag{6}$$

which remains constant, i.e.

$$E(t) = E(0), \quad \forall t \in [0,T].$$

The inequality (5) can be proved by several methods including multiplier techniques [6], microlocal analysis [1] and Carleman inequalities [13]. In the particular case of subset Γ_0 as above and $T > 2R$, the inequality (5) can be proved easily by the method of multipliers [6] that in the present paper we adapt to time-discrete equations.

Note, however, that the subsets Γ_0 of the boundary and the values of the minimal control time obtained in this way are not optimal. The obtention of optimal control subsets and times requires the use of methods of geometric optics (see [1]).

In this paper, we analyze time semi-discretization schemes for the systems (1) and (4). We are thus replacing the continuous dynamics (1) and (4) by time-discrete ones and analyze their controllability/observability properties. Here we take the point of view of numerical analysis and, therefore, we analyze the limit behavior as the time-step tends to zero.

More precisely, we set the time step h by $h = T/K$, where $K > 1$ is a given odd integer. Denote by y^k and u^k respectively the approximations of the solution y and the control u of (1) at time $t_k = kh$ for any $k = 0, \dots, K$. We then introduce the following trapezoidal time semi-discretization of (1):

$$\begin{cases} \dfrac{y^{k+1}+y^{k-1}-2y^k}{h^2} - \Delta\left(\dfrac{y^{k+1}+y^{k-1}}{2}\right) = 0, & \text{in } \Omega,\ k = 1,\dots,K-1, \\ y^k = u^k 1_{\Gamma_0}, & \text{on } \Gamma,\ k = 0,\dots,K, \\ y^0 = y_0, \quad y^1 = y_0 + hy_1, & \text{in } \Omega. \end{cases} \tag{7}$$

Here $(y_0, y_1) \in L^2(\Omega) \times H^{-1}(\Omega)$ are the data in the system (1). We refer to Theorem 1 below for the well-posedness of the system (1) by means of a transposition method.

The controllability problem for the system (7) is formulated as follows: *For any $(y_0, y_1) \in L^2(\Omega) \times H^{-1}(\Omega)$, to find a control $\{u^k \in L^2(\Gamma_0)\}_{k=1,\dots,K-1}$ such that the solution $\{y^k\}_{k=0,\dots,K}$ of (7) satisfies:*

$$y^{K-1} = y^K = 0 \quad \text{in } \Omega. \tag{8}$$

Note that (8) is equivalent to the condition $y^{K-1} = (y^K - y^{K-1})/h = 0$ that is a natural discrete version of (3).

As in the context of the above continuous wave equation, we also consider the uncontrolled system

$$\begin{cases} \dfrac{\varphi^{k+1}+\varphi^{k-1}-2\varphi^k}{h^2} - \Delta\left(\dfrac{\varphi^{k+1}+\varphi^{k-1}}{2}\right) = 0, & \text{in } \Omega,\ k = 1,\dots,K-1, \\ \varphi^k = 0, & \text{on } \Gamma,\ k = 0,\dots,K \\ \varphi^K = \varphi_0^h + h\varphi_1^h, \quad \varphi^{K-1} = \varphi_0^h, & \text{in } \Omega, \end{cases} \tag{9}$$

where $(\varphi_0^h, \varphi_1^h) \in (H_0^1(\Omega))^2$. In particular, to guarantee the convergence of the solutions of (9) towards those of (4), one considers convergent data such that

$$\begin{cases} \varphi_0^h \to \varphi_0 & \text{strongly in } H_0^1(\Omega), \\ \varphi_1^h \to \varphi_1 & \text{strongly in } L^2(\Omega), \\ h\varphi_1^h \to \varphi_1 & \text{is bounded in } H_0^1(\Omega), \end{cases} \qquad \text{as } K \to \infty \text{ (or } h \to 0). \tag{10}$$

Obviously, because of the density of $H_0^1(\Omega)$ in $L^2(\Omega)$, this choice is always possible.

Remark 1. Note that the choice of the values of φ^K and φ^{K-1} in (9) is motivated by the definition of the solution of the time-discrete non-homogenous problem (7) in the sense of transposition (see Definition 1).

The energy of the system (9) is given by

$$E_h^k \triangleq \frac{1}{2}\int_\Omega \left(\left| \frac{\varphi^{k+1} - \varphi^k}{h} \right|^2 + \frac{|\nabla \varphi^{k+1}|^2 + |\nabla \varphi^k|^2}{2} \right) dx, \quad k = 0, \dots, K-1,$$

which is a discrete counterpart of the continuous energy E in (6). It is easy to show that E_h^k is conserved in the discrete time variable $k = 0, \dots, K-1$. Consequently, the scheme under consideration is stable and its convergence (in the classical sense of numerical analysis) is guaranteed (in the finite-energy space $H_0^1(\Omega) \times L^2(\Omega)$ of the system (4)).

By means of classical duality arguments, it is easy to show that the above controllability property (8) is equivalent to the following boundary observability property for solutions $\{\varphi^k\}_{k=0,\dots,K}$ of (9):

$$E_h^0 \le Ch \sum_{k=1}^{K-1} \int_{\Gamma_0} \left| \frac{\partial}{\partial \nu} \left(\frac{\varphi^{k+1} + \varphi^{k-1}}{2} \right) \right|^2 d\Gamma_0, \quad \forall (\varphi_0^h, \varphi_1^h) \in (H_0^1(\Omega))^2. \tag{11}$$

As we mentioned above, the controllability/observability properties of numerical approximation schemes for the wave equation have been the object of intensive studies. However, most analytical results concern the case of space semi-discretizations (see [17] and the references cited therein). In practical applications, fully discrete schemes need to be used. The most typical example is the classical fully-discrete central scheme which converges under a suitable CFL condition [3, 4, 11]. However, in the present setting in which the space Laplacian Δ is kept continuous, without discretizing it, this scheme is unsuitable since it is unstable. Indeed, it is easy to see that the scheme

$$\frac{\varphi^{k+1} + \varphi^{k-1} - 2\varphi^k}{h^2} - \Delta \varphi^k = 0 \tag{12}$$

is unstable since $-\Delta$, with homogenous Dirichlet conditions, is a positive self-adjoint operator with an infinite sequence of eigenvalues $\{\mu_j^2\}_{j\ge1}$ tending to infinity. The stability of (12) would be equivalent to the stability of the scheme

$$\frac{\varphi^{k+1} + \varphi^{k-1} - 2\varphi^k}{h^2} + \mu_j^2 \varphi^k = 0$$

for all values of μ_j^2, $j \geq 1$. This stability property fails clearly, regardless how small h is, when μ_j^2 is large enough. Hence, we choose the trapezoidal scheme (9) for the time-discrete problem, which is stable (due to the property of conservation of energy), as mentioned before.

Let us now return to the analysis of (7) and (9). Noting that the spaces in which the solutions of these systems evolve are infinite dimensional while the number of time-steps is finite, it is easy to conclude that: *For any given* $h > 0$, *the inequality* (11) *fails and the system* (7) *is not exactly controllable.* Accordingly, to make the observability inequality possible, one has to restrict the class of solutions of the adjoint system (9) under consideration by filtering the high frequency components. Similarly, since the property of exact controllability of the system (7) fails, the final requirement (8) has to be relaxed by considering only low frequency projections of the solutions. Controlling such a projection can be viewed as a *partial* controllability problem. This filtering method has been applied successfully in the context of controllability of time discrete heat equations in [15] and space semi-discretization schemes for wave equations in [5, 16, 17].

In this paper, we sketch the discrete version of the classical multiplier approach developed in [14] which allows to derive the uniform observability estimate (with respect to the time step h) for the system (9) with initial data in a suitable filtered space, which, in turn, by duality, implies the partial controllability of (7), uniformly on h.

As in the continuous case, the multiplier technique applies mainly to the case when the controller/observer Γ_0 is given in (2) and some variants [9], but does not work when (T, Ω, Γ_0) is assumed to satisfy the sharp Geometric Control Condition (GCC) in [1]. As we shall see, the main advantage of our multiplier approach is that the filtering parameter we use has the optimal scaling in what concerns the frequency of observed/controlled solutions with respect to h.

The rest of the paper is organized as follows. In Section 2 we state the main results, i.e. the uniform controllability and observability of the systems (7) and (9) after filtering, respectively. In Section 3 we give a heuristic explanation of the necessity of the filtering analyzing the bicharacteristic rays and the group velocity. The key ingredients in the proof of the uniform observability results will be sketched in Sections 4 and 5. Finally, in Section 6, we shall briefly discuss some open problems and closely related issues.

2 Main Results

We begin with the well-posedness of the system (7). For this purpose, for any $\{f^k \in L^2(\Omega)\}_{k=1,\dots,K-1}$, and any $\{g^k \in H_0^1(\Omega)\}_{k=1,\dots,K}$ with $g^1 = g^K = 0$, we consider the following adjoint problem of the system (7):

$$\begin{cases} \dfrac{\zeta^{k+1}+\zeta^{k-1}-2\zeta^k}{h^2} - \Delta\left(\dfrac{\zeta^{k+1}+\zeta^{k-1}}{2}\right) & \\ \qquad = f_k + \dfrac{g^{k+1}-g^k}{h}, & \text{in } \Omega,\ k=1,\ldots,K-1, \\ \zeta^k = 0, & \text{on } \Gamma,\ k=0,\ldots,K, \\ \zeta^K = \zeta^{K-1} = 0, & \text{in } \Omega. \end{cases} \tag{13}$$

It is easy to see that (13) admits a unique solution $\{\zeta^k \in H_0^1(\Omega)\}_{k=0,\ldots,K}$. Moreover, this solution has the regularity

$$\frac{\partial}{\partial\nu}\left(\frac{\zeta^{k+1}+\zeta^{k-1}}{2}\right) \in L^2(\Gamma) \quad \text{for } k=1,\ldots,K-1.$$

Put

$$\mathcal{H} = \Big\{ \{y^k\}_{k=0,\ldots,K} \mid y^{i+1}+y^{i-1} \in L^2(\Omega) \text{ for } i=1,\ldots,K-1, \\ \frac{y^{j+1}-y^j}{h} + \frac{y^{j-1}-y^{j-2}}{h} \in H^{-1}(\Omega) \text{ for } j=2,\ldots,K-1 \Big\}. \tag{14}$$

We introduce the following:

Definition 1. *$\{y^k\}_{k=0,\ldots,K} \in \mathcal{H}$ is said to be a solution of (7), in the sense of transposition, if $y^0 = y_0$, $y^1 = y_0 + hy_1$, and for any $\{f^k \in L^2(\Omega)\}_{k=1,\ldots,K-1}$, and $\{g^k \in H_0^1(\Omega)\}_{k=1,\ldots,K}$ with $g^1 = g^K = 0$, it holds*

$$h\sum_{k=1}^{K-1}\int_\Omega f^k \frac{y^{k+1}+y^{k-1}}{2}dx - h\sum_{k=2}^{K-1}\left\langle g^k, \frac{y^{k+1}-y^k}{2h} + \frac{y^{k-1}-y^{k-2}}{2h}\right\rangle_{H_0^1(\Omega),H^{-1}(\Omega)} \\ = \langle \zeta^0, y_1\rangle_{H_0^1(\Omega),H^{-1}(\Omega)} - \int_\Omega \frac{\zeta^1-\zeta^0}{h}y_0 dx - h\sum_{k=1}^{K-1}\int_{\Gamma_0}\frac{\partial}{\partial\nu}\left(\frac{\zeta^{k+1}+\zeta^{k-1}}{2}\right)u^k d\Gamma_0, \tag{15}$$

where $\{\zeta^k \in H_0^1(\Omega)\}_{k=0,\ldots,K}$ is the solution of (13).

The above definition can be viewed as a discrete version of the classical transposition approach [6]. It is motivated by the following observation: When the control $\{u^k\}_{k=0,\ldots,K}$ and the initial data (y^0, y^1) are sufficiently smooth, multiplying both sides of (13) by $(y^{k+1}+y^{k-1})/2$, integrating the resulting identity in Ω and summing it for $k=1,\ldots,K-1$, one obtains (15).

The well-posedness of the system (7) is stated as follows:

Theorem 1. *Assume $(y_0, y_1) \in L^2(\Omega)\times H^{-1}(\Omega)$ and $\{u^k \in L^2(\Gamma_0)\}_{k=1,\ldots,K-1}$. Then the system (7) admits one and only one solution $\{y^k\}_{k=0,\ldots,K} \in \mathcal{H}$ in the sense of Definition 1. Moreover, $\left(y^{2\ell}, \frac{y^{2\ell+1}-y^{2\ell}}{h}\right) \in L^2(\Omega)\times H^{-1}(\Omega)$ for $\ell = 0, 1, \ldots, [\frac{K}{2}]$, and*

$$\max_{\ell=0,1,\ldots,[\frac{K}{2}]} \left\| \left(y^{2\ell}, \frac{y^{2\ell+1} - y^{2\ell}}{h} \right) \right\|^2_{L^2(\Omega)\times H^{-1}(\Omega)}$$
$$\leq C \left(\|(y_0, y_1)\|^2_{L^2(\Omega)\times H^{-1}(\Omega)} + h \sum_{k=1}^{K-1} \|u^k\|^2_{L^2(\Gamma_0)} \right). \quad (16)$$

We refer to [14] for the proof of Theorem 1 by means of a discrete multiplier approach.

Next, assume $\{\Phi_j\}_{j\geq 1} \subset H_0^1(\Omega)$ to be an orthonormal basis of $L^2(\Omega)$ consisting of the eigenvectors (with eigenvalues $\{\mu_j^2\}_{j\geq 1}$) of the Dirichlet Laplacian:

$$\begin{cases} -\Delta\Phi_j = \mu_j^2\Phi_j, & \text{in } \Omega \\ \Phi_j = 0, & \text{on } \Gamma. \end{cases}$$

For any $s > 0$, we set

$$\mathcal{C}_{1,s} = \left\{ f(x) \mid f(x) = \sum_{\mu_j^2 < s} a_j\Phi_j(x),\ a_j \in \mathbb{C} \right\} \subset H_0^1(\Omega), \quad (17)$$

$$\mathcal{C}_{0,s} = \left\{ g(x) \mid g(x) = \sum_{\mu_j^2 < s} b_j\Phi_j(x),\ b_j \in \mathbb{C} \right\} \subset L^2(\Omega), \quad (18)$$

and

$$\mathcal{C}_{-1,s} = \left\{ z(x) \mid z(x) = \sum_{\mu_j^2 < s} c_j\Phi_j(x),\ c_j \in \mathbb{C} \right\} \subset H^{-1}(\Omega), \quad (19)$$

subspaces of $H_0^1(\Omega)$, $L^2(\Omega)$ and $H^{-1}(\Omega)$, respectively, with the induced topologies. It is clear that $\bigcup_{k=1}^\infty \mathcal{C}_{1,k}$ is dense in $H_0^1(\Omega)$, and the same can be said for $\bigcup_{k=1}^\infty \mathcal{C}_{0,k}$ in $L^2(\Omega)$ and $\bigcup_{k=1}^\infty \mathcal{C}_{-1,k}$ in $H^{-1}(\Omega)$. Denote by $\pi_{1,s}$, $\pi_{0,s}$ and $\pi_{-1,s}$ the projection operators from $H_0^1(\Omega)$, $L^2(\Omega)$ and $H^{-1}(\Omega)$ to $\mathcal{C}_{1,s}$, $\mathcal{C}_{0,s}$ and $\mathcal{C}_{-1,s}$, respectively.

Our main results are stated as follows:

Theorem 2. *Let $T > 2R$. Then there exist three constants $h_0 > 0$, $\delta > 0$ and $C > 0$, depending only on T, R and the dimension d, such that for all $(\varphi_0, \varphi_1) \in \mathcal{C}_{1,\delta h^{-2}} \times \mathcal{C}_{0,\delta h^{-2}}$, the corresponding solution $\{\varphi^k\}_{k=0,\ldots,K}$ of (9) satisfies*

$$E_h^0 \leq Ch \sum_{k=1}^{K-1} \int_{\Gamma_0} \left| \frac{\partial}{\partial\nu} \left(\frac{\varphi^{k+1} + \varphi^{k-1}}{2} \right) \right|^2 d\Gamma_0, \quad \forall h \in (0, h_0]. \quad (20)$$

Remark 2. We refer to (38) for the exact form of δ, which depends only on d, T and R. In particular, it indicates that δ decreases as T decreases. This is natural since, as T decreases, less and less time-step iterations are involved in the system (9) and, consequently, less Fourier components of the solutions may be observed. Further, δ tends to zero as T tends to $2R$. This is natural too since our proof of (20) is based on the method of multipliers which works at the continuous level for all $T > 2R$ but that, at the time-discrete level, due to the added dispersive effects, may hardly work when T is very close to $2R$, except if the filtering is strong enough.

Remark 3. The problem considered in this paper could have been addressed, in $1-d$, using discrete Ingham inequalities as those in [8]. When doing that, one would get similar results. In [2] the problem of observability of time-discrete linear conservative systems is addressed in an abstract context including wave, plate and Schrödinger equations. The techniques employed in [2] are inspired in those in [10] based on resolvent estimates, which allow to derive, in a systematic way, observability results for time-discrete systems as consequences of those that are by now well-known for time-continuous ones. The results in [2] can be applied to the time-discrete wave equation considered in this article. The main drawback of the results in [2] is that the observability time one gets seems to be far from the expected optimal one. Another different approach, which gives weaker results, is viewing (by extension to continuous time) the solutions of (9) as perturbed solutions of the continuous conservative wave equation (4). Absorbing the remainder terms then requires stronger filtering than the multiplier method.

Remark 4. As shown in [14], the order h^{-2} of the filtering parameter (in Theorem 2) is optimal. This corresponds precisely to filtering numerical solutions whose wave length is of the order of the mesh-size h, for which resonance phenomena may arise. However, our analysis in the next section indicates that the inequality (20) may hold within the class $\mathcal{C}_{1,\delta h^{-2}} \times \mathcal{C}_{0,\delta h^{-2}}$ for any $\delta > 0$. This can be proved to hold by applying the abstract results in [2] to the present problem. The multiplier method we develop here needs to impose a smallness condition on δ. It is an interesting open problem to see if the multiplier method can be adapted to deal with arbitrarily large values of δ. But it is well known, even at the continuous level, that the method of multipliers is often unable to yield observability results that can be obtained by other ways.

As a consequence of the partial observability result in Theorem 2, by duality, we can derive the following uniform partial controllability result:

Theorem 3. *Let T, h_0 and δ be given as in Theorem 2. Then for any $h \in (0, h_0]$ and any $(y^0, \frac{y^1-y^0}{h}) \in L^2(\Omega) \times H^{-1}(\Omega)$, there exists a control $\{u^k \in L^2(\Gamma_0)\}_{k=0,\dots,K}$ such that the solution of (7) satisfies the following:*

(i) It holds

$$\pi_{0,\delta h^{-2}} y^{K-1} = \pi_{-1,\delta h^{-2}} \left(\frac{y^K - y^{K-1}}{h} \right) = 0 \quad in\ \Omega; \tag{21}$$

(ii) There exists a constant $C > 0$, independent of h, y^0 and y^1, such that

$$h \sum_{k=1}^{K-1} \int_{\Gamma_0} \left| \frac{u^{k+1} + u^{k-1}}{2} \right|^2 d\Gamma_0 \leq C \left\| \left(y^0, \frac{y^1 - y^0}{h} \right) \right\|^2_{L^2(\Omega) \times H^{-1}(\Omega)};$$

(iii) When $h \to 0$,

$$U_h \triangleq \sum_{k=1}^{K-1} u^k(x) 1_{[kh,(k+1)h)}(t) \longrightarrow u \quad strongly\ in\ L^2((0,T) \times \Gamma_0), \tag{22}$$

where u is a control of the system (1), fulfilling (3);

(iv) When $h \to 0$,

$$y_h \triangleq y^0 1_{\{0\}}(t) + \frac{1}{h} \sum_{k=0}^{K-1} \left[(t - kh) y^{k+1} - (t - (k+1)h)\, y^k \right] 1_{(kh,(k+1)h]}(t)$$

$$\longrightarrow y \quad strongly\ in\ C([0,T]; L^2(\Omega)) \cap H^1([0,T]; H^{-1}(\Omega)), \tag{23}$$

where y is the solution of the system (1) with the limit control u as above.

Remark 5. The above theorem contains two results: the uniform partial controllability and the convergence of the controls and states as $h \to 0$. The proof is standard. Indeed, the partial controllability statement follows from Theorem 2 and classical duality arguments [6]; while for the convergence result, one may use the approach developed in [17].

It is important to note that, in the limit, one can recover the controllability of (1) for all $T > 2R$, i.e. the same results as the multiplier method applied directly to the time-continuous wave equation yields, as we have shown in the last two properties of Theorem 3. Indeed, given any $T > 2R$, one can choose a sufficiently small δ such that Theorem 3 guarantees the controllability of the projections $\pi_{0,\delta h^{-2}}$ in time T. Since these projections involve the frequencies μ_j^2 such that $\mu_j^2 < \delta h^{-2}$, it is clear that, as $h \to 0$, this range of frequencies eventually covers the whole spectrum of the time-continuous wave equation. It is, however, important to underline that the filtering parameter δ has to be chosen depending on the value of T and that $\delta \to 0$ as T approaches $2R$, as indicated in Remark 2.

By duality, Theorem 3 is a consequence of Theorem 2. Hence, in the sequel we shall concentrate mainly on the proof of Theorem 2. To show Theorem 2, we shall develop a multiplier approach, which is a discrete analogue of the classical one for the time-continuous case [6]. There are two key ingredients when doing this. One is a basic identity for the solutions of (9) obtained by means of multipliers, which is a discrete version of the classical one on the time-continuous wave equation [6]. The other one is the construction of the filtering operator to guarantee the uniform observability of (9) after filtering. We shall explain them in more detail later in this paper.

3 Bicharacteristic Rays and Group Velocity

Before entering into the details of the proofs, we give an heuristic explanation of the necessity of the above filtering mechanism in terms of the group velocity of propagation of the solutions of the time-discrete system (see [12, 17]). For doing that we consider the time-discrete wave equation (9) in the whole space $\mathbb{R}^d$. Applying the Fourier transform (the continuous one in space and the discrete one in time), we deduce that the symbol of the time semi-discrete system (9) is

$$p_h(\tau,\xi) = -\frac{4\sin^2\frac{\tau h}{2}}{h^2} + |\xi|^2\cos(\tau h), \quad (\tau,\xi)\in\left[-\frac{\pi}{2h},\frac{\pi}{2h}\right]\times\mathbb{R}^d.$$

It is easy to see that, for all $\tau\in[-\pi(2h)^{-1},\pi(2h)^{-1}]$, $p_h(\tau,\xi)$ has two non-trivial roots $\xi^{\pm}\in\mathbb{R}^d$. The bicharacteristic rays are defined as the solutions of the following Hamiltonian system:

$$\begin{cases} \dfrac{dx(s)}{ds} = 2\xi\cos(\tau h), & \dfrac{dt(s)}{ds} = -\dfrac{2\sin(\tau h)}{h} - |\xi|^2 h\sin(\tau h), \\ \dfrac{d\xi(s)}{ds} = 0, & \dfrac{d\tau(s)}{ds} = 0. \end{cases}$$

As in the continuous case, the rays are straight lines. However, both the direction and the velocity of propagation of the rays in this time-discrete setting case are different from the time-continuous one.

Let us now, for instance, illustrate the existence of bicharacteristic rays whose projection on $\mathbb{R}^d$ propagates at a very low velocity or even does not move at all. For this, we fix any $x_0 = (x_{0,1},\ldots,x_{0,d})\in\Omega$ and choose the initial time $t_0 = 0$. Also, we choose the initial microlocal direction $(\tau_0,\xi_0) = (\tau_0,\xi_{0,1},\ldots,\xi_{0,d})$ to be a root of p_h, i.e.

$$|\xi_0|^2 = \frac{4\sin^2\frac{\tau_0 h}{2}}{h^2\cos(\tau_0 h)}, \quad \tau_0\in\left(-\frac{\pi}{2h},\frac{\pi}{2h}\right).$$

Note that the above condition is satisfied for $\xi_{0,1} = 2h^{-1}\sin\frac{\tau_0 h}{2}\cos^{-1/2}(\tau_0 h)$ and $\xi_{0,2} = \cdots = \xi_{0,d} = 0$, for instance. In this case we get

$$\frac{dx}{dt} = \frac{dx/ds}{dt/ds} = -\frac{\cos^{3/2}(\tau_0 h)}{\cos\frac{\tau_0 h}{2}}$$

and $dx_2(t)/dt = \cdots = dx_d(t)/dt = 0$. Thus, $x_j(t)$ for $j = 2,\ldots,d$ remain constant and

$$x_1(t) = x_{0,1} - t\cos^{3/2}(\tau_0 h)\cos^{-1}\frac{\tau_0 h}{2}$$

evolves with speed $-\cos^{3/2}(\tau_0 h)\cos^{-1}\frac{\tau_0 h}{2}$, which tends to 0 when $\tau_0 h\to\frac{\pi}{2}-$, or $\tau_0 h\to-\frac{\pi}{2}+$. This allows us to show that, as $h\to 0$, there exist rays that remain trapped on a neighborhood of x_0 for time intervals of arbitrarily large length. In order to guarantee the boundary observability, these rays have to be cut-off by filtering. This can be done by restricting the Fourier spectrum of the solution to the range $|\tau| \leq \frac{\rho\pi}{2h}$ with $0 < \rho < 1$. This corresponds to

$$|\xi|^2 \leq \frac{4\sin^2(\rho\pi/2)}{h^2\cos(\rho\pi/2)}, \tag{24}$$

for the root of the symbol p_h.

This is the same scaling of the filtering operators we imposed in Theorems 2 and 3, namely, $\mu_j^2 \leq \delta/h^2$. Note, however, that in (24), as $\rho \to 1$, the filtering parameter

$$\delta = \frac{4\sin^2(\rho\pi/2)}{\cos(\rho\pi/2)} \longrightarrow \infty.$$

Thus, in principle, as mentioned above, the analysis of the velocity of propagation of bicharacteristic rays does not seem to justify the need of letting the filtering parameter δ small enough as in Theorems 2 and 3. Thus, this last restriction seems to be imposed by the rigidity of the method of multipliers rather than by the underlying wave propagation phenomena.

We can reach similar conclusions by analyzing the behavior of the so-called group velocity. Indeed, following [12], in $1-d$ the group velocity has the form

$$C(\xi) = \frac{4}{(2+h^2\xi^2)\sqrt{4+h^2\xi^2}},$$

with the graphs as in Figure 1. Obviously, it tends to zero when $h^2\xi^2$ tends to infinity. This corresponds precisely to the high frequency bicharacteristic rays constructed above for which the velocity of propagation vanishes. Based on this analysis one can show that, whatever the filtering parameter δ is, uniform observability requires the observation time to be large enough with $T(\delta) \nearrow \infty$ as $\delta \nearrow \infty$. This may be done using an explicit construction of solutions concentrated along rays (see, for instance, [7]). The positive counterpart of this result guaranteeing that, for any value of the filtering parameter $\delta > 0$, uniform observability/controllability holds for sharp large enough values of time, is an interesting open problem whose complete solution will require the application of microlocal analysis tools. At this respect it is worth mentioning that, although the results in [2] can be applied for any $\delta > 0$, the value of the time they yield is larger than the one predicted by the analysis in this section.

4 A Key Identity via Multipliers

In this section we present the first key point of the proof of Theorem 2, i.e. an identity for the solutions of (9).

The desired identity is as follows:

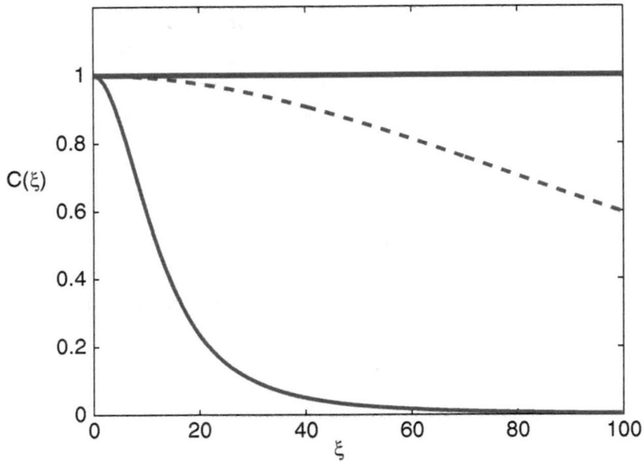

Fig. 1. The diagram of the group velocity $C(\xi)$. $h = 0.1$ (solid line) vs. $h = 0.01$ (dashed line). The thick horizontal segment corresponds to the theoretical group velocity $C(\xi) = 1$ (in the continuous case, i.e. for $h = 0$).

Lemma 1. *For any $h > 0$ and any solution $\{\varphi^k\}_{k=0,\dots,K}$ of (9), it holds*

$$\frac{h}{2}\sum_{k=0}^{K-1}\int_\Omega\left(\left|\frac{\varphi^{k+1}-\varphi^k}{h}\right|^2+\frac{|\nabla\varphi^{k+1}|^2+|\nabla\varphi^k|^2}{2}\right)dx+X+Y+Z$$
$$=\frac{h}{2}\sum_{k=1}^{K-1}\int_\Gamma (x-x_0)\cdot\nu\left|\frac{\partial}{\partial\nu}\left(\frac{\varphi^{k+1}+\varphi^{k-1}}{2}\right)\right|^2 d\Gamma, \quad (25)$$

where

$$X=\int_\Omega\left[(x-x_0)\cdot\nabla\left(\frac{\varphi^K+\varphi^{K-2}}{2}\right)+\frac{d-1}{2}\varphi^K\right]\frac{\varphi^K-\varphi^{K-1}}{h}dx$$
$$-\int_\Omega\left[(x-x_0)\cdot\nabla\left(\frac{\varphi^2+\varphi^0}{2}\right)+\frac{d-1}{2}\varphi^0\right]\frac{\varphi^1-\varphi^0}{h}dx, \quad (26)$$

$$Y=\frac{d}{2}\left[h^2\sum_{k=1}^{K-1}\int_\Omega\Delta\left(\frac{\varphi^{k+1}+\varphi^{k-1}}{2}\frac{\varphi^k-\varphi^{k-1}}{h}\right)dx\right.$$
$$\left.-h\int_\Omega\left|\frac{\varphi^K-\varphi^{K-1}}{h}\right|^2dx\right]$$
$$+\int_\Omega(x-x_0)\cdot\left[\nabla\left(\frac{\varphi^{K-1}-\varphi^{K-2}}{2}\right)\frac{\varphi^K-\varphi^{K-1}}{h}\right.$$
$$\left.+\nabla\left(\frac{\varphi^2-\varphi^1}{2}\right)\frac{\varphi^1-\varphi^0}{h}\right]dx, \quad (27)$$

$$
\begin{aligned}
Z = &\frac{(d-2)h}{8}\sum_{k=1}^{K-1}\int_{\Omega}\left|\nabla(\varphi^{k+1}-\varphi^{k-1})\right|^2 dx \\
&-\frac{(d-1)h}{4}\sum_{k=0}^{K-1}\int_{\Omega}\left|\nabla\left(\varphi^{k+1}-\varphi^{k}\right)\right|^2 dx \\
&-\frac{(d-1)h}{4}\int_{\Omega}\left(\nabla\varphi^{K}\cdot\nabla\varphi^{K-1}+\nabla\varphi^{1}\cdot\nabla\varphi^{0}\right) dx \\
&+\frac{(d-2)h}{4}\int_{\Omega}\left(|\nabla\varphi^{K-1}|^2+|\nabla\varphi^{1}|^2\right) dx. \qquad (28)
\end{aligned}
$$

Proof. Multiplying the first equation of (9) by $(x-x_0)\cdot\nabla(\varphi^{k+1}+\varphi^{k-1})/2$ (which is a discrete version of the classical multiplier $(x-x_0)\cdot\nabla\varphi$ for the wave equation), integrating it in Ω, summing it up from 1 to $K-1$ and using integration by parts, we obtain

$$
\begin{aligned}
&h\sum_{k=1}^{K-1}\int_{\Omega}(x-x_0)\cdot\nabla\left(\frac{\varphi^{k+1}+\varphi^{k-1}}{2}\right)\frac{\varphi^{k+1}+\varphi^{k-1}-2\varphi^{k}}{h^2}dx \\
&\quad = h\sum_{k=1}^{K-1}\int_{\Omega}(x-x_0)\cdot\nabla\left(\frac{\varphi^{k+1}+\varphi^{k-1}}{2}\right)\Delta\left(\frac{\varphi^{k+1}+\varphi^{k-1}}{2}\right)dx. \qquad (29)
\end{aligned}
$$

One can check that the left-hand side term of (29) coincides with

$$
\begin{aligned}
&\frac{d}{2}h\sum_{k=0}^{K-1}\int_{\Omega}\left|\frac{\varphi^{k+1}-\varphi^{k}}{h}\right|^2 dx + Y \\
&+\int_{\Omega}(x-x_0)\cdot\nabla\left[\left(\frac{\varphi^{K}+\varphi^{K-2}}{2}\right)\frac{\varphi^{K}-\varphi^{K-1}}{h}-\left(\frac{\varphi^{2}+\varphi^{0}}{2}\right)\frac{\varphi^{1}-\varphi^{0}}{h}\right]dx, \qquad (30)
\end{aligned}
$$

where Y is defined as in (27). We now use the classical multiplier identity for the Laplacian

$$
\int_{\Omega}(x-x_0)\cdot\nabla\psi\Delta\psi dx = \frac{1}{2}\int_{\Gamma}(x-x_0)\cdot\nu\left|\frac{\partial\psi}{\partial\nu}\right|^2 d\Gamma - \frac{2-d}{2}\int_{\Omega}|\nabla\psi|^2 dx, \quad (31)
$$

which holds for all $\psi\in H^2\cap H_0^1(\Omega)$ [6]. Then, using the identity $(a+b)^2 = 2(a^2+b^2)-(a-b)^2$ for any $a,b\in\mathbb{R}$, the right-hand side term of (29) may be written as

$$\frac{h}{2}\sum_{k=1}^{K-1}\int_\Gamma (x-x_0)\cdot\nu\left|\frac{\partial}{\partial\nu}\left(\frac{\varphi^{k+1}+\varphi^{k-1}}{2}\right)\right|^2 d\Gamma$$
$$+\frac{(d-2)h}{2}\left\{\sum_{k=0}^{K-1}\int_\Omega\frac{|\nabla\varphi^{k+1}|^2+|\nabla\varphi^k|^2}{2}dx\right.$$
$$\left.-\sum_{k=1}^{K-1}\int_\Omega\left|\nabla\left(\frac{\varphi^{k+1}-\varphi^{k-1}}{2}\right)\right|^2 dx-\frac{1}{2}\int_\Omega\left(|\nabla\varphi^{K-1}|^2+|\nabla\varphi^1|^2\right)dx\right\}. \quad (32)$$

On the other hand, multiplying the first equation of (9) by φ^k (which is a discrete version of the multiplier φ in the time-continuous setting, which allows establishing the identity of equipartition of energy), integrating it in Ω, summing it up for $k=1,\ldots,K-1$ and using integration by parts, as above, we obtain the following equipartition of energy identity:

$$h\sum_{k=0}^{K-1}\int_\Omega\left(\left|\frac{\varphi^{k+1}-\varphi^k}{h}\right|^2-\frac{|\nabla\varphi^{k+1}|^2+|\nabla\varphi^k|^2}{2}\right)dx$$
$$=-\frac{h}{2}\sum_{k=0}^{K-1}\int_\Omega\left|\nabla(\varphi^{k+1}-\varphi^k)\right|^2 dx-\frac{h}{2}\int_\Omega\left(\nabla\varphi^K\cdot\nabla\varphi^{K-1}+\nabla\varphi^1\cdot\nabla\varphi^0\right)dx$$
$$+\int_\Omega\left(\frac{\varphi^K-\varphi^{K-1}}{h}\varphi^K-\frac{\varphi^1-\varphi^0}{h}\varphi^0\right)dx. \quad (33)$$

By (29)–(33), recalling (26) and (28) respectively for X and Z, we arrive at the desired identity (25).

Remark 6. The identity (25) is a time-discrete analogue of the following well-known identity for the wave equation (9) obtained by multipliers [6]:

$$\frac{1}{2}\int_0^T\int_\Omega\left[|\varphi_t|^2+|\nabla\varphi|^2\right]dxdt+\mathcal{X}=\frac{1}{2}\int_0^T\int_\Gamma(x-x_0)\cdot\nu\left|\frac{\partial\varphi}{\partial\nu}\right|^2 d\Gamma dt, \quad (34)$$

where

$$\mathcal{X}=\int_\Omega\left[(x-x_0)\cdot\nabla\varphi+\frac{d-1}{2}\varphi\right]\varphi_t dx\Big|_{t=0}^{T}.$$

There are clear analogies between (25) and (34). In fact, the only major differences are that, in the discrete version (25), two extra remainder terms (Y and Z) appear, which are due to the time discretization. It is easy to see, formally, that Y and Z tend to zero as $h\to 0$. But this convergence does not hold uniformly for all solutions. Consequently, these added terms impose the need of using filtering of the high frequencies to obtain observability inequalities out of (25) and modify the observability time, as we shall see.

5 Filtering and Uniform Observability

In this section, we present the second key ingredient of the proof of Theorem 2, i.e. the choice of the filtering parameter which, combined with the identity in Lemma 1, leads to the desired uniform observability inequality in Theorem 2.

For this, we first derive the following result, which provides an estimate on the remainder term $X + Y + Z$ in Lemma 1 in terms of the energy:

Lemma 2. *Let K be an even integer, $s > 0$ and $T > 0$. Then, for any $(\varphi^0, \frac{\varphi^1-\varphi^0}{h}) \in \mathcal{C}_{1,s} \times \mathcal{C}_{0,s}$, for the corresponding solution $\{\varphi^k\}_{k=0,\dots,K}$ of (9), it holds*

$$X + Y + Z \geq -\left[2R + a_1 h + 3R\sqrt{s}h + T\left(\frac{d}{2}\sqrt{s}h + a_2 sh^2\right)\right] E_h^0, \tag{35}$$

where

$$a_1 = 3d - 2 + \max\left(\frac{d-1}{2}, 2\right), \quad a_2 = \min\left(1, (2-d)^+\right) + \frac{d-1}{2}. \tag{36}$$

Proof. For any $(\varphi^0, \frac{\varphi^1-\varphi^0}{h}) \in \mathcal{C}_{1,s} \times \mathcal{C}_{0,s}$, in view of the Fourier series decomposition of the corresponding solution $\{\varphi^k\}_{k=0,\dots,K}$ of (9), one sees that, for any k, we have

$$\begin{gathered}\int_\Omega |\nabla(\varphi^k - \varphi^{k-1})|^2 dx \leq s \int_\Omega |\varphi^k - \varphi^{k-1}|^2 dx,\\ \int_\Omega \left|\Delta\left(\frac{\varphi^{k+1}+\varphi^{k-1}}{2}\right)\right|^2 dx \leq s \int_\Omega \left|\nabla\left(\frac{\varphi^{k+1}+\varphi^{k-1}}{2}\right)\right|^2 dx.\end{gathered} \tag{37}$$

Recalling (26)–(28) and using (37), and noting $T = Kh$ and that the energy of the system (9) is conservative, we can show that

$$|X| \leq \left[2R + 2(d-1)h + Rh\sqrt{s}\right] E_h^0, \quad |Y| \leq h\left[d\left(\frac{\sqrt{s}T}{2} + 1\right) + 2R\sqrt{s}\right] E_h^0,$$
$$Z \geq -h\left\{\left[\min(1, (2-d)^+) + \frac{d-1}{2}\right] shT + \max\left(\frac{d-1}{2}, 2\right)\right\} E_h^0,$$

which gives (35).

Finally, Theorem 2 follows from Lemmas 1 and 2 immediately. Indeed, combining (25) and (35) and recalling the definition of Γ_0 in (2), we deduce that

$$\left\{T\left(1 - \frac{d}{2}\sqrt{s}h - a_2 sh^2\right) - \left[2R + a_1 h + 3R\sqrt{s}h\right]\right\} E_h^0$$
$$\leq \frac{R}{2} h \sum_{k=1}^{K-1} \int_{\Gamma_0} \left|\frac{\partial}{\partial\nu}\left(\frac{\varphi^{k+1}+\varphi^{k-1}}{2}\right)\right|^2 d\Gamma_0.$$

For this inequality to yield an estimate on E_h^0 we need to choose $s = \delta h^{-2}$ with h small enough such that

$$a_2\delta + \frac{d}{2}\sqrt{\delta} < 1,$$

or, more precisely,

$$0 < \sqrt{\delta} < \frac{4}{\sqrt{d^2 + 16a_2} + d}. \tag{38}$$

Once this is done, for $h \in (0, h_0)$, T has to be chosen such that

$$T > \frac{2R + a_1 h_0 + 3R\sqrt{\delta}}{1 - \frac{d}{2}\sqrt{\delta} - a_2\delta} \geq 2R. \tag{39}$$

Hence, (20) holds for $h \in (0, h_0]$.

Conversely, for any $T > 2R$ one can always choose h_0 and δ small enough so that (38) and (39) hold, guaranteeing the uniform observability inequality (20).

6 Further Comments and Open Problems

Fully Discrete Schemes

The analysis in this paper can be combined with previous works (see, for instance, [17]) concerning space semi-discretizations to deal with full discretization schemes. This has been done in [2] in a more abstract setting. But a complete analysis of this issue is still to be done.

Other Equations

The approach and results in this paper can be extended to other PDEs of conservative nature such as the Schrödinger, plate, Maxwell equations, and so on. There is a fruitful literature on the use of multiplier techniques for these models in the continuous setting (see, for instance, [6]). But, the analysis of the corresponding time-discrete systems, adapting the techniques developed in this paper, remains to be done.

Variable Coefficients and Nonlinear Problems

It is well-known that, in the continuous case, the multiplier approach can be applied to obtain the controllability/observability of the conservative PDEs with constant coefficients. As for the problems with variable coefficients and/or the nonlinear ones, one has to use microlocal analysis [1] and/or Carleman estimates [13] to get sharp results. In this time-discrete setting, it would be interesting to develop these other approaches to cover the same class of models as in the PDE setting. This is still to be done.

Acknowledgement. This work is supported by the Grant MTM2005-00714 of the Spanish MEC, the project MTM2008-03541 of the Spanish Ministry of Science and Innovation, the DOMINO Project CIT-370200-2005-10 in the PROFIT program of the MEC (Spain), the i-MATH project of Spanish MEC, the SIMUMAT projet of the CAM (Spain), the NSFC under grants 10831007 and 60821091, and the NSF of China under grant 10525105.

References

1. C. Bardos, G. Lebeau, and J. Rauch. Sharp sufficient conditions for the observation, control, and stabilization of waves from the boundary. *SIAM J. Control Optim.*, 30(5):1024–1065, 1992.
2. S. Ervedoza, C. Zheng, and E. Zuazua. On the observability of time-discrete linear conservative systems. *J. Funct. Anal.*, 254(12):3037–3078, 2008.
3. G. Glowinski. Ensuring well-posedness by analogy: Stokes problem and boundary control for the wave equation. *J. Comput. Phys.*, 103(2):189–221, 1992.
4. R. Glowinski, C. H. Li, and J.-L. Lions. A numerical approach to the exact boundary controllability of the wave equation. I. Dirichlet controls: description of the numerical methods. *Japan J. Appl. Math.*, 7(1):1–76, 1990.
5. J. A. Infante and E. Zuazua. Boundary observability for the space semi-discretizations of the 1-D wave equation. *M2AN Math. Model. Numer. Anal.*, 33(2):407–438, 1999.
6. J.-L. Lions. *Contrôlabilité exacte, perturbations et stabilisation de systèmes distribués, Vol. 1.* Masson, Paris, 1988.
7. F. Macià. The effect of group velocity in the numerical analysis of control problems for the wave equation. In *Mathematical and Numerical Aspects of Wave Propagation – WAVES 2003*, pages 195–200, Berlin, 2003. Springer.
8. M. Negreanu and E. Zuazua. Convergence of a multigrid method for the controllability of a 1-d wave equation. *C. R. Math. Acad. Sci. Paris*, 338(5):413–418, 2004.
9. A. Osses. A rotated multiplier applied to the controllability of waves, elasticity, and tangential Stokes control. *SIAM J. Control Optim.*, 40(3):777–800, 2001.
10. K. Ramdani, T. Takahashi, G. Tenenbaum, and M. Tucsnak. A spectral approach for the exact observability of infinite-dimensional systems with skew-adjoint generator. *J. Funct. Anal.*, 226(1):193–229, 2005.
11. J. W. Thomas. *Numerical partial differential equations: Finite difference methods.* Springer, Berlin, 1995.
12. L. N. Trefethen. Group velocity in finite difference schemes. *SIAM Rev.*, 24(2):113–136, 1982.
13. X. Zhang. Explicit observability estimate for the wave equation with potential and its application. *R. Soc. Lond. Proc. Ser. A Math. Phys. Eng. Sci.*, 456(1997):1101–1115, 2000.
14. X. Zhang, C. Zheng, and E. Zuazua. Time discrete wave equations: Boundary observability and control. *Discrete Contin. Dyn. Syst.*, 23(1–2):571–604, 2009.
15. C. Zheng. Controllability of the time discrete heat equation. *Asymptot. Anal.*, 59(3–4):139–177, 2008.
16. E. Zuazua. Boundary observability for the finite-difference space semi-discretizations of the 2-D wave equation in the square. *J. Math. Pures Appl.*, 78(5):523–563, 1999.
17. E. Zuazua. Propagation, observation, and control of waves approximated by finite difference methods. *SIAM Rev.*, 47(2):197–243, 2005.

Index

B
Banavar, J.R., 134
Bates, D.S., 214
Bensoussan, A., 9–23
Boi, S., 61, 65
Brennan, M.J., 214
Brenner, H., 128, 133
Burger, M., 61, 65

C
Camacho, J., 128, 133
Čanić, S., 41–57
Capasso, V., 59, 61
Cattaneo, C., 135
Chen, C.Y., 132

D
Davis, H.T., 139
Di Cesare, N., 27, 28

E
Ethier, S.N., 73

F
Fernández-Cara, E., 81–93
Fitzgibbon, W.E., 1–4
Flück, M., 169–181
Freundlich, H., 139
Frischat, G.H., 140
Fursikov, A.V., 84

G
Galdi, P., 128
Glowinski, R., 27, 213
Golub, G.H., 161
Gunzburger, M.D., 27

H
Hachiya, H., 148
Haslinger, J., 34
Has'minski, R.Z., 77
He, J., 27
Heston, S., 214
Hintermüller, M., 97–108
Hofer, T., 169–181
Hoppe, R.H.W., 97–108
Hu, H.H., 128

I
Imanuvilov, O.Yu., 84
Ito, K., 113–125

J
Jackson, R., 133
Janka, A., 169–181
Joseph, D.D., 127–128, 135, 137–139, 143

K
Kawarada, H., 147–160
Koplick, J., 134
Korteweg, D., 128, 134, 143
Kou, S.G., 213
Kurtz, T.G., 73

L
Landau, L.D., 133
Liao, T.Y., 137

Lifshitz, E.M., 133
Lions, J.-L., 213
Lowengrub, J., 140

M
Maddox, J., 135
Mäkinen, R.A.E, 34
Makridakis, S.G., 147
Mali, O., 183–197
Malone, P.C., 135
Masui, T., 148
Maxwell, J.C., 135
Maxworthy, T., 132
Meiburg, E., 132
Merton, R., 214
Meurant, G., 161–167
Mo, G., 133
Morale, D., 59, 61, 65
Munakata, S., 147
Mungall, J.E., 140

N
Neittaanmäki, P., 199–210

P
Paniagua, D., 41–57
Periaux, J.F., 1–4
Petitjeans, P., 132
Pironneau, O., 5–7, 29
Pojman, J.A., 140
Preziosi, L., 135, 138

Q
Quinke, G., 139

R
Rannacher, R., 214, 219
Rappaz, J., 169–181
Renardy, Y., 128
Repin, S., 183–197, 199–210
Rosenberger, F., 133
Runge, S., 140, 141

S
Saito, K., 147
Scholes, M., 213
Schwartz, E.S., 214
Sekine, F., 147
Serrin, J., 128
Smith, P.G., 139
Sokolowski, J., 32, 34
Strang, G., 220
Suito, H., 147–160

T
Tambača, J., 41–57
Toivanen, J., 213–225
Trémolières, R., 213
Truskinovsky, L., 140
Turi, J., 9–23

V
Van der Waals, M., 128, 134
Veretennikov, A.Y., 77

W
Wheatley, D.N., 135
Wheelwright, S.G., 147

Z
Zhang, Q., 113–125
Zhang, X., 229–244
Zheng, C., 229–244
Zheng, Y., 147–160
Zolésio, J.-P., 32, 34
Zoltowski, B., 140
Zuazua, E., 229–244

FSC
www.fsc.org
MIX
Papier | Fördert
gute Waldnutzung
FSC® C083411

Kolibri 360 GmbH
Ferdinand-Jühlke-Straße 7
99095 Erfurt, Deutschland
produktsicherheit@kolibri360.de